INTRODUCTION TO COASTAL ENGINEERING AND MANAGEMENT

ADVANCED SERIES ON OCEAN ENGINEERING

Series Editor-in-Chief
Philip L- F Liu (*Cornell University*)

Forthcoming titles:

Tsunami Run-Up
by Philip L- F Liu (Cornell Univ.), *Costas Synolakis* (Univ. Southern California),
Harry Yeh (Univ. Washington) *and Nobu Shuto* (Tohoku Univ.)

Beach Nourishment: Theory and Practice
by Robert G Dean (Univ. Florida)

Advanced Series on Ocean Engineering — Volume 16

INTRODUCTION TO COASTAL ENGINEERING AND MANAGEMENT

J. William Kamphuis

Queen's University, Canada

World Scientific

Singapore • New Jersey • London • Hong Kong

Published by

World Scientific Publishing Co. Pte. Ltd.

P O Box 128, Farrer Road, Singapore 912805

USA office: Suite 1B, 1060 Main Street, River Edge, NJ 07661

UK office: 57 Shelton Street, Covent Garden, London WC2H 9HE

British Library Cataloguing-in-Publication Data
A catalogue record for this book is available from the British Library.

ISBN 981-02-3830-4
ISBN 981-02-4417-7 (pbk)

Printed in Singapore.

to all students

Preface

What can I say? This book is really not about facts and formulas. It is about learning and understanding. It is about diligence and care, about stewardship of a precious resource. It was essentially 32 years in the making. It was developed from lecture notes for an introductory course and its stated purpose is *to bridge the gap* between an eager student who knows nothing about coastal engineering and management, and the available literature. My hope is that this book also finds its way on the bookshelves of the practitioners, as a handy reference to those "first things we all need to know".

This book distils things I learned from my professors, from reading, from interacting with colleagues, from practicing all over the world, from listening to stories, and from questions, comments and remarks of my students. My students asked me to write this book - that's why it's here.

My thanks to all who inspired me. My thanks also to the many who helped me – in particular: Mohamed Dabees, Steve Hughes, Tim Janssen, Han Ligteringen, Laura McHardy, Vicki Mitchell, Karim Rakha and Cathy Wagar. Without Queen's University and its Civil Engineering Department, this book would not have become reality. There I first learned the trade, particularly from Arthur Brebner and Bernard Le Méhauté and later Queen's paid me for the privilege to teach so many for so many years. I am also indebted to Delft University of Technology and Delft Hydraulics Laboratory who hosted me at the times that I needed to be away to write this book. I thank the National Sciences and Engineering Research Council of Canada for their continuous research support. And I thank my wife, Nelly, who provided the space and support for me to do this.

This book is about strategy, tactics and philosophy. It is not only about how we should design and manage, but also about design and management itself. It is also about enjoyment. Coastal problems are very complex. They allow us to put together elements of physics, oceanography, geology, geotechnical and structural design, and resource management. In the process, we rub shoulders with experts in each of these areas, and with biologists, chemists and environmentalists. We must also be familiar with the economic, legal and political frameworks, within which we practice. Because our art is young, we still approach our task with only a few rules. We have no coastal engineering design code. We have no precedents in our coastal management tasks. That means challenge, thinking, innovation and unfortunately it may mean mistakes. I enjoy such a challenge, I hope you do.

There is much to do. People still die because of natural disasters. Much of the coastal work to date has been ill-conceived, ill-designed or poorly constructed and needs to be redone. We are faced with the largest migration of people in history. This migration has become a true invasion of the coast, putting tremendous pressure on a scarce natural resource. We are dealing with a mega shift in priorities as we convert industrial areas, rail yards and loading docks of the previous era into residential and recreational settings. We are also asked to integrate. Projects must fit into systems. Physical coastal systems must fit into biological, environmental, legal and sociological systems. Finally, we know so much in theory and at pilot scale, but the translation of this knowledge into prototype reality is so very difficult.

The information in this book goes beyond the printed text. **Bold letters** and the symbol (®) identify computer programs, tables and examples that are available in electronic form on the website that accompanies this text (http://www.wspc.com.sg/others/software/4064).

I have provided a basic tool. The tool is incomplete. It only discusses some of the topics needed in our trade. There is much literature for you to expand into. Good luck on your further journey.

Kingston, June 2000.

Contents

Notation

A   =   horizontal orbital amplitude of the wave,

      A_B = horizontal orbital amplitude at the bottom,

      A_S = horizontal orbital amplitude at the surface,

      A_o = horizontal orbital amplitude in deep water,

    =   slope in regression analysis of transformed co-ordinates,

    =   area

      A_a = surface area (per unit length) of the armor layer of a breakwater,

      A_e = erosion area in the profile around still water level,

      A_s = surface area,

A_i =   coefficient in **ONELINE**,

A_n =   frequency function in wave frequency analysis;

$|A_n|$ =   amplitude spectrum,

A_p =   beach profile coefficient,

a   =   wave amplitude,

    =   parameter of Jonswap spectrum,

a_b =   berm elevation of a berm breakwater,

a_i =   amplitude of tidal constituent,

a_n =   Fourier coefficient,

B   =   vertical orbital amplitude of the wave,

      B_B = vertical orbital amplitude at the bottom,

      B_S = vertical orbital amplitude at the surface,

      B_o = vertical orbital amplitude in deep water,

    =   intercept in regression analysis of transformed co-ordinates,

    =   width

      B_b = berm width in a berm breakwater,

      B_c = width of the soil column affected by a caisson and rock berm,

      B_t = width of the berm seaward of a caisson,

      B_v = width of the caisson of a vertical breakwater,

$\quad\quad$ = $\quad$ (as a subscript) at the bottom,

B_e $\quad$ = $\quad$ effective basin length to calculate seiche,

B_i $\quad$ = $\quad$ coefficient in **ONELINE**,

b $\quad$ = $\quad$ distance between adjacent wave rays,

$\quad\quad$ = $\quad$ (as a subscript) at breaking,

b_b $\quad$ = $\quad$ height of a berm breakwater,

b_n $\quad$ = $\quad$ Fourier coefficient,

C $\quad$ = $\quad$ velocity of propagation; phase speed (= L/T),

$\quad\quad$ C_G $\quad$ = $\quad$ group velocity,

$\quad\quad$ $\mathbf{C_G}$ $\quad$ = $\quad$ group velocity vector,

$\quad\quad$ C_o $\quad$ = $\quad$ velocity of propagation in deep water,

C $\quad$ = $\quad$ coefficient

$\quad\quad$ C_c $\quad$ = $\quad$ calibration coefficient,

$\quad\quad$ C_d $\quad$ = $\quad$ damping coefficient,

$\quad\quad$ C_H $\quad$ = $\quad$ wave height coefficient,

$\quad\quad$ C_{sw} $\quad$ = $\quad$ design coefficient for standing wave,

$\quad\quad$ C_u $\quad$ = $\quad$ design coefficient for wave uplift,

$\quad\quad$ C_w $\quad$ = $\quad$ design coefficient for maximum water level,

$\quad\quad$ C_1 $\quad$ = $\quad$ design coefficient for p_1,

$\quad\quad$ C_3 $\quad$ = $\quad$ design coefficient for p_3,

C_E $\quad$ = $\quad$ earth's center of rotation,

C_{ME} $\quad$ = $\quad$ center of rotation of the earth-moon system,

C_n $\quad$ = $\quad$ Fourier coefficient,

C_Q $\quad$ = $\quad$ ratio of actual and potential sediment transport rate,

CDF $\quad$ = $\quad$ cumulative distribution function,

C_H $\quad$ = $\quad$ crest width of a berm breakwater,

c_i $\quad$ = $\quad$ various coefficients defined locally and only valid locally,

ch $\quad$ = $\quad$ (as a subscript) characteristic,

c_H $\quad$ = $\quad$ modification factor for effective depth at a structure,

D $\quad$ = $\quad$ depth of water including storm surge (=d+S),

$\underline{D}$ $\quad$ = $\quad$ diffusion coefficient [$=Q/(\alpha_b\, d_p)$],

D_a $\quad$ = $\quad$ nominal armor size,

D_{50} $\quad$ = $\quad$ median grain or rock size size,

$\quad\quad$ D_{15} $\quad$ = $\quad$ 15% of the grain sizes are smaller than this size,

$\quad\quad$ D_{85} $\quad$ = $\quad$ 85% of the grain sizes are smaller than this size,

d $\quad$ = $\quad$ depth of water,

$\quad\quad$ d^* $\quad$ = $\quad$ dimensionless depth (=gd/U),

$\quad\quad$ d_B $\quad$ = $\quad$ berm height under vertical breakwater caisson,

$\quad\quad$ d_b $\quad$ = $\quad$ depth of water at breaking,

$\quad\quad$ d_c $\quad$ = $\quad$ closure depth; seaward limit of the active beach profile,

d_d = dune height,

d_p = profile depth,

d_s = depth of water at the structure,

d_s' = modified depth of water at the structure,

d_t = depth at the top of the berm under a vertical breakwater,

d_v = depth at the bottom of a vertical breakwater caisson,

d_v' = d_v for standing wave ($=d_v+\Delta_H$),

d_{5H} = depth 5 wave heights seaward of a structure,

df = frequency increment,

df' = resolution of the wave spectrum,

E = wave energy density,

E_o = wave energy density in deep water,

E( ) = wave energy density spectrum,

e = porosity of the armor layer,

F = fetch length,

F_{eff} = effective fetch length for limited storm duration,

F^* = dimensionless fetch length ($=gF/U^2$)

= Force

F_b = buoyancy force,

F_d = dynamic wave force,

F_h = horizontal force,

F_H = hydrostatic force from the harbor side,

F_{ice} = ice force,

F_m = vertical force from the mass of a caisson,

F_s = hydrostatic force from a standing wave,

F_u = wave-generated uplift force,

F_v = vertical force,

F_w = horizontal force from waves and water level,

F_v = vertical force,

F_n = Fourier transform,

F(u) = function of u,

f = frequency ($= 1/T$),

f_c = highest frequency to be considered in a wave analysis,

f_N = Nyquist frequency,

f_n = Fourier frequency,

f_p = peak frequency of the wave spectrum,

= (as a subscript) derived by frequency analysis,

f_f = friction coefficient between a structure and its sub-base,

f_r = resistance function,

f_s = load function,

f_v = freeboard above still water,

f_v' = freeboard above mean wave level,

G = reduced variate for Gumbel distribution,

= failure surface,

$G(\theta)$ = directional spreading function for directional wave spectrum,

g = gravitational acceleration,

= (as a subscript) with respect to failure surface,

H = wave height,

$\overline{H}$ = mean wave height,

H_b = breaking wave height,

H_{ch} = characteristic wave height,

H_d = wave height determined by zero down-crossing method

H_{des} = design wave height,

H_I = incident wave height,

H_{max} = maximum wave height,

H_{min} = minimum wave height,

H_{mo} = zero moment wave height,

H_{mo}^* = dimensionless zero moment wave height ($=gH_{mo}/U^2$),

H_{mode} = most probable wave height,

H_o = deep water wave height,

H_o' = deep water wave height without refraction,

H_Q = wave height with a probability of exceedence Q,

$\overline{H_Q}$ = average of all the waves larger than H_Q,

H_R = reflected wave height,

H_{rms} = root mean square wave height,

H_s = significant wave height,

H_{sb} = significant breaking wave height,

H_{srb} = significant wave height due to shoaling, refraction and breaking,

H_{T_R} = wave height for return period T_R

H_T = transmitted wave height,

H_t = threshold wave height,

H_u = wave height determined by zero up-crossing method

$H_{0.1}$ = wave height that is exceeded 10% of the time,

$\overline{H_{0.1}}$ = average of the highest 10% of the waves,

$H_{0.01}$ = wave height that is exceeded 1% of the time,

$\overline{H_{0.01}}$ = average of the highest 1% of the waves,

$H_{0.5}$ = median wave height,

$\overline{H_{1/3}}$ = average of the highest 1/3 of the waves ($=H_s$),

h	=	rise in water level,
h_v	=	height of a caisson,
h_w	=	maximum water level reached by waves against a caisson,
I	=	number of tidal constituents,
	=	(as a subscript) incident,
I_c	=	the moment of inertia of the soil column under a rock berm,
I_s	=	bulk sediment transport rate,
i	=	index,
	=	ranking of data point in extreme value analysis,
j	=	(as a subscript) index referring to time,
	=	(as a subscript) index referring to ensemble,
K	=	number of realizations in an ensemble,
K_b	=	berm breakwater design factor,
K_D	=	damage coefficient in rubble mound breakwater design,
K_d	=	diffraction coefficient,
K_M	=	armor mass factor $[=\Delta_a (K_D \cot \theta / \rho_a)^{1/3}]$,
K_{max}	=	maximum wave height factor $[=(H_b)_{max}/H_{sb}]$,
K_p	=	pressure response factor,
K_R	=	reflection coefficient $(=H_R/H_I)$,
K_r	=	refraction coefficient,
K_S	=	spring constant,
K_s	=	shoaling coefficient $(=H/H_o')$,
KE	=	kinetic energy density of the wave,
k	=	wave number $(= 2\pi/L)$,

$\qquad \mathbf{k} \quad$ = wave number vector,

$\qquad k_o \quad$ = wave number in deep water,

	=	bottom roughness,
	=	(as a subscript) index referring to realization,
k_a	=	armor shape factor,
L	=	wave length,

$\qquad L_{bp}$ = the breaking wave length with peak period,

$\qquad L_d$ = the wave length at depth d,

$\qquad L_o$ = wave length in deep water,

$\qquad L_{op}$ = L_o related to the peak frequency of the wave spectrum,

	=	model type – long term and large area,
M	=	mass,

$\qquad M_A$ = mass of accelerometer,

$\qquad M_a$ = mass of armor unit for a rubble mound breakwater,

$\qquad M_s$ = mass of a vertical breakwater structure,

	=	model type – medium term and medium area,

M   =   overturning moment; same subscripts as for forces (above),
$\tilde{M}_o$ = moment about the landward corner of a caisson,
$\tilde{M}_c$ = moment about center of a soil column,
M_f =   number of frequency increments used to average a wave spectrum,
MWL=   mean wave level,
M_Φ =   mean grain size,
M2   =   semi-diurnal tide constituent for the moon,
m   =   beach slope,
m_b = beach slope in the breaking zone,
$\overline{m}$ = average beach slope,
      =   moment of a spectrum ,
      =   harmonic of a seiche,
      =   (as a subscript) model,
N   =   number of samples,
      =   number of points in extreme value analysis,
N_a =   number of armor units per unit length of a rubble mound breakwater,
N_b =   geotechnical indicator; "blow count",
N_L =   project design life,
N_s =   stability number,
N_w =   number of waves,
n   =   energy flux parameter (=C_G/C),
      =   porosity,
      =   general model scale,
n_x = model scale of x (= x_p/x_m),
      =   (subsript) index referring to frequency component,
      =   (as a subscript) index referring to moment of the spectrum,
n_a =   nominal number of layers of armor,
O( ) =   order: terms of order greater than ...,
o   =   (as a subscript) deep water,
P   =   cumulative probability of non-exceedence,
P(r) =   cumulative distribution function for resistance,
$\overline{P}$ =   wave power averaged over a wave period,
      =   the energy flux or wave power between wave rays,
P_o = P in deep water,
$\overline{P}'$ = the average wave power per unit length of beach,
P_a = the alongshore component of wave power,
P_{ab} = P_a in the breaking zone,
P_{asb} = P_{ab} for significant wave height of irregular waves,
P_b =   overall porosity of a breakwater,

P_E = encounter probability,

P_F = probability of failure of the design condition,

P_L = P_F during the lifetime of a project,

P_r = fraction of rounded stones,

PE = potential energy density of the wave,

p = pressure,

$\quad$ p_b = pressure from buoyant force,

$\quad$ p_d = dynamic pressure at still water,

$\quad$ p_h = pressure generated by a standing wave,

$\quad$ p_{sw} = standing wave pressure,

$\quad$ p_u = wave-generated uplift pressure,

$\quad$ p_u = wave-generated uplift pressure,

$\quad$ p_v = pressure at the bottom of a vertical caisson,

$\quad$ p_1 = pressure at still water level (or mean wave level),

$\quad$ p_2 = pressure at the top of a caisson,

$\quad$ p_3 = pressure at the bottom of a caisson,

p( ) = probability density function,

$\quad$ p(r) = probability density function for resistance,

$\quad$ p(s) = probability density function for loading,

Q = sediment transport rate,

$\quad$ Q_a = actual sediment transport rate,

$\quad$ Q_{by} = bypassing sediment transport rate,

$\quad$ Q_c = bulk potential sediment transport rate by CERC formula,

$\quad$ Q_g = sediment transport rate through a groin field,

$\quad$ Q_{gross} = gross sediment transport rate,

$\quad$ Q_i = sediment transport rate at section (i),

$\quad$ Q_i^* = new sediment transport rate at section (i),

$\quad$ Q_k = bulk potential sediment transport rate by Kamphuis formula,

$\quad$ Q_{net} = net sediment transport rate,

$\quad$ Q_p = potential sediment transport rate,

$\quad$ Q_u = sediment transport rate outside a groin field,

$\quad$ = probability of exceedence,

Q_f = fluid discharge,

q = surcharge on soil from rock berm,

$\quad$ = collection of sediment transport parameters,

q_o = cross-shore gain of sediment transport,

R = Rayleigh reduced variable,

$\quad$ = resistance or strength of a structure,

$\quad$ = recession,

R_{ch} = characteristic resistance,

R_i = grouping of known terms in **ONELINE**,

$R_{2\%}$ = runup exceeded by 2% of the waves,

r = correlation coefficient,

r_a = armor layer thickness,

r_f = friction factor on the front slope of a rubble mound breakwater,

S = storm surge,

  = design loading,

    S_{ch} = characteristic loading,

  = model type – short term and small area,

S_a = damage to armor layer of a rubble mound breakwater,

S_e = effective length of structure,

S_i = grouping of known terms in **ONELINE**,

S_t = surface tension,

S_s = structure length,

S ( ) = wave variance spectral density function; wave spectrum,

    S_J = Jonswap spectrum,

    S_{PM} = Pierson-Moskowitz spectrum,

SWL = still water level,

S2 = semi-diurnal tide constituent for the sun,

s = sample standard deviation,

  = wave direction (**7**),

  = direction along the shoreline (**14**),

S_m = mean wave steepness, related to mean wave period,

T = wave period,

    $\bar{T}$ = average period,

    T_p = period corresponding to the peak of the spectrum ($=1/f_p$),

    T_p^* = dimensionless peak wave period ($=gT_p/U$),

    T_1 = period using the zero and first moment of the spectrum,

    T_2 = period using the zero and second moment of the spectrum,

    T_m = period of oscillation of the mth harmonic of a seiche,

T_R = return period,

t = time,

    t_i = initial time,

    t_F = final time,

    t_{full} = time for a structure to fill with sediment

  = storm duration ,

    t^* = dimensionless storm duration ($=gt/U$),

t_r = length of record,

U = wind speed,

    U_L = wind speed over land,

U_t = effective wind speed representative of duration t,
U_w = wind speed over the water,
U_z = wind speed at z m above the ground or water surface,
U_1 = maximum hourly average wind speed over duration t,
U_{10} = wind speed at 10 m above the ground or water surface,
= depth-integrated velocity in the x-direction,
= mass transport velocity,
$\quad U_o$ = mass transport in deep water,
$\quad U_B$ = mass transport velocity at the bottom,
u = horizontal component of wave orbital velocity,
$\quad u_B$ = u at the bottom,
$\quad \hat{u}$ = maximum value u,
= sediment transport variable $(=[y/\sqrt{(4\underline{D}t)}]$,
V = depth integrated velocity in the y-direction,
V_L = longshore current velocity,
v = velocity in the y-direction,
W = reduced variate in the Weibull distribution,
W_n = width of nourishment,
w = vertical component of wave orbital velocity,
$\quad \hat{w}$ = maximum value of w,
w_f = fall velocity,
X = transformed x-axis,
x = horizontal direction (variously defined),
= direction over which storm surge is calculated,
= horizontal distance parallel to direction of wave propagation (2),
= cross-shore horizontal direction offshore of the still water line,
$\quad x_c$ = distance from still water line to closure depth,
$\quad x_i$ = location of shoreline,
$\quad x_i^*$ = new location of shoreline,
$\quad x_s$ = shoreline location against structure,
$\quad x_o$ = shoreline location at a structure,
Y = transformed y-axis
y = horizontal direction perpendicular to the x-direction,
= alongshore horizontal direction,
y_o = parameter used to calculate wave speed $(=2\pi d/Lo)$,
Z = standard normal variate $(=\Phi^{-1}(P))$,
Z_r = number of standard deviations that R is removed from its mean,
Z_s = number of standard deviations that S is removed from its mean,
z = upward vertical direction (datum is variously defined),
= vertical distance above still water level (2),

		= instantaneous water surface above an arbitrary datum (3),
z_a	=	vertical movement of accelerometer,
z_m	=	vertical movement of vibrating mass in accelerometer,
z_r	=	relative movement (z_a-z_m),
α	=	parameter in Weibull distribution; (4),
	=	Spectrum constant (3),

    α_P = Phillips constant,
    α_{PM} = Pearson Moskowitz constant,
    α_J = Jonswap constant,
    α_M = Mitsuyasu constant,

	=	angle of wave incidence with respect to the x-axis,

    α_b = angle of wave incidence at breaking,
    α_{bd} = diffracted breaking wave angle,
    α_e = effective angle,
    α_m = morphology wave angle,
    α_S = angle with respect to a structure,
    α_s = angle of the shoreline,
    α_o = angle of wave incidence in deep water,

α_i	=	phase of tidal constituent,
α_Φ	=	skewness of Φ grain size distribution,
β	=	parameter in Pierson-Moskowitz spectrum,
	=	parameter in Weibull and Gumbel distributions,
	=	wind direction change,
	=	reliability index,
Γ	=	Gamma function,
	=	(global) factor of safety (=$\gamma_r \gamma_s$),
Γ'	=	factor of safety used in calculations with uncertainties,
γ	=	Euler's constant = .5772 ...(2),
	=	overshoot parameter in the Jonswap spectrum (3),
	=	parameter in Weibull and Gumbel distributions; (4),
	=	partial design coefficients (8, 9),

    γ_e = partial safety coefficient for the equation,
    γ_{ice} = partial safety coefficient for ice forces,
    γ_r = performance factor; partial safety coefficient for resistance,
    γ_{RHS} = partial coefficient for the right hand side of the equation,
    γ_s = load factor, partial safety coefficient for loading,
    γ_O = the partial safety coefficient for overturning,
    γ_S = partial safety coefficient for sliding,

γ_w = partial load factor for waves,

γ_b = breaker index $(=H_b/d_b)$,

γ_{sb} = breaker index for significant wave,

Δ = vector operator,

Δ_a = relative underwater density of the armor $\{(=\rho_a-\rho)/\rho\}$,

Δ_H = mean wave level – mean level between wave crest and trough,

Δf = smallest frequency in a wave record of length t_R $(=1/t_R)$,

= frequency increment,

Δh = barometric pressure surge,

Δp = change in atmospheric pressure,

Δt = sampling interval,

= time step,

Δx = distance step in the x- direction,

Δy = distance step in the y- direction,

δ = Jonswap spectrum parameter;

ε = spectrum bandwidth parameter,

η = instantaneous water surface elevation above SWL (2),

= instantaneous water surface elevation above mean water level (3),

η_T = tidal water level,

θ = wave direction in directional spectrum (3),

= wind direction (5),

= seaward slope of a rubble mound breakwater (8, 9),

= wave direction with respect to shadow line (14),

θ_n = phase spectrum

= random phase angle in random phase model,

λ = number of events per year;

μ = mean value,

μ_g = mean value of the failure condition,

μ_r = mean value of the resistance,

μ_s = mean value of the loading,

$\mu[\ ]$ = expected value (mean value),

= dynamic viscosity of the fluid,

ν = kinematic viscosity of the fluid $(=\mu/\rho)$,

ξ = surf similarity parameter,

ξ_m = surf similarity parameter related to mean wave period,

ξ_p = surf similarity parameter related to the peak wave period,

ρ = density of water,

ρ_a = density of armor units,

ρ_s = density of sediment,

σ =   standard deviation,

    σ_g = standard deviation of the failure function,

    σ_r = standard deviation of the resistance,

    σ_s = standard deviation of the loading,

  =   standard deviation of water surface position,

    σ_f = σ determined from frequency analysis,

    σ_z = σ determined from zero crossing analysis,

    σ_ω = σ calculated using frequency ω,

  =   maximum allowable soil pressure on a sandy bottom,

    σ_b = σ under a structure of width B_v,

    σ_c = σ on the column of soil under a vertical breakwater,

    σ_d = additional allowable pressure due to surcharge,

    σ_u = underwater maximum allowable soil pressure,

σ' =   uncertainty (coefficient of variation),

    σ_H' = σ' for wave height,

    σ_r' = σ' for the resistance

    σ_s' = σ' for the loading,

    σ_T' = σ' for wave period,

    σ_α' = σ' for wave angle,

σ_Φ =   standard deviation for Φ grain size distribution,

τ =   shear stress,

Φ =   cumulative standard normal probability (calculates Φ from.z),

  =   grain size parameter [$=-\log_2(D)$],

  =   wave spectrum parameter

    Φ_{PM} = Pierson Moskowitz filter,

    Φ_P = Phillips function,

    Φ_J = Jonswap enhancement function; developing seas filter,

    Φ_d = depth limitation function,

Φ^{-1} =   inverse calculation of Φ (calculates z from Φ),

ϕ =   angle between the wind direction and the x-axis,

  =   parameter in implicit finite difference method,

ω =   wave angular frequency ($=2\pi/T$),

ω_i =   angular frequency of a tidal constituent,

ω_{ME} =   angular velocity of the earth-moon system,

ζ =   constant in surge calculation ($=3.2 \times 10^{-6}$),

ζ_i =   design parameter for vertical breakwaters,

1. Introduction

1.1 Introduction

Coastal Engineering and management are very old and at the same time very new professions. They have a long history, leading to high sophistication in more developed areas of the world. Yet they are virtually non-existent in newly developing countries. Historically, humans have always wanted to protect themselves from flooding to the extent that their tools permitted. Peoples living in the estuaries and deltas of the world's rivers, in particular, faced difficult coastal management problems, as history of Middle Eastern civilizations shows. They lived on land with little vertical relief that needed periodic flooding by the river water in order for the soil to remain fertile and for crops to grow. Yet major floods resulting from storm-generated, high water levels and waves threatened life and limb.

Herein lies the contradiction that is the basis for our work. How can you live near the coast, take advantage of its great abundance and yet survive? In the case of our ancestors: How could they encourage and experience minor floods, necessary for survival, while not being killed by major floods?

Flooding and its consequences have been dealt with in many ingenious ways. One common solution was to construct high areas to which the people could flee in case of flooding. Pliny in 47 AD already describes such Dutch *terps* or mounds, of which eventually over 1200 were built. The construction of such safe areas was a major feat in coastal engineering, but imagine trying to prevent the waves from eroding such a safe area. With no mechanical earth moving equipment, the physical size of such safe areas was small. Any erosion by floodwaters and waves of such a limited area would be dangerous. There was also no rock available in delta areas to serve as a hard perimeter protection around the outside of such a mound.

1

Simple methods of providing safe areas are still common in developing countries where scarce resources are channeled toward production of basic foodstuffs necessary for survival, rather than toward esoteric coastal protection structures. Yet, in highly populated, low-lying deltas, such safe areas are often too small and too difficult to reach in time for large numbers of people, resulting in periodic disasters involving the drowning of hundreds or even thousands of people.

More elaborate means are used in countries where greater economic resources are available for personal safety. The Netherlands, for example, uses every type of protection to prevent possible flooding of 2/3 of that country. Driving through the flat countryside there, it is still possible to see the old safe mounds. These usually have a church on it, which served as shelter and the only *pointe fixe* in an otherwise endless area of wetland and water. Further toward the sea, there are dikes, seawalls and revetments (structures built parallel to shore), groins (structures perpendicular to shore) and immense masses of sand, artificially placed against the shore by large dredges to protect the hinterland by extensive beach-dune systems. Yet, in spite of such investments in coastal protection, the basic conflict remains. As recently as 1953, the sea won another battle in the war for control of the Dutch shore zone when a combination of waves, high tides and high water levels swept up by very strong winds (storm surge) created very extensive damage and cost 1835 lives.

Another example of the precariousness of the coastal zone is the barrier island system along the East Coast of the United States and the Gulf of Mexico. Some of these islands are only a meter or so above high water. The waves, winds and tides move the sand from the seaward side of the islands to the backside, eroding the seaside, accreting the backside and literally rolling those islands toward the mainland. Even extensive coastal protection will not keep these islands in place. Structures can only provide short-term protection for relatively calm conditions, but under severe conditions the sea wins another battle and people on these low-lying islands must evacuate to safer areas. The residents do not run to locally built mounds of earth in this case, but drive to higher ground along congested roads.

Coastal engineering and management engages the sea in a war over control of the shore zone. *We* can win a few battles, but *the sea* will also to win some. The conflict exists at every land-sea interface, regardless of social, political or economic conditions. Some people suggest that we should not protect against the sea, but that is not practical. The lives and livelihood of millions of people depend on the safe use of the shore zone for production of food, transportation by land and water, accommodation and recreation.

With the present interest in the environment, many of the weapons used in the past to fight this war are now considered inappropriate. For example, hard shore protection structures, such as groins and seawalls (Ch. 15) are in many cases not considered acceptable. We can, however, implement more environmentally friendly solutions such as artificial nourishment with sand, retreat to more defensible shoreline positions and natural shore systems such as wetlands, mangroves and fallen trees.

Coastal management traditionally involved providing adequate and safe transportation facilities, and will continue to be involved in design and construction of harbors and marinas. However, modern coastal management involves much more than transportation and protection from the sea. Issues such as water quality, dispersion of pollutants and the proper management of the complete coastal ecosystem have become important. In fact, the actual design of shore structures is now only a small aspect of coastal management.

The present chapter presents an overview. It is largely philosophical and sets the stage for the other chapters, which will deal with specific aspects of our mission as coastal managers or engineers. First of all, we need some definitions for coastal management and engineering. Historically the two concepts were synonymous. Management of the coast was provision of safety and military advantage, mainly through building engineered structures. It is only recently that the two are viewed separately. Management involves such concepts as guidance, control, steering and stewardship. Coastal management is essentially the management of conflicting uses of highly populated coastal areas (Ch. 10).

Coastal engineering, on the other hand involves design and centers on three keywords: synthesis, simplification and systems.

1.2 Synthesis

Most technical papers and lectures related to coastal engineering deal with the scientific appraisal of the coastal zone. They explain what goes on in this very complex region. Such explanations present an *analysis* of the physical phenomena. An engineer must solve a particular coastal problem by *synthesis* of many such scientific concepts and available data. Even a minor, small-scale coastal design involves the simultaneous consideration of different physical phenomena. For example: consider a storm water drainage pipe that is periodically blocked by beach sand after sustained wave action. To propose a solution to this relatively simple problem involves many management considerations about the environment, social issues, etc. For the design aspects alone, it is necessary to put together (synthesize) at least the items in Table 1.1. The terms in Table 1.1 will be explained in later chapters. The point here is that there are many facets to even simple coastal design.

Table 1.1   Design Considerations for a Simple Design

1.   Wind Climate (speed and direction)
       - long-term statistical data
       - data for major storms

2.   Wave Climate
       - long-term statistical data
       - data for major storms

3.   Water Levels
       - tides
       - surges
       - fluctuations (annual, decades and long-term)

4.   Wave Transformation
       - wave shoaling
       - wave refraction
       - wave diffraction
       - wave breaking
       - wave reflection
       - wave attenuation

5.   Beach Parameters
       - profiles and profile variability
       - grain sizes and distributions

6.   Currents
7.   Ice

8.   Sediment Transport Relationships
       - alongshore
       - cross-shore

9.   Wave Forces on Structures
10.   Diffusion and Dispersion
11.   Environmental Impact

1.3 Simplification

Science and engineering research define many of the concepts in Table 1.1 in detail and with considerable accuracy. For instance, the wave climate can be expressed by directional wave spectra - a sophisticated tool that describes wave energy as a function of wave frequency and wave direction (Ch. 3). Large and complex computer programs are available for detailed computation of wave transformation by refraction, diffraction, attenuation and reflection (Ch. 7). Complex theories exist to compute sediment transport (Ch. 11, 12 and 13). Diffusion can also be calculated with mathematical models. However, even the best representations are simplifications of reality.

The large number of items in Table 1.1 that need to be taken into account necessarily leads to simplification. This is true for major projects. Their design involves large and costly structures, possible loss of life and large potential for damage if failure occurs. Even though we use the most sophisticated design methods available, there will be always be simplifications. What about small projects? Do we need (or even want) the most sophisticated information to re-design a storm water drain? For most engineering designs, relatively simple expressions for the various concepts of Table 1.1 are sufficient and indeed preferable to provide solutions within a budget. This represents further simplification. The bulk of this text focuses specifically on the simpler concepts. These are the concepts normally used and judged to be sufficient to accomplish most studies and routine designs.

1.4 Systems

The third keyword, systems, recalls ideas such as system analysis, system design, etc. A system is may be defined as all the inputs, outputs and interactions that affect a physical process or event. The boundaries of the system are defined so that the number of input-outputs is a minimum. Because some input-output always takes place across the system boundaries, it is normally necessary to visualize a hierarchy of systems, nested within each other. Thus the system for a blocked storm water outfall may be a part of a larger system (a section of beach). This system may contain within it several sub-systems such as nearshore circulation cells, etc., while being part of an even larger system, such as the California Coast, Lake Baikal or the Bay of Fundy, which in turn is part of ...

Appropriate design considers only as many systems as necessary. To arrive at a technically and environmentally satisfactory design, we must understand each element and its interactions, the inputs and outputs, and how they affect the system and neighboring systems.

Consider sediment transport. Figure 1.1 indicates some of the inputs, outputs and elements of the simplest of coastal zone sub-systems, a short beach section between two structures, placed more or less perpendicular to the shore. Note that the system boundaries are drawn far enough seaward and landward as well as along the two structures for the input-output to be a minimum. This sub-system may be adjacent to similar sub-systems or to totally different sub-systems such as a small tidal inlet, a harbor entrance, etc. (Fig. 1.2).

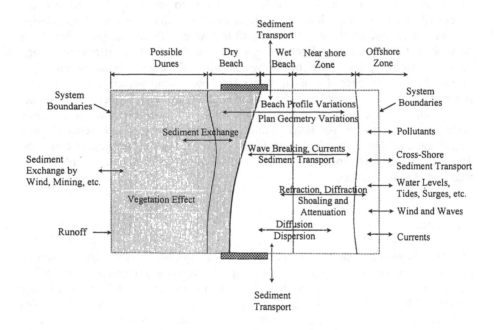

Figure 1.1 Simple Coastal Subsystem

If sufficient input-output takes place across the sub-system boundaries to cause measurable changes to adjacent systems, the sub-system must be considered part of the system in Fig. 1.2. Thus, any system is part of a larger system if it is not completely self-contained and interacts with adjacent systems. In engineering design it is necessary to consider all systems that affect the design and all the systems that are affected by the design. Thus for coastal sediment transport or morphology in the sub-system of Fig. 1.1, it is clearly necessary to take into account the complete system in Fig. 1.2. But should any super-system that encompasses the system of Fig. 1.2 and other similar systems be considered? That depends on the flows of water and sediment across the landward and seaward boundaries and past the two headlands of the system in Fig. 1.2. When the headlands contain virtually all sediment and when no river flow or sediment flow can be diverted to adjacent systems, Fig. 1.2 can be considered as a complete and isolated coastal system (littoral cell).

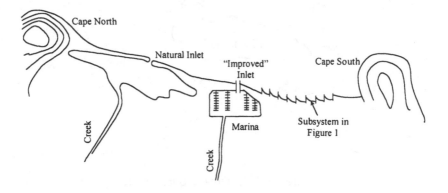

Figure 1.2 Coastal System

In the past, mistakes were made by not considering the proper system boundaries or by not considering a super-system, when necessary. For example, the origin of the sediment along the California coast is mainly the sediment brought to the coast by major rivers. When large power and water-supply dams were built along these rivers, the authorities did not take into account that the sand trapped behind the dams should continue to travel downstream to feed the California beaches. This mistake is perhaps understandable, since the dams are many hundreds of kilometers from the beaches. Power generation and water supply were of great economic concern at the time, while coasts and beaches were merely of recreational value and not as important. Yet this mistake resulted in a large negative impact on the health of the California coast.

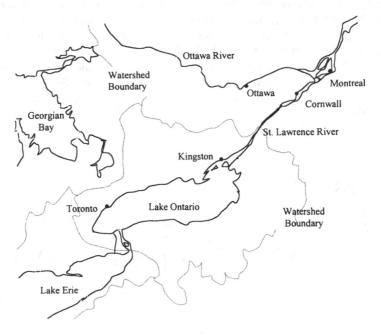

Figure 1.3 Lake Ontario Super-System

As a second example, coastal erosion along Lake Ontario is integrally connected to the water levels in the lake. These water levels are a function of the supply of water from the Lake Ontario watershed, the upstream inflow from Lake Erie and the downstream discharge through the St Lawrence River. The Ottawa River is never closer than 150 km to Lake Ontario (Fig. 1.3) and has no apparent connection to Lake Ontario. However, the combined discharge of the St. Lawrence and the Ottawa River systems determine flooding or low water levels in Montreal. Thus when there is a high runoff on the Ottawa River, the outlet control for the St. Lawrence-Lake Ontario system, located at Cornwall, can only pass a relatively small discharge down the St. Lawrence toward Montreal. A number of consecutive wet years will affect both the lower Ottawa River and the Great Lakes – St Lawrence drainage basins. At such times the high flood peaks on the Ottawa River will prevent adequate drainage of the floodwaters from Lake Ontario. The St. Lawrence River is then backed up, raising the water levels along the Lake Ontario shores excessively, causing unusual flooding, resulting in extensive shore erosion.

Similarly, several dry years will result in extreme low waters on Lake Ontario and low flows along the Ottawa River. In order to keep water levels in Montreal high enough for shipping purposes, extra water must be supplied from the lake at such times. Thus the super-system for coastal design in Lake Ontario includes the regulation of the Ottawa River drainage basin, many kilometers away and seemingly not connected to the St. Lawrence river system. And that is only the physical part of the story. Added to this are many management issues, such as conflicting social and economic interests of navigation, power generation, agriculture and recreation; political boundaries and jurisdictions, and private property ownership.

The system to be considered for any design is not absolute by definition. The system in Fig. 1.2 may be a self-contained system with respect to sediment transport but pollutants can easily pass by the headlands. Ecology and habitat normally need to consider larger systems than fluid mechanics and sediment transport. This leads to the definition of the ecological system (ecosystem). It is the system that fully contains the ecological problem under consideration. In theory, one would need to consider the complete universe, or at least the whole earth; but for engineering design, practical limitations govern the choice of the systems.

An example of an ecosystem is the Greater Toronto Bio-region (Crombie, 1992), shown in Fig. 1.4. Any ecologically sensitive design within this region must take into account the complete bio-region which extends from the open waters of Lake Ontario to the watershed at the Niagara Escarpment and the Oak Ridges Moraine. Any animal living anywhere in the system may depend, for its food and habitat, on plant and animal life throughout the whole watershed. Even that is not sufficient. There are other inputs and outputs to be considered, such as air-borne transportation of nutrients and pollutants. We need also be concerned, for example, with birds that migrate 10,000 km.

1.5 Jargon and Terminology

Engineers and managers are often accused of using jargon (yield strength, shear stress, water quality, zoning, wave spectra). Indeed we spend much time learning these concepts, so that we all understand each other and so that a particular definition means the same to everyone, even though such a definition may itself be quite arbitrary. Such expressions are really *terminology*. Most disciplines, such as geomorphologists, dentists and airline pilots understand each other through different, but equally necessary and important terminology.

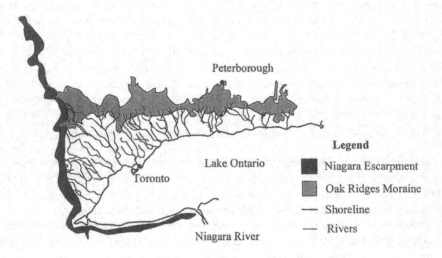

Figure 1.4 Greater Toronto Bio-Region (after Crombie, 1992)

There is also jargon - terms invented without much reason or necessity. Among those we find pretentious terms such as "subaqueous and subaerial", simply meaning below or above water. There are many terms that are not as straightforward and mean something different for each user. Such unclear, undefined or unnecessary terms, which are nevertheless in common usage, are *jargon*. This tendency to invent words is especially prevalent in the relatively new stream of environmental consciousness. And so we need to deal with terms such as ecodesign, bio-region and even the word environment itself, that have crept into daily usage, without having a single, clear definition. As managers and engineers we need to work closely with people from different backgrounds and therefore it is certainly necessary for us to understand their terminology. When they resort to jargon, we need to determine what the particular person means that specific term.

1.6 Engineering Time

Coastal engineering (and management) are normally based on considerations that seldom involve less than one year or more than a few hundred years. Thus engineering time lies somewhere between the monthly bills and geologic time. A thorough understanding of the origin of our shorelines is essential to the proper execution of our work. A detailed review of the Quaternary (last three million or so years) history of the site may be needed, but certainly the Holocene period of the last 10,000 years must be understood. This is particularly true in areas where much of the local geomorphological history depends heavily on recent glaciation. The associated crustal uplift and the extensive accretion, erosion and transportation of sediment are very important considerations in any design. Yet, all the research by geologists and geomorphologists can seldom define details over the last 1000 years that are desperately needed for proper design.

At the other end of the knowledge spectrum lies short-term research. The literature contains many papers about: "Look what happened to my beach during my test period", where "test period" could be anything from a few hours to one summer because that's nice and warm and coincides with research-intensive time at universities. Watch out for conclusions drawn from such tests.

Historical design input relating to engineering time is not easy to find. Beyond the published literature, we depend on hydrographic and topographic charts, maps, air photos, archival research, old surveys, photographs and sketches. And we must listen to the locals as they give us their (non-technical) version of the processes. They are the ones who have actually seen the site under stress by waves, currents, ice and water level fluctuations. There is a large difference in the experience of someone who lived through a flood or hurricane, and of an engineer who has never actually encountered such a disaster but learned about it from the literature. These two knowledge bases are complementary, however, and should both be used.

1.7 Handy References

To review coastal engineering and management beyond this introductory text, there is the ubiquitous Shore Protection Manual (CERC, 1984) prepared by and for the U.S. Corps of Engineers, the agency responsible for all coastal work in the United States. The manual has gone through a number of editions since it first appeared as "Shore Protection, Planning and Design" in the 1950s. It contains a wealth of experience, and packages recent research materials in usable form for managers and engineers. This publication is being re-written to bring it into the 21st Century, integrating text with software, etc. Shore Protection Manual and some other "manuals" have the appearance of being design codes, but they are not! Because of their apparent simplicity, such manuals also tend to make coastal engineers out of people with little or no previous coastal experience. Any engineer, upon discovering such a manual, can become (over)confident about this apparently "simple subject".

Table 1.2   Partial Reference List

Coastal Engineering:
- Abbott and Price (1994), *"Estuarial and Harbour Engineers Reference Book"*,
- CERC (1984), *"Shore Protection Manual"*,
- Dean and Dalrymple (1984), *"Water Wave Mechanics for Engineers"*,
- Fredsoe and Deigaard (1992), *"Mechanics of Coastal Sediment Transport"*
- Goda (1985), *"Random Seas and Design of Maritime Structures"*,
- Herbich (1989), *"Handbook of Coastal and Ocean Engineering"*,
- Horikawa (1978), *"Coastal Engineering"*,
- Horikawa (1988), *"Nearshore Dynamics of Coastal Processes"*,
- Ippen (1966), *"Coastline and Estuary Hydrodynamics"*,
- Komar (1983), *"Handbook of Coastal Processes and Erosion"*,
- Le Méhauté and Hanes (1990), *"The Sea"*,
- Muir Wood and Fleming (1981), *"Coastal Hydraulics"*,
- Sawaragi (1995), *"Coastal Engineering, Waves, Beaches and Wave-Structure Interaction"*,
- Silvester (1974), *"Coastal Engineering"*,
- Sorensen (1993), *"Basic Wave Mechanics for Coastal and Ocean Engineers"*,
- Wiegel (1964), *"Oceanographical Engineering"*.

Geomorphology:
- Bird (1984), *"Coasts" (3rd Ed.)*,
- Bird (1985), *"Coastline Changes, a Global Review"*,
- Bird (1993), *"Submerging Coasts : the Effects of a Rising Sea Level on Coastal environments"*, ",
- King (1972), *"Beaches and Coasts"*,
- Komar (1998). *"Beach Processes and Sedimentation" (2nd Ed.)*,

Management:
- Carter (1988), *"Coastal Environments"*,

Table 1.2 gives a partial list of reference texts. In general, these publications are more advanced than the present text, but you will find it helpful to browse through them. The complete references are given at the back of the book. The list only shows one reference under management, because little has been written to address coastal management specifically. The other references, however, contain much information relevant to coastal management. A delightful paperback book by Bascom (1964) entitled *Waves and Beaches* is probably the most palatable introduction to the subjects. The most notable technical journals on coastal engineering are the *Journal of the Waterways, Port, Coastal and Ocean Division of the American Society of Civil Engineers (ASCE)* and *Coastal Engineering,* a journal published by Elsevier. Other publications, relevant to both engineering and management are *Journal of Coastal Research, Journal of Geophysical Research* and *Shore and Beach.* Finally, many conferences are held on the subject. The foremost of these is the *International Conference on Coastal Engineering (ICCE)* held every two years under the auspices of ASCE. More specifically for managers is the *Coastal Zone* series of conferences held by ASCE. A number of specialty conferences are also organized and published, mostly by ASCE, the Institution of Civil Engineers (UK).

1.8 Data Requirements

It is obvious that for coastal design and management we need data that fit the requirements of our design (synthesis) and the concepts of simplification, systems and engineering time. Most coastal data are difficult to measure, which means that they contain large inherent uncertainties. Such uncertainties mean that even the best designs and solutions will be approximate. Uncertainties are discussed throughout this text. There are ten basic data sets that are required for almost all coastal designs and these are summarized in Table 1.3.

For designs involving structural stability, it is necessary to obtain short-term wave data, particularly about extreme conditions. Measured wave records are normally analyzed as wave spectra - distributions of wave energy with frequency and possibly direction (Ch. 3). Long-term distributions of wave heights, periods and directions are important to determine long-term evolution of coastal processes and extreme values to design stability of structures (Ch. 4). To obtain measured long-term values requires continuous operation of rugged, but highly sophisticated equipment. (Ch. 13), but that is expensive and uses a relatively new technology. Hence, long term data bases for most project designs are obtained by hindcasting. They are calculated from previously recorded wind observations (Ch.5). Projects involving water quality and habitat require both long-term and short-term wave data.

Table 1.3 Data Requirements

---

1. **Wave Data**
   – Short-term wave spectra (measured or hindcast)
   – Long-term distributions of wave height, period and direction (usually hindcast)
2. **Meteorological Data**
   – Wind (speed, direction and frequency of occurrence)
   – Barometric pressure
   – Storms (tracks, frequencies)
   – Extreme values
3. **Water Level Data**
   – Tides, seiche and storm surge
   – Seasonal and annual fluctuations
   – Longer term fluctuations (decades)
   – Fluctuations on a geologic time scale (sea level rise, isostatic rebound)
   – Water level fluctuations due to climate change
4. **Current Data**
   – Tidal, wind-driven and wave-driven currents
5. **Hydrographic Data**
   – Sufficient resolution in time and space
   – Above water, through the breaker zone and in deeper water
6. **Sediment Transport and Morphology Data**
   – Rates
   – Directions
   – Erosion - accretion
7. **Environmental Data**
   – Water quality
   – Habitat
8. **Sociological Data**
   – Land use
   – Economic impact
9. **Historical Data**
   – Extreme water levels (high and low)
   – Major erosion and accretion events
   – Old charts and paintings, maps, photographs and air photos.
10. **Materials Data**
   – Availability, quality and cost

Meteorological data such as barometric pressure variations, frequency, size and tracks of weather systems, and wind velocities and directions are valuable for wave hindcasting. They also define variations in water level by seiches, storm surge, etc. (Ch. 6). Very large depressions and hurricanes delineate maximum values of stress.

Normally wind speed and direction are recorded hourly. Directional accuracy is usually not very good. Ideally the wind data should be collected close to the site and over the water, but wind data are most often collected on land, at airports. This means measured wind speeds and directions need to be corrected, since over-water wind speeds may be up to 50% stronger and their direction may be up to 30° different from the overland values. The quality of wave hindcasts deteriorates further if geographic features such as hills, rivers or lakes exist that cause the winds to lift or funnel. Particularly local thermal breezes modify the general wind climate substantially. They may generate waves that occur consistently over the water for much of the open-water season.

The end result of a wave hindcast, using time series of crudely corrected hourly wind data, is an hourly time series of crudely calculated wave heights, periods, directions. Unless sophisticated wave hindcast techniques are used, wave direction is usually assumed to be the same as the wind direction. Such retrospective time series of waves may be very different from what actually took place, because of the approximations involved in the computation and hence all hindcasts must be carefully *calibrated* with available measured wave data. Even sophisticated hindcast procedures need at least some measured waves as an essential input into any design. If good measured wave data are not available for calibration and the hindcast is based on "experience" alone, it must be treated with caution.

Structures and beach-dune systems are especially susceptible to damage by high water levels. Ecosystems units such as wetlands may even be severely affected by small fluctuations in water levels. Thus historical and predicted water level fluctuations are important data to be included in any coastal design. Information on seiches, storm surges and tidal fluctuations is needed. On lakes and reservoirs, extensive data on lake levels are needed (Power and water supply reservoirs can fluctuate tens of meters regularly and over a short time). Geological estimates of past water levels, rates of sea level rise, isostatic rebound (crustal movement resulting from the release of ice pressure that occurred during the latest glaciation) and rates of land emergence and submergence should also be collected. Future water levels should be carefully estimated, especially in the light of long-term sea level rise and global warming (Ch. 6).

Currents are also an important aspect of many designs, particularly those involving the environment, water quality and habitat. Currents are very site-specific, which means that an on-site current measurement program should be an integral part of the data collection for a site where currents are important. Remote sensing, using radar and other imagery obtained by aircraft or satellites, will improve future data quality and coverage, particularly for large current patterns.

Offshore hydrographic data should be accurate enough to permit the correct calculation of wave transformation (refraction, diffraction and shoaling) for the site (Ch. 7). Inshore hydrographic data should be able to define beach profiles, longshore bars, sand waves, etc. Offshore hydrographic data are available from charts and field sheets, which are usually of sufficient quality. The inshore hydrographic data are the most difficult to obtain. It is easy where or when the water is warm and the wave action is limited. A survey crew can wade into the water to a depth of about 1.5 m and a small boat can survey in water as shallow as 1 m. However, when the water is cold or the waves are high, this is not possible and thus inshore hydrographic data often displays a gap in the observations for cold water, stormy seasons. A complicating factor is the frequency required for inshore hydrographic surveys for proper design. Because the inshore is very dynamic - inshore profiles can change rapidly - it is difficult to understand many beach-related problems without frequent inshore hydrographic surveys. Remote sensing and modern survey methods will bring about some future improvement with bathymetry data, but also these methods have difficulty measuring near the sea-land interface.

Sediment transport and coastal morphology data are important, but difficult to measure or calculate with any accuracy (Ch. 11 and 13). Environmental parameters such as water quality and habitat are also difficult to ascertain, and they can change rapidly and unpredictably. Yet permits to proceed with a project often depend on being able to demonstrate no-net-loss of habitat or biomass, which means that extensive base data need to be collected, prior to construction.

Socio-economic data are also important. It makes no sense, for example, to consider re-constructing a storm water outfall in the middle of a tourist beach or near a hotel. At many locations, historic socio-economic activity may have resulted in a major distortion of the physical environment. Carefully managed and landscaped shorelines, for example, may originally have been mangroves. Major subsidence of the landmass may be the result of pumping of water or natural gas from below the shore zone.

Data on historical development of a site are needed. Understanding the future requires a thorough understanding of the past. The data will focus on extreme events and historical changes. Historical data become especially important when calibrating models that simulate long-term coastal development (Ch. 13 and 14). The required data are not on a geological time scale, but concern developments over the past few hundred years or so, on an engineering time scale. Local residents can often supply such data. Care should be exercised in using such data, but biases can often be removed by talking with many people and sorting through many old plans, documents and photographs. In some cases, one may be fortunate enough to find historical surveys, since most areas of erosion were likely to be eroding also 20 or 50 years ago. Many excellent coastal surveys have been performed since 1800. Historical development may also be pieced together from air photos. Many areas have been flown at regular intervals since 1930. Finally, old charts, maps and even paintings and personal diaries are helpful, if interpreted carefully. In addition to the history of the physical environment, it is usually necessary to construct an environmental and socio-economic history of the site so that proper management decisions can be made against those backgrounds.

Finally, information on availability, quality and cost of necessary construction materials is essential. The acquisition cost of rock, concrete, nourishment sand and many other required materials can easily vary by a factor ten. Cost depends on many factors, such as availability, location of quarries, available transportation infrastructure, necessary transportation routes and loading restrictions, cost of on-site stockpiling, proximity of dredging equipment and relation of the project to other projects requiring the same materials at the same time. Finally, for many projects, cost is a function of the ecological windows of opportunity that determine when quarrying, dredging and construction can take place. These windows involve weather patterns and wave action, but also habitat and spawning and nesting periods for animals such as fish, turtles and birds.

1.9 Coastal Design

To provide some perspective, we present some ideas about the coastal design process in this introductory chapter. Coastal engineering is a field for which there are *no design codes*. Some standard procedures exist but solutions are generally site specific. Thus every project becomes a unique challenge. Input conditions cannot be defined with sufficient accuracy and the "strength of materials" is uncertain. Therefore, normal design, as one might design a bridge for example, is not possible. Design by full-scale trial and error is socially and economically also not acceptable. Hence, coastal projects are normally designed using *models*, which are essentially trial and error tools (Ch. 13). The two basic types of models are physical (or hydraulic) models and numerical (or computer) models.

The ideal coastal design procedure is shown in Fig. 13.1. Design must always be guided by extensive coastal *knowledge*, which is the combination of theory and experience. Theory can be obtained from textbooks such as this. Experience is more difficult to obtain. It is certainly not foolhardy application of handbook methods by someone with little real prior experience in the coastal environment. Yet, many firms who have done little previous coastal work feel quite confident to produce coastal designs. They assume that experience from other fields is transferable. But most designs in other fields are based on codes with specified requirements for loading, strength, probabilities of failure, and associated risks, based on many years of experience and research. For the coastal zone there are no such codes. Its formal history is too short and its projects are too varied. It is too seldom that standard practice can be repeated, as when one designs a building. There is no accepted practice, a few standard formulas and no standard solutions. That is why experience is such an important ingredient of the coastal design process.

Local experience is vital. Only a person who lives at the site believes the required design wave heights, sizes of structures, possible forces, the potential for sediment transport or the rapid rates of erosion that are possible in the shore zone. Anyone who has not actually experienced a coastal site subjected to large storm waves, or high flood levels finds it difficult to believe the magnitude of the forces, the possible damage and the necessity for apparently monumental structures. Few people have such experience, since major storms and flood events may be separated by decades, even centuries.

Without the combination of extensive on-site experience and general coastal engineering experience, there is a tendency to reduce design values because:
−   the resulting design is judged to be too conservative (based on "experience" in other related fields),
−   the authors of handbooks and textbooks are expected to be conservative,
−   the resulting design is too costly,
−   the resulting structure is not esthetically appealing.

There are other pitfalls with experience. Glossy, simplistic pamphlets explain coastal processes, how to protect property, etc. These can lead even professionals to think coastal design is simple. A number of inexpensive so-called "solutions" to coastal problems are always on the market. Some are said to have been "thoroughly tested". Such testing has often been done in other environments such as along rivers and then simply transferred to the (much more hostile) coastal environment. Others are said to have been "proven to collect beach material", or "have stood up for a number of years to hostile coastal environments". Investigate such claims carefully. For example, almost any coastal structure is stable and can collect sand along the Great Lakes when the lake levels are on a long-term falling trend, even when it has been subjected to adverse wave conditions.

1.10 Concluding Remarks

You have read a bit of philosophy about synthesis, systems, simplification, terminology, time, etc. You have seen some basic data requirements and we have introduced you to coastal design. You have also been made aware of some of the limitations of our trade and that experience with coastal environments is very important in a design process where there is little in the way of design guidelines or codes. What else can we say by way of introduction? As you work through this text, you will hope for more accurate methods and "textbook". There are none!

Disappointing? Of course not! Coastal management and engineering are exciting because:
- much research is still needed, even on the very basics,
- large doses of ingenuity, inventiveness and intuition (the original engineering skills) are required,
- you will work closely with and learn from many related disciplines, such as geologists, biologists, geographers, planners and lawyers.

Besides, in what other profession can you lie on the beach, swim, dive or cruise in a boat and tell the boss you are working?

2. Water Waves

2.1 Introduction

2.1.1 Description of Waves

Water waves are fluctuations of the water level, accompanied by local currents, accelerations and pressure fluctuations. Their simplest form is sinusoidal (Fig. 2.1) and we will use it here to define the most basic wave properties. The high water levels are the wave crests, the low levels are the wave troughs. The vertical distance between a crest and a trough is the wave height H. The distance over which the wave pattern repeats itself is the wave length L. The waves propagate with a velocity C, and the time that is required for a wave to pass a particular location is the wave period T. The inverse of the wave period is the wave frequency f.

The subject of water waves covers phenomena ranging from capillary waves that have very short wave periods (order 0.1 sec.) to tides, tsunamis (earthquake generated waves) and seiches (basin oscillations), where the wave periods are expressed in minutes or hours. Waves also vary in height from a few millimeters for capillary waves to tens of meters for the long waves. A classification by wave frequency of the various types of waves is given in Fig. 2.2. In the middle of the range are *gravity* or *wind-generated waves*. They are the focus of this chapter and have periods from 1 to 30 sec. and wave heights that are seldom greater than 10 m and mostly of the order of 1 m. They are generated by wind against the gravitational force that wants to restore the still water level. Because of their prevalence, these waves account for most of the total available wave energy.

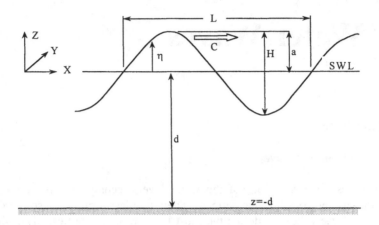

Figure 2.1 Basic Wave Nomenclature

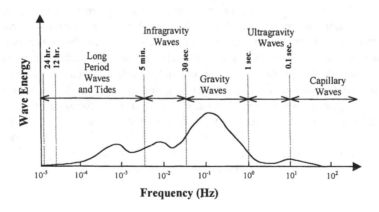

Figure 2.2 Wave Classification by Frequency (after Kinsman, 1965)

The actual shape of a water surface subjected to wind does not look like Fig. 2.1, but is so complex that it almost defies description. Even when the first puffs of wind impact a flat water surface the resulting distortions present non-linearities that make rigorous analysis impossible. When the first ripples generated by these puffs are subsequently strengthened by the wind and interact with each other, the stage has been set for what is known as a confused sea. The waves will continue to grow ever more complex and we have to make sense out of this confusion. As a result, we use simplified concepts in design. This chapter will establish a bridge from the confusing and complex sea state to theoretical expressions that are simple and can be used for design purposes.

2.1.2 Wind and Waves

For theoretical analysis of wave generation, the reader is referred to Dean and Dalrymple (1984), Dingemans (1997), Horikawa (1978), Ippen (1966), Kinsman (1965), Sarpkaya and Isaacson (1981), who discuss various theoretical models and give references to original papers. In general, wind speed and wave activity are closely related. There are other important variables to consider such as depth of water, duration of the storm and fetch (the distance the wind blows over the water to generate the waves). These will all be discussed in more detail in Ch. 5. For the moment we will look only at the effect of wind and assume water depth, wind duration and fetch are unlimited. The resulting waves are called *fully developed sea*. These conditions are approximated in the deep, open sea.

The relationship between wind and waves in the open sea is so predictable that sailors have for centuries drawn a close parallel between wind and waves. The Beaufort Scale in Table 2.1 is a formalized relationship between sea state and wind speed, and we can use it to obtain an estimate of waves in the open sea when wind speed is known. In fact, sailors have used this concept in reverse for centuries to estimate wind velocity from the sea state. Even today, windsurfers and dingy sailors know that the wind speed is 10 knots when the first whitecaps appear and the crew of a yacht knows that spindrift is formed by gale force winds. Remarkable series of photographs relating sea state to wind velocity may be found in Neumann and Pierson (1966).

Table 2.1 Beaufort Scale Of Wind And Sea State[1]

Beaufort Wind Force	Wind Speed (knots)[2]	Description of Wind	Description of Sea	Approx H_s (m)	Approx T (sec)
0	0-1	Calm	Sea is like a mirror.	0	1
1	1-3	Light airs	Ripples are formed.	0.025	2
2	4-6	Light breeze	Small wavelets. Still short but more pronounced; crests have a glassy appearance, but do not break	0.1	3
3	7-10	Gentle breeze	Large wavelets. Crests begin to break. Perhaps scattered white caps.	0.4	4
4	11-17	Moderate breeze	Small waves, becoming larger. Fairly frequent white capping.	1	5
5	17-21	Fresh breeze	Moderate waves, taking a more pronounced long form. Many white caps are formed (chance of some spray).	2	6
6	22-27	Strong breeze	Large waves begin to form. The white foam crests are more extensive everywhere (probably some spray).	4	8
7	28-33	Moderate gale	Sea heaps up and white foam from breaking waves begins to be blown in streaks along the direction of the wind (spindrift).	7	10
8	34-40	Fresh gale	Moderately high waves of greater length. Edges of crests break into spindrift. The foam is blown in well-marked streaks along the direction of the wind. Spray affects visibility.	11	13
9	41-47	Strong gale	High waves. Dense streaks of foam along the direction of the wind. Sea begins to roll. Visibility is affected.	18	16
10	48-55	Whole gale[3]	Very high waves with long overhanging crests. The resulting foam is in great patches and is blown in dense white streaks along the direction of the wind. On the whole, the surface of the sea takes a white appearance. The rolling of the sea becomes heavy and shocklike. Visibility is affected.	25	18
11	56-63	Storm[3]	Exceptionally high waves (small and medium sized ships might for a long time be lost to view behind the waves). The sea is completely covered with long white patches of foam lying along the direction of the wind. Visibility is affected.	35[4]	20[4]
12	64-71	Hurricane[3]	Air filled with foam and spray. Sea completely white with driving spray; visibility very seriously affected.	40[4]	22[4]

[1] Fully developed sea - unlimited fetch and duration.
[2] 1 knot $\cong$ 1.8 km/hr $\cong$ 0.5 m/s
[3] Required durations and fetches are seldom attained to generate fully developed sea.
[4] Really only a 30-40 m deep interface between sea and air.

2.1.3 Sea and Swell

Waves generated locally by wind are known as *sea*. It consists of waves of many different wave heights and periods as shown in the time series in Fig. 2.3. These waves propagate more or less in the wind direction. In fact, as discussed in Ch. 5, waves are formed by a crossing pattern of two wave trains propagating at a small angle away from the wind direction as shown in Fig. 2.4. Local peaks in the water level occur where the two wave trains add and lower water levels exist where they subtract, resulting in the irregular wave pattern of Fig. 2.3 at any particular location.

On large bodies of water, the waves will travel beyond the area in which they are generated. For example, waves generated by a storm off the coast of Newfoundland may travel in an easterly direction and eventually arrive in Portugal. While the waves travel such long distances, the energy of the individual waves is dissipated by internal friction and wave energy is transferred from the higher frequencies to lower frequencies. The resulting waves arriving in Portugal will be more orderly than the initial sea generated off Newfoundland, with longer wave periods (10-20 sec), smaller wave heights and more pronounced wave grouping (discussed in Fig. 2.10). Waves, generated some distance away are called *swell*. The difference between sea and swell is shown in Fig. 2.5.

On most coasts, sea and swell occur simultaneously. The exceptions are enclosed bodies of water such as lakes, reservoirs and inland seas, where swell cannot arrive from long distances away. Even large enclosed water bodies such as the Great Lakes essentially only experience locally generated sea.

Silvester (1974) discusses locations of high wind velocities and shows that there is a global wind pattern. High winds occur predominantly between 40°-60° North and South latitude. Elementary meteorology tells us that warm air flowing toward the poles and cold air flowing away from the poles meet there to form the Polar Front (Fig. 2.6). The earth's rotation then causes depression-type storms that move along this front. These storms occur throughout the year and at short intervals and wave conditions can be expected to vary from hour to hour, throughout the year and in unpredictable patterns, because this polar front shifts in North-South direction as a result of small pressure and temperature changes[1]. The polar front is, therefore, where most of the sea is generated – not including waves generated by tropical storm activity such as hurricanes or typhoons.

---

1. These shifts are a combination of short term, annual and longer term changes (such as El Niño).

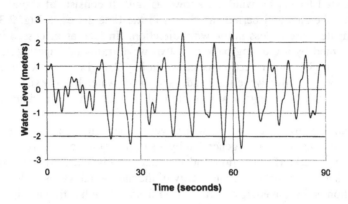

Figure 2.3 Record of Locally Generated Sea

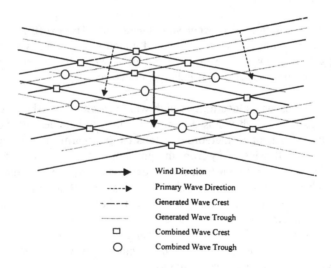

Figure 2.4 Crossing Pattern of Waves

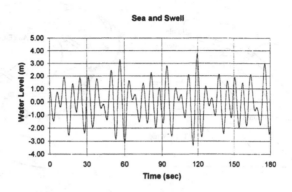

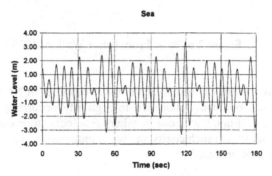

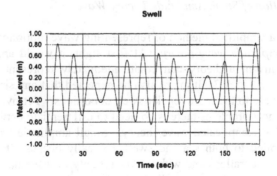

Figure 2.5 Sea and Swell Separated

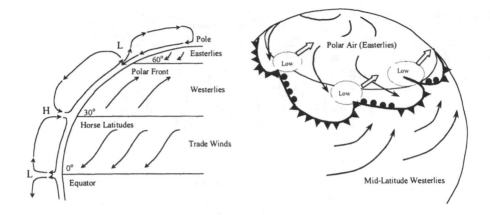

Figure 2.6 Polar Front (after Walker, 1973)

Owing to the polar front, the heavily developed coastlines of North America, Europe and Japan are predominantly subjected to regularly occurring storms and locally generated sea. Swell-dominated coasts, where few local storms occur, may be found closer to the equator. On such coasts, the wave parameters may be virtually constant for weeks or months on end.

2.1.4 Introduction of Small Amplitude Theory Wave

In this chapter a simplified method of representing wave motion will be introduced, called *small amplitude wave theory*. At first sight, it would appear to be almost impossible to invent a theory that could adequately represent locally generated confused sea as in Fig. 2.3. It might also be expected that any simple theory would be more applicable to the more regular swell conditions. Yet over the years, it was found that for most problems there is no need to differentiate between sea and swell or to use a more complicated wave theory. Small amplitude wave theory can be confidently applied to both sea and swell, basically because it is consistent with other design considerations and with the uncertainty in wave data[2]. The data upon which designs are based are normally scant, incomplete, and approximate, and the

---

2. The uncertainties in wave data and design will be discussed in detail throughout this book.

design methods with which the wave theory is used are relatively crude. More complex wave theories have been developed, but they are normally used only for research and very complex designs.

The basis for small amplitude wave theory is the sinusoidal wave, shown in Fig. 2.1. We use a right hand system of coordinates, with its origin at still water level (SWL), defined as the water surface that would exist in the absence of any wave action. The x-axis is horizontal and parallel to the direction of wave propagation and we assume that there is no variation in the y direction, perpendicular to the x axis. The z-axis is vertically up and therefore the position of the bottom is at $z = -d$ and the water surface is centered around $z = 0$. The sinusoidal water surface may be described by

$$\eta = a \cos(kx - \omega t) = a \cos\left(\frac{2\pi x}{L} - \frac{2\pi t}{T}\right) \qquad (2.1)$$

where a is the amplitude of the wave, x is distance in the direction of wave propagation, t is time, k is the wave number, ω is the angular wave frequency, L is the wave length, T is the wave period and

$$k = \frac{2\pi}{L}; \qquad \omega = \frac{2\pi}{T} \qquad (2.2)$$

The maximum vertical distance between crest and trough of the wave is the wave height, H(=2a). Since in an actual wave train, such as in Fig. 2.3, the wave heights and lengths are not all the same, more precise definitions are introduced in Ch. 3. The ratio of wave height to wave length (H/L) is called wave steepness. The wave form moves forward and the velocity of propagation

$$C = \frac{L}{T} \qquad (2.3)$$

Further terms and definitions are introduced as required in the following sections.

Mean wave level is defined as the level midway between wave crest and trough. In small amplitude wave theory (Fig. 2.1), it is the same as SWL, but for higher order wave theories it will be above SWL. Waves are described as long-crested or short-crested, which refers to the length of the wave crest in the y direction. Swell is normally long crested (the wave is recognizable as a single crest over a hundred meters or so). Sea is normally short crested, forming local peaks as shown in Fig. 2.4. Finally, waves are said to be in deep water when $d/L > 0.5$ and in shallow water when $d/L < 0.05$. Between these conditions, the water depth is called transitional. This terminology for water depth will be further explained in Section 2.2.1.

2.2 Wave Theories

Historically wind-generated waves have been described by several different theoretical developments. The most common approach uses the equations of motion and continuity for a frictionless (ideal) fluid and subjects them to the appropriate boundary conditions. The result is Stokes Wave Theory. This theory may be found in the standard texts mentioned earlier. If a further assumption is made that the wave height is infinitesimally small compared to the other defining lengths, such as wave length and water depth, the result is Small Amplitude Wave Theory, first discussed by Airy (1845). Complete developments may again be found in the above texts. Small Amplitude Wave Theory constitutes the first order of approximation of the Stokes theory (Stokes, 1847). As the wave amplitudes become larger, higher orders of approximation to the Stokes theory can be used to describe the finite amplitude waves more precisely.

There are other theoretical approaches. Rotational Wave Theory developed by Gerstner (1809) is a simple approach that is seldom used today. Cnoidal Wave Theory is derived from the Korteweg deVries equations first developed in 1895 and described in detail by Korteweg and de Vries (1985), Svendsen (1974), Isobe (1985) and others. Solitary Wave Theory, valid for very shallow water, was developed by Boussinesq (1872), McCowan (1891, 1894) and others and made popular by Munk (1949) as an approximation for waves close to breaking.

From the assumption of frictionless fluid it would appear that Stokes wave theory would be most useful in deep water. Small amplitude theory would then be best used for small waves in deep water while higher order approximations of the Stokes theory could be used for the larger waves in deep water. In shallower water we need to apply Cnoidal wave theory. It accounts for distortion of the wave shape by interference from the bottom. In the limit, as the wave reaches very shallow water and is about to break, solitary wave theory can be used. The three theoretical approaches are continuous. Cnoidal theory becomes Stokes theory in deep water and solitary theory in shallow water. The applicability of the various wave theories is summarized in Fig. 2.7, adapted from Le Méhauté(1976).

Many simplifying assumptions are necessary to develop the above closed form (or analytical) solutions. To describe waves more precisely, there are also numerical formulations. In these, it is possible to specify, for example, non-sinusoidal wave forms. Such formulations are discussed in Sarpkaya and Isaacson (1981). The best-known numerical theory is the Stream Function Theory of Dean (1974). Much interest also exists in the Boussinesq equations. These equations are related to the Korteweg deVries equations and simulate waves in shallow water. With recent research, it has been possible to extend the range of applicability of these equations to deeper water. Thus, although computationally intensive, they form a powerful non-linear, numerical wave theory that can calculate waves accurately throughout a large range of depths - Dingemans (1997, Vol. 2).

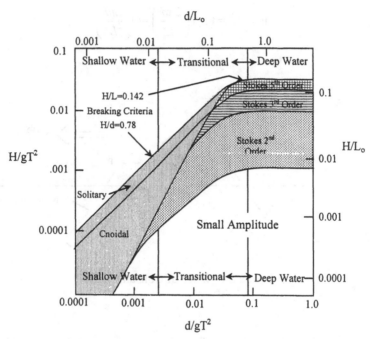

Figure 2.7 Applicability of Various Wave Theories (after Le Méhauté, 1976)

2.3 Small Amplitude Wave Theory

The Small Amplitude Wave Theory expressions are summarized in Table 2.2. The last three equations (14, 15,and 16) give additional useful expressions derived from

other wave theories.   Equation [1][3] of Table 2.2 describes the water surface
fluctuation as shown in Fig. 2.1. Equation [2] calculates the velocity of propagation,
C, assuming the wave retains a constant form as shown in Fig. 2.8. The 'tanh' term in
Eq. [2] has two asymptotic values. For large depths (kd is large)

$$\tanh kd = \tanh\left(\frac{2\pi d}{L}\right) \rightarrow 1 \qquad (2.4)$$

For small depths

$$\tanh kd = \tanh\left(\frac{2\pi d}{L}\right) \rightarrow \left(\frac{2\pi d}{L}\right) \qquad (2.5)$$

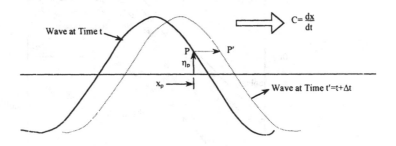

Figure 2.8 Wave of Constant Form

Thus, it is possible to give *deep* and *shallow* water asymptotic values for C as in
Table 2.2. It has been customary to define deep water as d/L>0.5 (tanh kd = 0.996)
and shallow water is usually defined as d/L<0.05 (kd = 0.312, while tanh kd =
0.302).

---

3. Square brackets refer to equations in Table 2.2

Table 2.2: Common Expressions For Progressive Waves

	Deep Water $(d/L>0.5)$	Shallow Water $(d/L>0.5)$
1. Water Surface [m]	$\eta = \dfrac{H}{2}\cos(kx-\omega t)$	
2. Velocity of Propagation [m/s] (Dispersion Equation)	$C = \dfrac{L}{T} = \dfrac{\omega}{k} = \dfrac{gT}{2\pi}\tanh kd$ $= \sqrt{\dfrac{gL}{2\pi}\tanh kd}$ $C_o = \dfrac{gT}{2\pi}$	$C = \sqrt{gd}$
3. Wave Length [m]	$L = CT = \dfrac{gT^2}{2\pi}\tanh kd$ $L_o = \dfrac{gT^2}{2\pi}$	
4. Horizontal Component of Orbital Velocity [m/s]	$u = \dfrac{\pi H}{T}\dfrac{\cosh k(z+d)}{\sinh kd}\cos(kx-\omega t)$ $u_o = \dfrac{\pi H_o}{T}e^{k_o z}\cos(k_o x-\omega t)$	
5. Vertical Component of Orbital Velocity [m/s]	$w = \dfrac{\pi H}{T}\dfrac{\sinh k(z+d)}{\sinh kd}\sin(kx-\omega t)$ $w_o = \dfrac{\pi H_o}{T}e^{k_o z}\sin(k_o x-\omega t)$	
6. Horizontal Orbital Semi-Axis [m]	$A = \dfrac{H}{2}\dfrac{\cosh k(z+d)}{\sinh kd}$ $A_o = \dfrac{H_o}{2}e^{k_o z}$	
7. Vertical Orbital Semi-Axis [m]	$B = \dfrac{H}{2}\dfrac{\sinh k(z+d)}{\sinh kd}$ $B_o = A_o$	

		Deep Water (d/L>0.5)	Shallow Water (d/L>0.5)
8. Pressure [m of water]	$\dfrac{p}{\rho g} = -z + K_p \eta$		
9. Pressure Response Factor	$K_p = \dfrac{\cosh k(z+d)}{\cosh kd}$	$K_p = e^{k_o z}$	
10. Energy Density [j/m²]	$E = \dfrac{1}{8}\rho g H^2; \quad KE = PE = \dfrac{E}{2}$		
11. Wave Power [w/m]	$P = EC_G$	$P_o = \dfrac{EC_o}{2}$	$P = EC$
12. Group Velocity [m/s]	$C_G = nC$	$(C_G)_o = \dfrac{C_o}{2}$	$C_G = C$
13. Group Velocity Parameter	$n = \dfrac{1}{2}\left[1 + \dfrac{2kd}{\sinh 2kd}\right]$	$n_o = \dfrac{1}{2}$	$n = 1$
14. Mass Transport at Bottom [m/s]	$U_B = \dfrac{5}{4}\dfrac{a^2 k\omega}{\sinh^2 kd}$		
15. Wave Breaking Criterion	$\left(\dfrac{H}{L}\right)_{max} = 0.142 \ \tanh \ kd$	$\left(\dfrac{H}{L}\right)_{max} = 0.142$	$\left(\dfrac{H}{d}\right)_{max} = 0.78$
16. MWL – SWL [m]	$\Delta = \dfrac{H^2 k}{8}\coth kd$	$\Delta = \dfrac{H^2 k}{8}$	

Table 2.3 Wave Table

d/L₀	tanh kd	d/L	kd	sinh kd	cosh kd	n
0.000	0.000	0.0000	0.000	0.000	1.00	1.000
0.002	0.112	0.0179	0.112	0.113	1.01	0.996
0.004	0.158	0.0253	0.159	0.160	1.01	0.992
0.006	0.193	0.0311	0.195	0.197	1.02	0.988
0.008	0.222	0.0360	0.226	0.228	1.03	0.983
0.010	0.248	0.0403	0.253	0.256	1.03	0.979
0.015	0.302	0.0496	0.312	0.317	1.05	0.969
0.020	0.347	0.0576	0.362	0.370	1.07	0.959
0.025	0.386	0.0648	0.407	0.418	1.08	0.949
0.030	0.420	0.0713	0.448	0.463	1.10	0.939
0.035	0.452	0.0775	0.487	0.506	1.12	0.929
0.040	0.480	0.0833	0.523	0.548	1.14	0.919
0.045	0.507	0.0888	0.558	0.588	1.16	0.910
0.050	0.531	0.0942	0.592	0.627	1.18	0.900
0.055	0.554	0.0993	0.624	0.665	1.20	0.891
0.060	0.575	0.104	0.655	0.703	1.22	0.880
0.065	0.595	0.109	0.686	0.741	1.24	0.872
0.070	0.614	0.114	0.716	0.779	1.26	0.863
0.075	0.632	0.119	0.745	0.816	1.29	0.853
0.080	0.649	0.123	0.774	0.854	1.31	0.845
0.085	0.665	0.128	0.803	0.892	1.34	0.836
0.090	0.681	0.132	0.831	0.929	1.37	0.827
0.095	0.695	0.137	0.858	0.968	1.39	0.819
0.10	0.709	0.141	0.886	1.01	1.42	0.810
0.11	0.735	0.150	0.940	1.08	1.48	0.794
0.12	0.759	0.158	0.994	1.17	1.54	0.778
0.13	0.780	0.167	1.05	1.25	1.60	0.762
0.14	0.800	0.175	1.10	1.33	1.67	0.747
0.15	0.818	0.183	1.15	1.42	1.74	0.733
0.16	0.835	0.192	1.20	1.52	1.82	0.718
0.17	0.850	0.200	1.26	1.61	1.90	0.705
0.18	0.864	0.208	1.31	1.72	1.99	0.692
0.20	0.888	0.225	1.41	1.94	2.18	0.668
0.21	0.899	0.234	1.47	2.05	2.28	0.656
0.22	0.909	0.242	1.52	2.18	2.40	0.646
0.23	0.918	0.251	1.57	2.31	2.52	0.635
0.24	0.926	0.259	1.63	2.45	2.65	0.626
0.25	0.933	0.268	1.68	2.60	2.78	0.616
0.26	0.940	0.277	1.74	2.75	2.93	0.608
0.27	0.946	0.285	1.79	2.92	3.09	0.599
0.28	0.952	0.294	1.85	3.10	3.25	0.592
0.29	0.957	0.303	1.90	3.28	3.43	0.585
0.30	0.961	0.312	1.96	3.48	3.62	0.578
0.31	0.965	0.321	2.02	3.69	3.83	0.571
0.32	0.969	0.330	2.08	3.92	4.05	0.566
0.33	0.972	0.339	2.13	4.16	4.28	0.560
0.34	0.975	0.349	2.19	4.41	4.53	0.555
0.35	0.978	0.358	2.25	4.68	4.79	0.550
0.36	0.980	0.367	2.31	4.97	5.07	0.546
0.37	0.983	0.377	2.37	5.28	5.37	0.542
0.38	0.984	0.386	2.43	5.61	5.70	0.538
0.39	0.986	0.395	2.48	5.96	6.04	0.535
0.40	0.988	0.405	2.54	6.33	6.41	0.531
0.41	0.989	0.415	2.60	6.72	6.80	0.529
0.42	0.990	0.424	2.66	7.15	7.22	0.526
0.43	0.991	0.434	2.73	7.60	7.66	0.523
0.44	0.992	0.443	2.79	8.07	8.14	0.521
0.45	0.993	0.453	2.85	8.59	8.64	0.519
0.46	0.994	0.463	2.91	9.13	9.18	0.517
0.47	0.995	0.472	2.97	9.71	9.76	0.516
0.48	0.995	0.482	3.03	10.3	10.4	0.514
0.49	0.996	0.492	3.09	11.0	11.0	0.513
0.50	0.996	0.502	3.15	11.7	11.7	0.512
0.75	1.000	0.746	4.69	54.5	54.5	0.501
1.0	1.000	0.981	6.16	269.5	269.5	0.500

2.3.1 Wave Tables

To solve Eq. [2] and all the other equations in Table 2.2, it is necessary to know the wave length, L, which may be calculated using Eq. [3]. However, Eq. [3] is implicit and can only be solved numerically. Tables of solutions have been prepared that yield L as well as other important wave characteristics. Such tables are known as wave tables and have been published in CERC (1984) and Wiegel (1964). An abbreviated set of wave tables is presented in Table 2.3. To use the wave tables, we first calculate the deep water approximation of wave length as given by Eq. [3].

$$L_o = \frac{gT^2}{2\pi} \qquad (2.6)$$

Then we use the depth of water d to calculate d/L_o and from there we can evaluate all the remaining wave parameters from Table 2.3. An example is given at the end of the next section.

2.3.2 Small Amplitude Expressions

Waves propagate with velocity C, but the individual water particles do not propagate; they move in particle orbits as shown in Fig. 2.9. For small amplitude wave theory, such particle orbits are elliptical and if the water is 'deep', they become circular. Their size decreases with depth. Horizontal and vertical orbital velocity components, u and w, and the horizontal and vertical orbital amplitudes, A and B, are given in Eqs. [4] to [7].

The pressure fluctuations at any point below the water surface are related to the water level fluctuations at the surface. If the wave were infinitely long, the water level would be horizontal at any time, there would be no particle motion and the pressure fluctuations would be hydrostatic and equal to ρgH, where ρ is the fluid density and g is the gravitational acceleration. For waves of limited length the pressure fluctuations are smaller than (ρgH). The ratio of the actual pressure fluctuations to (ρgH), is called the pressure response factor, K_p, and it is a function of wave length (or wave period) and depth below the surface. For longer waves or close to the water surface, the pressure response factor approaches 1. For shorter waves or far below the water surface, the pressure response factor approaches 0. Equations. [8] and [9] quantify the pressure response.

Equation [10] expresses wave energy per unit surface area, or energy density E, in joules/m^2. It is made up of half potential energy and half kinetic energy. Equation

[11] gives wave power, P, arriving at any location. Its units are watts/m of wave crest.

Equation. [2] indicates that longer period waves travel faster than shorter period waves. A real wave train, as in Fig. 2.3, contains many different wave periods, and therefore it would stretch out (disperse) as it traveled. The longest waves would lead and run further and further ahead with time and distance, while the shortest waves would lag further behind. Hence Eq. [2] is called the *dispersion equation*. Equation [2] also means that away from their immediate, generating area, waves of roughly the same period tend to travel together. From basic physics we know that waves of almost the same period interfere to form beats or wave groups. The theoretical expression for the interference pattern of waves of almost the same period is shown in Fig. 2.10.

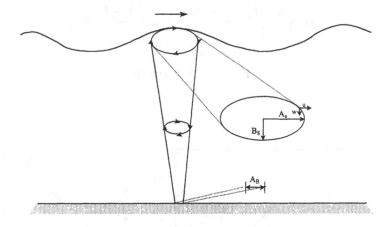

Figure 2.9 Particle Orbits

The resulting wave consists of two waves that are superimposed – one related to the average values of k and ω of the two interfering waves and another, much longer wave, called the wave group, related to the differences in k and ω. There are two wave speeds involved (ω/k) - one for the short waves $C=(\omega_1+\omega_2)/(k_1+k_2)$ and another for the wave group $G_G=(\omega_1-\omega_2)/(k_1-k_2)$. The speed of the wave group is related to C by the factor n, as given in Eq. [13]. In deep water $n \to \frac{1}{2}$ and in shallow water $n \to 1$. Thus $C_G<C$, but in very shallow water $C_G \to C$. Figure 2.10 shows that the wave group consists of a series of individual waves that increase in size and then decrease.

This gives rise to the adage "every 7th wave is a big one". Because $C > C_G$, the individual waves travel through the group. At the back of the group they are small. Then they increase in size as they travel through the group, decrease in size past the centre of the group and eventually disappear at the front of the group.

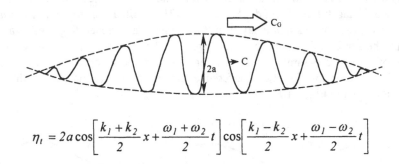

$$\eta_t = 2a \cos\left[\frac{k_1 + k_2}{2}x + \frac{\omega_1 + \omega_2}{2}t\right]\cos\left[\frac{k_1 - k_2}{2}x + \frac{\omega_1 - \omega_2}{2}t\right]$$

Figure 2.10 Wave Group

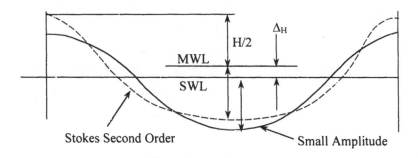

Figure 2.11 Wave Shape and Mean Wave Level

Eqs. [14], [15] and [16] are expressions derived from other theories, often used as simple extensions of small amplitude theory. According to higher order theory, the particle orbits of Fig. 2.9 are not closed. There is a small net movement of the water particles in the direction of wave propagation called Mass Transport. The mass transport velocity at the bottom, U_B, which is important for calculating sediment transport, is given in Eq. [14]. Three wave breaking criteria are given in Eq. [15]. There is a limit to the steepness of the wave. In shallow water, this reduces to a limit of the ratio (H/d), known as the breaking index. Solitary wave theory defines this limit as 0.78, a value supported by experimental observation. Finally, although the wave crest and trough are equidistant (H/2) from the still water level in small amplitude wave theory, higher order wave theory estimates that the wave crests are higher and the wave troughs become shallower than in small amplitude theory (Fig. 2.11). This creates a difference between mean wave level midway between crest and trough and still water level, as given by Eq [16].

Example 2.1 Use of the Wave Table

We will now calculate the wave characteristics for a wave of period T = 8 sec and a wave height H = 1.5 m in a depth of water d = 6 m. We use small amplitude wave theory (Table 2.2) and the wave table (Table 2.3).

It is first necessary to calculate the deep water wave length and relative depth

$$L_o = \frac{gT^2}{2\pi} = 1.56T^2 = 1.56(64) = 100m; \quad \frac{d}{L_o} = \frac{6}{100} = 0.060 \qquad (2.7)$$

The wave table (Table 2.3) now yields the following

$$\frac{d}{L} = 0.104; \quad \tanh kd = 0.575;$$
$$\sinh kd = 0.703; \quad \cosh kd = 1.22; \quad n = 0.881 \qquad (2.8)$$

From the value of $\frac{d}{L}$, the wave length in 6 m of water and wave number, k, may now be calculated

$$L = \frac{d}{0.104} = 57.5m; \quad k = \frac{2\pi}{L} = 0.109\,m^{-1} \qquad (2.9)$$

From these, the following parameters may be computed; ρ is assumed to be 1035 kg/m^3 for sea water.

$$C = \frac{L}{T} = 7.2 \ m/s; \quad C_G = nC = 0.881(7.2) = 6.35 \ m/s;$$

$$E = \frac{\rho g H^2}{8} = 2854 \ j/m^2; \quad P = EC_G = 18,124 \ w/m \ of \ wave \ crest \tag{2.10}$$

At the bottom:

$$z = -d; \quad k(z + d) = 0; \quad \sinh k(z + d) = 0; \quad \cosh k(z + d) = 1.0 \tag{2.11}$$

and the horizontal component of orbital velocity is

$$\begin{aligned} u_B &= \frac{\pi H}{T} \frac{1}{\sinh kd} \cos(kx - \omega t) \\ &= \frac{\pi(1.5)}{8} \frac{1}{.703} \cos(kx - \omega t) = 0.84 \cos(kx - \omega t) \end{aligned} \tag{2.12}$$

Thus, at the bottom, u_B has a maximum value $\hat{u}_B = 0.84$ m/s and the vertical velocity component of orbital motion at the bottom is zero. The amplitude of the orbital motion at the bottom is

$$A_B = \frac{H}{2 \sinh kd} = \frac{1.5}{2(0.703)} = 1.07 \ m. \tag{2.13}$$

and the orbital diameter is $2A_B = 2.14$ m. The pressure response factor K_p at the bottom is

$$(K_p)_B = \frac{1}{\cosh kd} = \frac{1}{1.221} = 0.82 \tag{2.14}$$

which means that the maximum pressure fluctuation is

$$K_p H = 0.82 \ (1.5) = 1.23 \ (\text{m of water}) \tag{2.15}$$

or for sea water

$$\rho g(K_p H) = 1035 \ (9.81)(0.82)(1.5) = 12,788 \ Pa = 12.8 \ kPa \tag{2.16}$$

At a distance of 4 m below the water surface: $z = -4$m and $k(z+d) = 0.218$. This gives $\sinh k(z+d) = 0.220$ and $\cosh k(z+d) = 1.024$, $\hat{u} = 0.84(1.024) = 0.86$ m/s, $\hat{w} = 0.84(0.220) = 0.18$ m/s, $A = 1.07(1.024) = 1.10$ m/s, $B = 1.07(.220) = 0.24$ m/s. Finally, $K_p = 0.82(1.024) = 0.84$.

2.3.3 Calculation by Computer

The above calculation using the wave table is fine if you have only a few calculations to make. What about if you need to make many calculations and use a computer? In that case, L or C may be calculated using a root finding technique such as Newton-Raphson, but such a technique requires iteration. To speed up such computations, approximations may be used, such as the one proposed by Hunt (1979)

$$\frac{C^2}{gd} = \left[y_o + (1 + 0.6522y_o + 0.4622y_o^2 + 0.0864y_o^4 + 0.0675y_o^5)^{-1} \right]^{-1} \quad (2.17)$$

where

$$y_o = \frac{2\pi d}{L_o} \quad (2.18)$$

Fig. 2.12 presents a spreadsheet calculation for Example 2.1, using Eq. 2.17.

2.4 Reflected Waves

When a wave reaches a rigid, impermeable vertical wall it is completely reflected. After some time, under controlled conditions, the reflected waves and the incident waves together form a standing wave. The wave form no longer moves forward in space. A theoretical expression for such a standing wave (Fig. 2.13) may be obtained by superposition of the equations for an incident and a reflected wave. The small amplitude expressions for a standing wave are given in Table 2.4. A maximum wave height (antinode) is present at the structure and at every half wave length away from the structure. A zero wave height (node) is located L/4 from the wall and then at every half wave length. The maximum wave height is twice the height of the original incident wave. The dashed lines in Fig. 2.13 indicate streamlines and the Eqs. [3] and [4] of Table 2.4 show that the horizontal component of particle velocity, u, is maximum below the nodes and zero below the antinodes; the vertical velocity, w, is maximum at the antinodes and zero at the nodes.

			Surface	At z=-4.0	Bottom
		Wave Table Calculator			
Title:		Example 2.1			
INPUT AREA:	Insert the values in the boxes		Surface	At z=-4.0	Bottom
Gravity	g	(m/s2)	9.806	9.806	9.806
Fluid Density	Rho	(kg/m3)	1035	1035	1035
Kinematic Viscosity	Nu	(m2/s)	0.000001	0.000001	0.000001
Incident Wave Period		(s)	8.00	8.00	8.00
Incident Wave Height		(m)	1.50	1.50	1.50
Incident Water Depth		(m)	6.00	6.00	6.00
Depth of Interest (-)	(-)z	(m)	0.00	-4.00	-6.00
WAVE TABLE CALCULATION:					
Depth of water	d	(m)	6.000	6.000	6.000
Wave Period	T	(s)	8.000	8.000	8.000
Wave Height	H	(m)	1.500	1.500	1.500
Depth of Interest (-)	(-)z !	(m)	0.000	-4.000	-6.000
Deep Water Wave Length (Table 2.2 - Eq 3)	Lo	(m)	99.840	99.840	99.840
Deep Water Wave Speed (Table 2.2 - Eq 2)	Co	(m/s)	12.480	12.480	12.480
	d/Lo		0.060	0.060	0.060
Deep Water Argument [2*Pi*d/Lo]			0.378	0.378	0.378
Wave Speed (Eq 2.17)	C	(m/s)	7.189	7.189	7.189
Wave Length (=CT)	L	(m)	57.513	57.513	57.513
Argument (kd=2*Pi*d/L)			0.655	0 655	0.655
sinh (kd)			0.703	0.703	0.703
cosh (kd)			1.223	1.223	1.223
tanh (kd)			0.575	0.575	0.575
Group Velocity Parameter (Table 2.2 - Eq. 13)	n		0.881	0.881	0.881
Group Velocity (=nC)	Cg	(m/s)	6.334	6.334	6.334
Depth below Elevation (-z)	D=d+z	(m)	6.000	2.000	0.000
Argument at depth (-z) (=kD)			0.655	0.218	0.000
sinh (kD)			0.703	0.220	0.000
cosh (kD)			1.223	1.024	1.000
Semi Major Axis (Table 2.2 - Eq. 6)	A	(m)	1.304	1.092	1.066
Semi-Minor Axis (Table 2.2 - Eq. 7)	B	(m)	0.750	0.235	0.000
Maximum Horizontal Velocity (Table 2.2 - Eq. 4)	<u>	(m/s)	1.024	0.857	0.837
Maximum Vertical Velocity (Table 2.2 - Eq. 5)	<v>	(m/s)	0.589	0.184	0.000
Pressure Response Factor (Table 2.2 - Eq 9)	Kp		1.000	0.838	0.818
Maximum Pressure Fluctuation (=Kp*H)	Del p	(m)	1.500	1.256	1.227
Energy Density (Table 2.2 - Eq. 10)	E	(kj/m2)	2.854	2.854	2.854
Wave Power (Table 2.2 - Eq. 11)	P	(kw/m)	18.081	18.081	18.081
Mass Transport Velocity (Table 2.2 - Eq. 14)	Ub	(m/s)	0.122	0.122	0.122

Figure 2.12 Spreadsheet Calculation of Wave Parameters

Table 2.4 Common Expressions for Reflected Waves

	Complete Reflection	Partial Reflection
1. Water Surface [m]	$\eta = H \cos kx \cos \omega t$	$\eta = \frac{1}{2}(H_I - H_R)\cos(kx + \omega t)$ $+ H_R \cos kx \cos \omega t$
2. Nodes [m]	$x_{NODE} = \frac{L}{4}, \frac{3L}{4}$	
3. Horizontal Component of Orbital Velocity [m/s]	$u = \frac{2\pi H}{T} \frac{\cos h\, k(z+d)}{\cosh kd} \sin kx \sin \omega t$	
4. Vertical Component of Orbital Velocity (m/s)	$w = -\frac{2\pi H}{T} \frac{\sinh k(z+d)}{\cosh kd} \cos kx \cos \omega t$	
5. Pressure Response Factor	$K_p = \frac{\cosh k(z+d)}{\cosh kd}$	
6. Reflection Coefficient	$K_R = 1$	$K_R = \frac{H_R}{H_I} = \frac{(H_{max} - H_{min})}{(H_{max} + H_{min})}$
7. Incident Wave Height [m]	$H_I = H$	$H_I = \frac{(H_{max} + H_{min})}{2}$
8. MWL - SWL [m]	$\Delta_H = \frac{H^2 k}{2} \coth kd$	

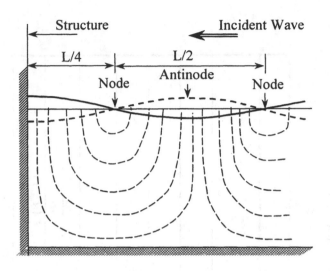

Figure 2.13 Standing Waves

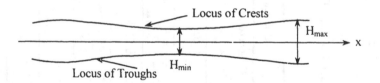

Figure 2. 14 Envelope of Partial Wave Reflection

Partial wave reflection will result if the reflecting surface is sloping, flexible or porous. The reflected wave is then smaller than the incident wave, which yields a standing wave that varies in wave height with distance, as shown in Fig. 2.14. The partial antinodes (H_{max}) are less than twice the incident wave height, while the partial nodes (H_{min}) are greater than zero. The resulting wave envelope can be used to estimate the reflection coefficient and the incident wave height. For simple sinusoidal waves the relationships are given in Eq. [6] and [7] of Table 2.4. The envelope can be defined by a number of wave probes that measure waves simultaneously at different locations over half a wave length. For real waves such as in Fig. 2.3, three wave probes can define the envelope well enough to be able to determine the reflection coefficient, which will be a function of wave frequency (Mansard and Funke, 1980).

2.5 Wave Measurement

In order to understand the coastal environment, probably the most important parameter to determine is the wave climate - the waves that are present at a location over the long term (years) and over the short term (storms and individual waves). The measurement of such waves is the topic of this section. Analysis of waves over the short term will be discussed in Ch. 3 and long-term wave analysis is in Ch. 4.

2.5.1 Wave Direction

If a single sensor is placed to measure waves, the record resembles Fig. 2.3. It gives information of water level fluctuations with respect to time, but it gives no idea of the direction in which the waves were traveling when the recording was made. Such a wave record is simply the sum total of all wave components arriving from all directions. If the direction of approach of the various components must be known, more sensors need to be placed and the record of each sensor needs to be related to the others.

It is easy to visualize that for the array of sensors in Fig. 2.15, long crested waves traveling from the North, (Direction N) would give identical records on all sensors. The records of C and D would be exactly in phase while the records of A and B would lead C and D in time. Conversely, long crested waves traveling from the west, (Direction W) would again give identical records on all four sensors. Those on A and B would be exactly in phase while C would lead and D would lag behind A and B. Thus from inspection of the records and comparison of phase differences between them, the direction of wave travel may be obtained. Even if the direction of travel were such that none of the records occur at the same time, a little arithmetic can determine wave direction. The intuitive method of comparison of separate wave recordings, as described above, may be generalized using cross-correlation techniques to produce estimates of wave direction. Since waves are normally not long-crested, the wave recordings should be taken as closely together as possible to ensure that all measurements are of the same short-crested wave. One common technique actually uses one pressure sensor and two orthogonally mounted current meters, placed at the same location and close enough to the surface to obtain strong signals. Using equations such as in Table 2.2, η can be computed from p, u and v, where v is the velocity measured in the y direction. Cross-correlation of the three signals then makes it possible to calculate wave direction from these three simultaneous measurements.

Remote sensing with radar or air and satellite photography can also be used to determine wave direction. When wave direction is not measured, it is generally inferred from the wind direction, as described in Ch. 5, in combination with refraction, diffraction and reflection patterns of the waves, described in Ch. 7. This is a rather inaccurate business, but often it is the only source of directional information.

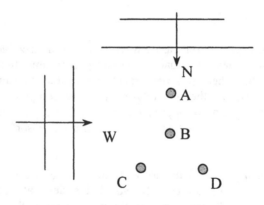

Figure 2. 15 Wave Direction and Array of Four Wave Gauges

2.5.2 Equipment

A number of different types of wave sensors have been used. Originally wave staffs were used. Electrical contacts were placed along a vertical staff of sufficient length to measure the largest waves. Each contact was individually wired so that immersion of a contact changed the resistance of a circuit. Thus the resistance represented the water level. Problems with corrosion at the contacts and with mounting the large and unwieldy staffs resulted in a search for alternative methods of wave measurement.

In Section 2.3 it was seen that subsurface pressure is related to water level fluctuation by the pressure response factor, K_p. Thus it should be possible to install a pressure sensitive device under water and translate its record into a record of water level fluctuation. However, K_p is a function of depth of immersion as well as wave length. If the pressure recorder is placed in deep water it will only measure longer waves such as tides, seiches and tsunamis (earthquake generated waves) and this is a common method of measuring long period fluctuations in the open ocean. A recorder placed close to the surface will distort all the frequency components to some extent and will still filter out the high frequency waves of very short wave length. Figure 2.16 shows this effect for pressure recordings, at 2 and 10 m depth, of the sea and swell record of Fig. 2.5. Thus, for the measurement of short waves, it is important to install a pressure sensor relatively close to the surface so that also higher frequency (shorter wave length) components can be recorded. Each individual wave frequency will have a different K_p which means that the pressure record must be converted into a water level record, using a transfer function (K_p vs f). One obvious advantage of a pressure gauge is that it can be left in place during the winter when ice would damage other types of gauges.

A commonly used wave measurement device is the accelerometer, shown schematically in Fig. 2.17. A mass m vibrates inside the accelerometer that floats on the water. The motion of the mass may be described by

$$M_A \frac{d^2 z_m}{dt^2} + C_d \frac{d(z_m - z_a)}{dt} + K_S(z_m - z_a) = 0 \qquad (2.19)$$

or

$$M_A \frac{d^2 z_r}{dt^2} + C_d \frac{dz_r}{dt} + K_S z_r = -M_A \frac{d^2 z_a}{dt^2} \qquad (2.20)$$

where K_S is a spring constant, C_d a damping coefficient, z_m is the movement of the vibrating mass, z_a the movement of the whole accelerometer and $z_r = z_m - z_a$, the relative displacement inside the instrument. Thus by measuring the velocities and accelerations inside the accelerometer, the acceleration of the accelerometer itself may be obtained. For a floating buoy accelerometer, the motion imparted to the accelerometer will be equal to the wave motion if the buoy is small enough compared to the waves to be measured. As long as the mass-spring system remains vertical, the vertical acceleration of the accelerometer and hence of the waves can be obtained as a function of time by Eq. 2.20. The vertical water surface displacement (z_a) may then be calculated by double integration.

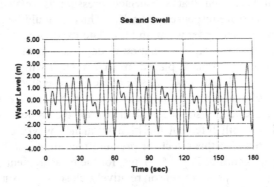

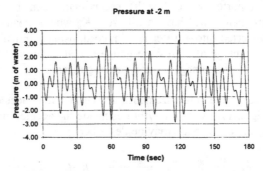

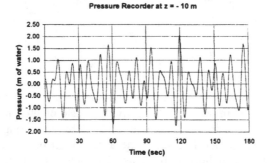

Figure 2.16 Pressure Record

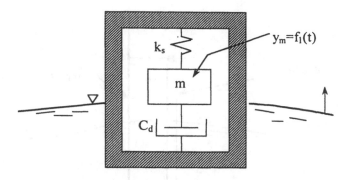

Figure 2.17   Accelerometer

Stereo photogrammetry has also been used to determine sea state. This method is effective when determining short term sea states, but the costs of recording and analyzing at regular intervals over a long time are prohibitive. The same is true about remote sensing methods using aircraft. The cost of repetitive flights is high and flights can only take place in fair weather. Satellite remote sensing looks promising. Satellites cover large areas at regular intervals and neither the flights nor some of the imagery are weather dependent. The resolution is good enough to yield approximate values of wave heights and directions, and is now used to forecast sea states at coastal zones and in shipping lanes.

2.5.3 Laboratory Sensors

Wave sensors in the laboratory may be classified as resistance and capacitance gauges. The laboratory resistance gauge consists of two conducting wires that are stretched on a frame and the rising and falling water levels close the resistance circuit yielding an output resistance proportional to the water level (Fig. 2.18). In a capacitance probe a conducting wire is coated by an insulator of uniform thickness. The conductor and the water form two plates of a capacitor while the insulator forms the dielectric. As the water level changes, the area of the capacitor changes and thus capacitance is a function of water level. Both resistance and capacitance laboratory wave probes are sensitive to dirt, temperature and meniscus effects, and need to be cleaned and calibrated often.

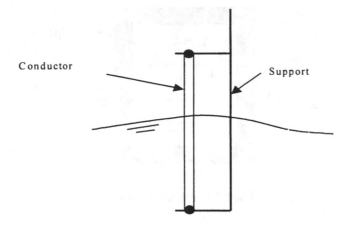

Figure 2.18 Laboratory Capacitance Wave Gauge

2.6 Summary

This introduction to water waves may be summarized as follows:
- wind-generated waves are complex
- they may be distinguished as irregular sea (in the generating area) and more organized swell (away from the generating area).
- simplification is needed to be able to quantify the wave action and its effects.
- small amplitude theory as outlined in Sections 2.3 and 2.4 is adequate for most design.

3. Short-Term Wave Analysis

3.1 Introduction

In analysis of wave data, it is important to distinguish between Short-Term and Long-Term wave analysis. *Short-Term* analysis refers to analysis of waves that occur within one wave train[1] or within one storm (Fig 2.3 and 3.1). *Long-Term* analysis refers to the derivation of statistical distributions that cover many years. To emphasize the difference between the two, they have been arranged into two separate chapters. Short-term wave analysis is discussed here and long-term wave analysis in Ch. 4.

It was stated in Ch. 2 that the complex sea surface appears to defy scientific analysis. A number of simplifying assumptions must be made to describe short-term recordings of the water surface and research has shown that a number of excellent approximations can be made (Goda, 1970, 1985). Because there are many sizes of waves in any wave record we will need to resort to statistical analysis.

We define z as the instantaneous water level related to a datum, and η as the difference between the instantaneous water and the mean water level. The values of z and η are functions of location (x, y) and time (t). A water level record such as shown in Fig. 3.1 therefore represents the process z(t) at a specific location. Water level records are normally not continuous, because they are recorded digitally. Thus z is only sampled at sampling intervals of Δt. A record of length t_R then consists of N samples z_j taken at times $j\Delta t$, where $1 \leq j \leq N$.

---

1. Series of waves.

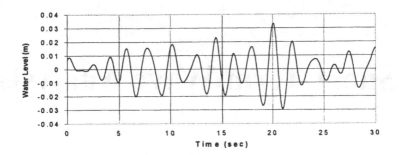

Figure 3.1 Water Level Record

The water level record in Fig. 3.1 is one realization of the process z(t). We will call this $z_1(t)$. To understand the relevant terminology, imagine a basin of water with a wave generator at one end. We start up the wave generator steered by a certain drive signal, and after 5 minutes we measure the water level in the middle of the basin for 30 seconds. That produces a short-term record z(t) as shown in Fig. 3.1. Now we shut off the generator and wait for the water to become quiet. Then we start the generator up again (with the same drive signal) and after 5 minutes we measure another 30 second record at the same location. The second record is a second realization, $z_2(t)$ of the same process. We could repeat this many times to produce an infinite number of realizations (records) of the process z(t), which in this case represents water surface fluctuations in the middle of a basin, after 5 minutes of wave generation. Three realizations of this process are shown in Fig. 3.2. The complete set of K realizations $z_k(t)$ is called an Ensemble.

We can take all the values of z at $t=j\Delta t$ in the k realizations and calculate statistical parameters such as ensemble mean $\bar{z}_j$ and ensemble standard deviation σ_j, where

$$\bar{z}_j = \frac{1}{K}\sum_{k=1}^{K} z_{k,j} \quad \text{and} \quad \sigma_j = \sqrt{\frac{1}{K}\sum_{k=1}^{K}(z_{k,j}-\bar{z}_j)^2} \qquad (3.1)$$

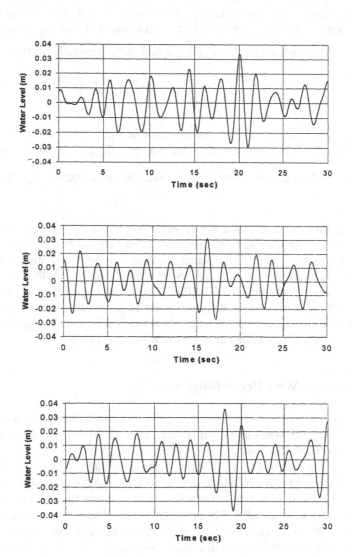

Fig. 3.2 Ensemble of Three Realizations of a Stationary and Ergodic Progress

Ensemble skewness and kurtosis can also be determined. If none of these ensemble parameters vary in time, the process is called *stationary;* if only the ensemble mean and standard deviation are constant, the system is said to be weakly stationary. If the time average for each realization $\overline{z_k(t)}$ is equal to the ensemble average $\overline{\overline{z}}_k$, where

$$\overline{z_k(t)} = \frac{1}{N}\sum_{j=1}^{N} z_j \quad \text{and} \quad \overline{\overline{z}}_k = \frac{1}{K}\frac{1}{N}\sum_{k=1}^{K}\sum_{j=1}^{N} z_{k,j} \tag{3.2}$$

the process is called *ergodic.* For the ensemble in Fig. 3.2, the process is stationary and ergodic. It is stationary because, for example, $\overline{z}_k$ and σ_j, do not increase with time and ergodic because $\overline{z_k(t)}$ for each realization is equal to $\overline{\overline{z}}_k$, the ensemble average.

What does this mean in practice? A water level record is always only a single realization of the process to be studied. We have no other realizations. We cannot turn off and later "re-play" the same situation as we could in the wave basin example. Therefore any record is only an approximation of the process. Weak stationarity can only be inferred from this single wave record if $\overline{z}$ and σ do not vary with time (there is no trend in the mean water level and the wave heights). Finally, with only one realization, we can never show that the process is ergodic; we simply must assume ergodicity as also discussed in Kinsman (1965).

3.2 Short-Term Wave Height Distribution

To determine wave heights, it is necessary to use η, the difference between the water level and the mean water level. It is usual to think of η as a superposition of an infinite number of small waves, each generated by its own wind eddies at different locations and at different times. The resulting sea surface is, therefore, the sum of a large number of statistically independent processes, and common sense would tell us that it is impossible to predict the exact value of η at any time or location. In other words, η is a random variable. The probability that η has a certain value is called the Probability Density Function (PDF), $p(\eta)$. The Central Limit Theorem states that the PDF for a sum of many independent variables is *Gaussian,* which means that $p(\eta)$ can be described by the *normal distribution*. The overall behavior of $p(\eta)$ may be summarized by its mean, $\overline{\eta}$, standard deviation, σ, and possibly some additional statistical parameters such as skewness and kurtosis. Most often a two-parameter normal distribution is used, defining p(η) by $\overline{\eta}$ and σ only. But by definition, $\overline{\eta} = 0$ and hence

$$p(\eta) = \frac{1}{\sigma\sqrt{2\pi}} \exp\left[\frac{-\eta^2}{2\sigma^2}\right] \tag{3.3}$$

where σ is the standard deviation of the process, $\eta(t)$. It is equal to the square root of the variance of η.

$$\sigma^2 = \overline{\eta^2} = \lim_{t_R \to \infty} \frac{1}{t_R} \int_{t=0}^{t=t_R} \eta^2 \, dt = \frac{1}{N} \sum_{j=1}^{N} \eta_j^2 \tag{3.4}$$

If the wave frequencies all occur within a narrow frequency band, (if the wave periods do not vary greatly) it may be shown theoretically (Longuet-Higgins, 1952; Cartwright and Longuet-Higgins, 1956; Benjamin and Cornell, 1970) that the PDF of the maximum instantaneous water levels is:

$$p(\eta_{max}) = \frac{\eta_{max}}{\sigma^2} \exp\left[\frac{-\eta_{max}^2}{2\sigma^2}\right] \tag{3.5}$$

If it is assumed that for waves of a narrow frequency band the wave height H is equal to $2\eta_{max}$, then the PDF for H becomes:

$$p(H) = \frac{1}{4} \frac{H}{\sigma^2} \exp\left[\frac{-H^2}{8\sigma^2}\right] \tag{3.6}$$

Equations 3.5 and 3.6 are known as the Rayleigh Distribution. This distribution is shown in Fig. 3.3.

To determine the Cumulative Distribution Function (CDF) of wave heights, the Rayleigh distribution is integrated to yield the probability that any individual wave of height H' is not higher than a specified wave height H

$$P(H' < H) = \int_0^H p(H) \, dH = 1 - \exp\left[\frac{-H^2}{8\sigma^2}\right] \tag{3.7}$$

The Probability of Exceedence, the probability that any individual wave of height H' is greater than a specified wave height H may be obtained as

$$Q(H' > H) = 1 - P(H' < H) = \exp\left[\frac{-H^2}{8\sigma^2}\right] \tag{3.8}$$

The functions P and Q are also shown in Fig. 3.3. Research has shown that for practically all locations the wave height distribution is reasonably close to a Rayleigh distribution. One exception is in shallow water when the waves are about to break.

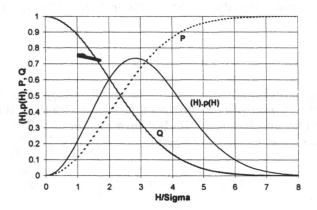

Figure 3.3 Rayleigh Distribution

The wave height with a probability of exceedence Q, may be determined from Eq. 3.8 as

$$H_Q = \sqrt{8\sigma^2 \, (-\ln Q)} = \sqrt{8\sigma^2 \, \ln\left(\frac{1}{Q}\right)} = 2\sigma \sqrt{2 \ln\left(\frac{1}{Q}\right)} \qquad (3.9)$$

To determine $\overline{H_Q}$, the average height of all the waves that are larger than H_Q in a record or a storm

$$\overline{H_Q} = \frac{\int_{H_Q}^{\infty} H \, p(H) \, dH}{Q} \qquad (3.10)$$

Equation 3.9 and the numerical evaluation of Eq. 3.10 yield Table 3.1, in which a number of common wave height definitions are related to σ. Of all these definitions,

Significant Wave Height (H$_s$) is the most important[2]. It is defined as the average of the highest 1/3 of the waves in a wave train, $\overline{H}_{1/3}$. In terms of significant wave height, four commonly used relationships based on the Rayleigh distribution are

$$\overline{H}_{0.1} = 1.27H_s; \quad \overline{H}_{0.01} = 1.67H_s$$
$$\overline{H} = 0.63H_s; \quad H_{rms} = 0.707H_s \tag{3.11}$$

Table 3.1 Commonly Used Wave Height Parameters

Symbol	Description	Value
$\overline{H}_{.01}$	Average of highest 1% of the waves	6.67σ
$H_{.01}$	Height, exceeded by 1% of the waves	6.07σ
$\overline{H}_{02}$	Average of the highest 2% of the waves	6.23σ
$H_{.02}$	Height, exceeded by 2% of the waves	5.59σ
$\overline{H}_{0.1}$	Average of highest 10% of the waves	5.09σ
$H_{0.1}$	Height exceeded by 10% of the waves	4.29σ
$H_s = \overline{H}_{1/3}$	Significant wave height (Average height of the highest 1/3 of the waves)	4.0σ
$\overline{H}$	Average wave height	$\sqrt{2\pi}\,\sigma$
$H_{0.5}$	Median wave height	2.35σ
H_{mode}	Most probable wave height	2.0σ
H_{rms}	$\dfrac{\sqrt{H_1^2 + H_2^2 + H_3^2 + \ldots}}{N}$	$2\sqrt{2}\,\sigma$

The meaning of average wave height is self-explanatory. The modal or most probable wave height is the wave height with the greatest probability of occurrence. The median wave height has 50% probability, i.e., half the waves in the wave train

---

2. This wave height definition was historically chosen as "significant" because it comes closest to the traditional estimates of average wave height by experienced observers before we had instruments to measure wave heights.

are higher and half the waves are lower than this wave height. The rms wave height is the constant wave height that represents the total energy of the whole wave height distribution.

The probability of exceedence for the average wave heights may now be calculated from their values in Table 3.1, using Eq. 3.8.   The results of this calculation are shown in Table 3.2.

Table 3.2 Probabilities of Exceedence of Average Wave Heights

H	$\bar{H}_{0.01}$	$\bar{H}_{0.1}$	H_s	$\bar{H}$	H_{mode}
Q(H'>H)	0.004	0.039	0.136	0.456	0.606

The expected value of the maximum wave in a wave train of N_w waves could be estimated by setting $Q = 1/N_w$ in Eq. 3.9. A more accurate estimate is

$$\mu\left[\frac{H_{max}}{2\sigma}\right] = \sqrt{2 \ln N_w} + \frac{\gamma}{\sqrt{2 \ln N_w}} + O[(\ln N_w)^{-3/2}] \qquad (3.12)$$

where $\mu(x)$ denotes "expected value" of x, γ is the Euler constant (=0.5772) and $O(x)$ denotes terms of order greater than x, i.e. small terms.

Example 3.1 Calculation of Short-Term Wave Heights

To analyze a wave record it must be stationary. Hence, it is normal to record waves for relatively short time durations (10 to 20 minutes). A longer record would not be stationary because wind and water level variations would change the waves. Thus it is usual to record, for example, 15 minutes every three hours. It is subsequently assumed that the 15 min. record is representative of the complete three hour recording interval.

Suppose the analysis of such a record yields

$$\bar{T} = 10 \sec \quad and \quad \sigma = 1.0 \ m \qquad (3.13)$$

We want to calculate significant wave height H_s, average wave height $\overline{H}$, average of the highest 1% of the waves $\overline{H}_{0.01}$, and the maximum wave height in the record.

From Table 3.1

$$H_s = 4\sigma = 4.0 \, m$$

$$\overline{H} = \sqrt{2\pi}\sigma = 2.5 \, m$$

$$\overline{H}_{0.01} = 6.67\sigma = 6.7 \, m$$

With $\overline{T} = 10 \sec$, the average number of waves in the 15 min record $N_w = 90$ and Eq. 3.12 yields

$$\mu\left[\frac{H_{max}}{2\sigma}\right] = \sqrt{2\ln(90)} + \frac{0.5772}{\sqrt{2\ln(90)}} = 3.00 + 0.19 = 3.19 \, m \qquad (3.14)$$

or the expected value of $H_{max}=(3.19)(2)(1.0)=6.4$ m. This calculated value of H_{max} can be verified against the actual record.

If the record is representative of a 3 hour recording interval, then $\overline{T}$, H_s, $\overline{H}$ and $\overline{H}_{0.01}$ for the 3 hours would be the same as above. However, H_{max} would be larger than 6.4 m. For the 3 hour recording interval, $N_w = 1080$ and Eq. 3.14 yields $H_{max,3hrs} = 7.8$ m.

3.3 Wave Period Distribution

In the above discussion the frequency bandwidth for the waves was assumed to be small (the wave periods are more or less the same) and in practice, wave period variability is often ignored. One attempt to define wave period distribution has been made by Bretschneider (1959) who postulated that the squares of the wave periods form a Rayleigh distribution. His expression for the PDF for wave periods is

$$p(T) = 2.7\frac{T^3}{\overline{T}^4}\exp\left[-0.675\left(\frac{T}{\overline{T}}\right)^4\right] \qquad (3.15)$$

From this, by integration, the expression for probability of exceedence of a certain wave period becomes

$$Q(T' > T) = \int_T^\infty p(T)dt = \exp\left[-0.675\left(\frac{T}{\overline{\overline{T}}}\right)^4\right] \qquad (3.16)$$

Wave periods are related to wave heights particularly in a growing, locally-generated sea, where high wave heights are always accompanied by long wave periods. A joint distribution of wave heights and periods can be postulated and that is normally assumed to be a joint Rayleigh distribution.

3.4 Time Domain Analysis of a Wave Record

Wave recordings are time series of water levels that typically look like Fig. 3.4. They are discrete time series z(t), sampled at N short intervals of Δt. The water level recording must first be converted into a discrete time series $\eta(t)$, the fluctuation about mean water level by subtracting the mean water level from the record.

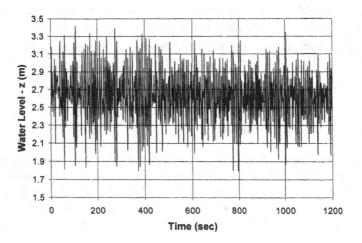

Figure 3.4: Water Level Record

Although the record is assumed to be stationary, there may be a small change in mean water level with time, as is the case in Fig 3.4. This could be, for example a result of tides. Because the record is short (20 min), the water level fluctuation is assumed to be a linear function of time, $\bar{z} = a + bt$. It is determined from the record by regression analysis and then subtracted, so that $\eta = z - \bar{z} = z - (a + bt)$. Figure 3.5 presents the η time series for Fig. 3.4. The bottom graph for the first 120 seconds shows more detail.

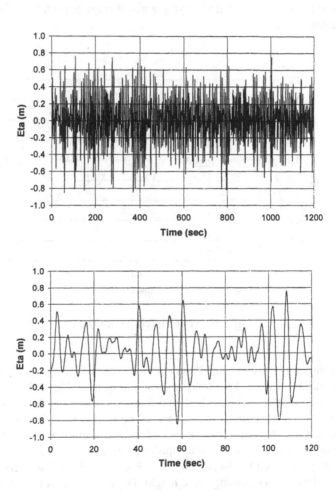

Fig. 3.5 Wave Record for Fig. 3.4

As an *initial analysis* of a record, we could simply calculate σ, using Eq 3.4. The wave analysis program **WAVAN**®was used and σ was found to be 0.28 m for the record in Fig. 3.5. If a Rayleigh distribution of wave heights is assumed (Tables 3.1 and 3.2). we can determine the values for the important wave heights for Fig 3.5.

$$H_s = 1.12\,m, \quad \overline{H} = 0.70\,m, \quad H_{rms} = 0.79\,m \quad and \quad \overline{H}_{0.1} = 1.43\,m \quad (3.17)$$

This initial analysis does not tell us anything about wave period, the other important wave parameter.

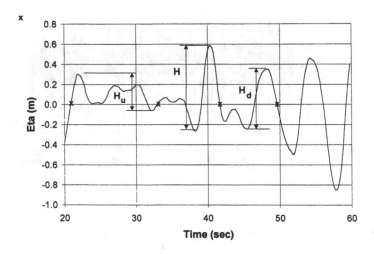

Fig. 3.6 Wave Height Definitions

The actual distributions of H and T may be obtained from the record by analysis of individual waves. Figure 3.6 shows a short segment of Fig. 3.5. First, the term *wave* must be defined. The earliest definition for *wave height* H is the vertical distance between a wave crest and the preceding trough (Fig. 3.6) where crest and trough are defined as a local maximum and a local minimum in the record. That definition would result in all the small ripples being identified as waves. How many waves are there between t = 20 and 60 sec in Fig. 3.5? To define wave height more realistically, zero down-crossing wave height, H_d, is defined as the vertical distance between the maximum and minimum water levels that lie between two subsequent zero down-crossings (in which η crosses zero on the way down). Similarly zero up-crossing wave height, H_u, is the difference between maximum and minimum water levels between two subsequent zero up-crossings. These definitions are also shown in Fig. 3.6. They disregard the small ripples that do not cross the mean water level. Example 3.2 presents the *zero crossing analysis* of the wave record of Fig. 3.5 and compares the results with the values in Eq. 3.17.

Example 3.2 Zero Crossing Analysis of Figure 3.5

The zero up-and down crossing wave heights in Fig. 3.5 were determined using **WAVAN**® . The two estimates were virtually identical. The following wave statistics were obtained by averaging the up-and down crossing results

$$H_s = 1.05\,m, \quad \overline{H} = 0.68\,m, \quad H_{rms} = 0.76\,m \quad and \quad \overline{H}_{0.1} = 1.30\,m \qquad (3.18)$$

When these calculated values are compared with Eq. 3.17 it is seen that Eq. 3.17 overpredicts for this record.

The wave heights were also grouped into 10 bins and the histogram of the wave heights is shown in Fig. 3.7.a. This distribution can be compared with the Rayleigh distribution. It would also be possible to plot the cumulative distribution function (P, as in Eq. 3.7) or the probability of exceedence (Q, as in Eq. 3.8). However, since the wave height distribution is expected to be Rayleigh, it is best to compare wave heights directly with this theoretically expected distribution. Equation 3.8 may be re-written as

$$-\ln(Q) = \left(\frac{H}{2\sqrt{2}\sigma}\right)^2 \qquad (3.19)$$

Thus,

$$R = \{-\ln(Q)\}^{1/2} = \frac{H}{2\sqrt{2}\sigma} \quad or \quad H = (2\sqrt{2}\sigma)R \qquad (3.20)$$

where R is called the Rayleigh parameter. A true Rayleigh distribution would plot as a straight line with zero intercept and a slope of $2\sqrt{2}\sigma$ on a graph of H vs R. Values of R were calculated for each wave height bin and H vs R was plotted in Fig. 3.7b.

The solid line in Fig. 3.7b represents the initial analysis, combining σ=0.28 m with Tables 3.1 and 3.2. This analysis clearly overpredicts the actually measured values of H.

The best fit line through the measured values has a slope of 0.69 and from this, using Eq. 3.20, another estimate for σ may be computed - $\sigma_z = 0.69/(2\sqrt{2}) = 0.24m$. This version is called σ_z since it was determined by zero-crossing analysis. An estimate of average wave period may be obtained by dividing the record length (t_R=1200 seconds) by the number of waves (N_w=197) to find $\overline{T}$=6.1 seconds.

Finally H_{max} was found to be 1.56 m from the record. This can be compared to the theoretical value of H_{max}=1.92m, obtained from Eq. 3.12, using σ=0.28 m; σ_z=0.24 m gives H_{max}=1.65 m.

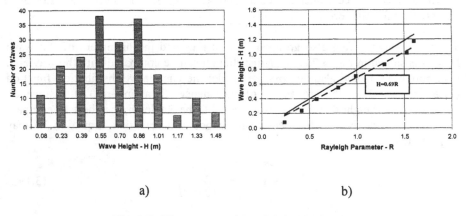

a)   b)

Fig. 3.7 Histogram and Rayleigh Distribution

3.5 Frequency Domain Analysis of a Wave Record

A completely different type of analysis, based on wave frequencies, is called *wave spectrum analysis.* We use the statistical assumptions that the wave record is both stationary and ergodic. Although these assumptions are necessary to perform a statistically correct analysis, in practice we have no choice but to assume that records are short enough to be both stationary and ergodic.

To express the time signal η in terms of frequency, we can use a Fourier series summation for each value of η_j

$$\eta_j = a_o + \sum_{n=1}^{\infty} a_n \cos(2\pi f_n t_j) + \sum_{n=1}^{\infty} b_n \sin(2\pi f_n t_j) \qquad (3.21)$$

Using Euler's relationship

$$e^{i\psi} = \cos\psi + i\sin\psi \qquad (3.22)$$

Eq. 3.21 becomes

$$\eta_j = \sum_{n=-\infty}^{\infty} C_n e^{i[2\pi(f_n t_j)]} \qquad (3.23)$$

where C_n is a complex coefficient

$$C_o = a_o; \quad C_n = \frac{1}{2}(a_n - ib_n); \quad C_{-n} = \frac{1}{2}(a_n + ib_n) = C_n{}^* \qquad (3.24)$$

Equation 3.23 expresses the time signal, η, in terms of discrete frequency components and is known as a Discrete Fourier Transform (DFT). The series in Eqs. 3.21 and 3.23 are infinite. However, a wave record, such as in Fig 3.5 is neither infinitely long, nor continuous. The water level is sampled only at N specific times, Δt apart. There are N values of (η_j) at times $t_j = j\Delta t$, where $1 \le j \le N$. As a result, the frequency domain is also not continuous. The smallest frequency that can be defined from a record of length t_R is $f_{min} = 1/t_R$. To provide the most accurate representation in the frequency domain (best resolution) we will use this smallest possible frequency as the frequency increment. Therefore $\Delta f = 1/t_R$ and we define $f_n = n\Delta f$. The highest frequency that can be defined from a time series with increments Δt is the Nyquist frequency

$$f_N = \frac{1}{2\Delta t} = \frac{N}{2t_r} = \frac{N}{2}\Delta f \qquad (3.25)$$

This results in the finite discrete Fourier transform (FDFT)

$$\eta_j = \sum_{n=-\frac{N}{2}+1}^{\frac{N}{2}} F_n e^{i[2\pi(f_n t_j)]} \qquad (3.26)$$

Because $F_n = F_{-n}{}^*$, Eq. 3.26 may also be written as:

$$\eta_j = \sum_{n=0}^{N-1} F_n e^{i[2\pi(f_n t_j)]}$$

(3. 27)

The inverse of Eq. 3.27 is

$$F_n = \frac{1}{N} \sum_{j=1}^{N} \eta_j e^{-i[2\pi(f_n t_j)]}$$

(3.28)

Equation 3.26 and 3.27 form a FDFT pair that permits us to switch between the time and frequency domains. Equation 3.26 allows us to calculate the complex frequency function F_n from a real time function η_j and Eq. 3.27 permits calculation of the real time function η_j from the complex frequency function F_n.

The complex variable F_n may also be expressed as

$$F_n = |F_n| e^{-i\theta_n}$$

(3.29)

where

$$|F_n| = \frac{1}{2} \sqrt{a_n^2 + b_n^2} \quad and \quad \theta_n = \tan^{-1}\left(-\frac{b_n}{a_n}\right)$$

(3.30)

Substitution of Eq. 3.30 into Eq. 3.29 results in

$$\eta_j = \sum_{n=0}^{N-1} |F_n| e^{i[2\pi(f_n t_j)-\theta_n]}$$

(3.31)

In practice waves have only positive frequencies, only frequencies lower than f_N can be defined, and η_j is real. Therefore, using Eq. 3.22, we can rewrite Eq. 3.32 as

$$\eta_j = \sum_{n=0}^{N-1} |F_n| \cos(2\pi ft - \theta_n) \doteq$$

$$\sum_{n=0}^{N/2} 2|F_n| \cos(2\pi ft - \theta_n) = \sum_{n=0}^{N/2} |A_n| \cos(2\pi ft - \theta_n)$$

(3.32)

where

$$| A_n | = 2 | F_n |$$ (3.33)

Here $|A_n|$ is called the amplitude spectrum and θ_n the phase spectrum. In standard wave spectrum analysis only the amplitude spectrum $|A_n|$ is calculated. In effect, θ_n is assumed to be a random variable $-\pi < \theta_n < \pi$ resulting in the *random phase model*. This unfortunate assumption loses all phase relationships between the N terms, which means, for example, that wave groups are not reproduced when calculating η_j from $|A_n|$ using with random θ_n. Resonance and reflection patterns are also not reproduced.

Parseval's theorem can be used to calculate σ from the amplitude spectrum $|A_n|$ because

$$\sigma^2 = \overline{\eta^2} = \sum_{n=-\infty}^{\infty} |C_n|^2 = \sum_{n=0}^{N-1} |F_n|^2 = \sum_{n=0}^{N/2} \frac{1}{2} |A_n|^2$$ (3.34)

Thus the variance at any frequency can be expressed as

$$S(f_n)df = \frac{1}{2} |A_n|^2 \quad or \quad S(f_n) = \frac{1}{2df} |A_n|^2$$ (3.35)

where S(f) is known as the wave variance spectral density function or *wave spectrum*. Variance is a statistical term. In physical terms, wave energy density is

$$E = \rho g \sigma^2$$ (3.36)

and hence wave energy distribution as a function of frequency is

$$E(f) = \rho g \, S(f)$$ (3.37)

Wave spectra for Fig. 3.5 were computed using **WAVAN®** and are shown in Fig. 3.8. Because we always have only one realization of the process and the record length (t_R) is finite, resulting in finite increments of frequency (Δf), the calculated value of S(f) is always an estimate of the true S(f).

The wave spectrum, for the record in Fig. 3.5 as produced by Eqs. 3.32 to 3.35 is shown in Fig. 3.8a. This spectrum gives the maximum possible resolution and distinguishes between frequencies that are $\Delta f = 1/t_R = 1/1200 = 0.00083$ Hz apart. It contains many closely spaced spikes of wave energy. Physically, such very local energy concentrations are not possible. They are a result of the uncertainties in our estimates and therefore the wave spectrum is smoothed. That can be done by averaging S(f) over frequency ranges longer than Δf so that

$$S(f) = \frac{1}{df'} \sum_{m=-M/2}^{m=M/2} |A_{n+m}|^2 \quad \text{where} \quad df' = M_f \Delta f \qquad (3.38)$$

where df' is the resolution of the spectrum and M_f denotes how many values of Δf are averaged. The results for M_f=6 and 12 (df'=0.005 and 0.01 Hz) are shown in Figure 3.8b and c. Smoothing produces a more regular spectrum. The amount of smoothing to be used depends on the purpose of the analysis. If a general impression of a wave field is needed, Fig 3.8c is most useful. If specific frequencies need to be identified, then less smoothing such as in Fig 3.8b may be more appropriate. Another method used for smoothing the spectrum is to divide the record into shorter sections, compute the spectrum for each section and then average the results. Fig. 3.8d, presents the average of 4 spectra, each for ¼ t_R. The resolution is now 4/t_R=0.0033, and averaging the four spectra introduces some further smoothing.

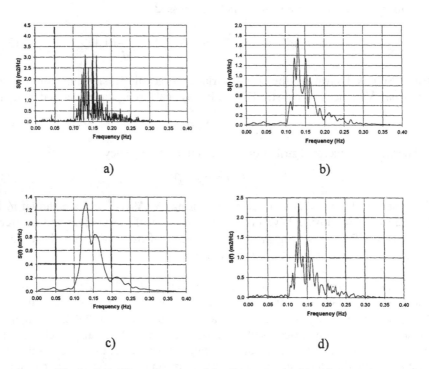

Figure 3.8 Wave Spectra of the Wave Record in Figure 3.5

The value of σ, as computed from the frequency analysis by integrating the spectrum of σ is denoted as σ_f because it was derived by frequency analysis. Because Eqs. 3.4 and 3.34 both integrate η^2, σ_f for all four spectra in Fig. 3.8, as well as σ found using Eq. 3.4, are the same and equal to 0.28.

Only the basic principles of wave spectrum are presented here. There are other methods of computing the wave spectra and dealing with smoothing of the spectra. Further details may be found in the literature, such as Bendat and Piersol (1966) and ASCE (1974) and Janssen (1999).

Frequencies that exceed the Nyquist frequency (Eq. 3.25) cannot be defined as separate frequencies. The energy in these high frequency waves is superimposed on the spectrum by a process known as *aliasing*. Figure 3.9a demonstrates aliasing in the time domain. A wave of frequency 1.1 Hz is sampled at $\Delta t=1.0$, for which $f_N=0.5$ Hz. It is seen that the sampled signal (square points) does not have a frequency of 1.1 Hz, but of 0.1 Hz. The energy of such a wave component would therefore become added at 0.1 Hz in a wave spectrum. Aliasing in the frequency domain is depicted in Fig. 3.9b.

Aliasing can be prevented by filtering frequencies greater than f_N out of the signal. Alternately, if f_c is the highest frequency that must be computed correctly (without aliasing), then it is reasonable to assume that $f_c=f_N/2$. That defines the necessary sampling interval for the record as $\Delta t \leq 1/(2f_N)=1/(4f_c)$. For wind waves, if we wish to define the spectrum correctly for all frequencies $f<2Hz$, Δt should be less than 1/8 sec.

3.6 Parameters Derived from the Wave Spectrum

The moments of the wave spectrum are defined as

$$m_h = \int_{f=0}^{f=\infty} f^h S(f)\, df \qquad (3.39)$$

The zero moment (n=0) is therefore the area under the spectrum

$$m_0 = \int_{f=0}^{f=\infty} S(f)\, df = \sigma_f^2 \qquad (3.40)$$

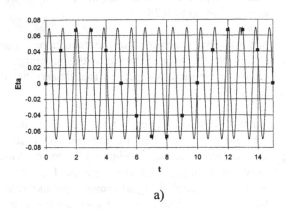

a)

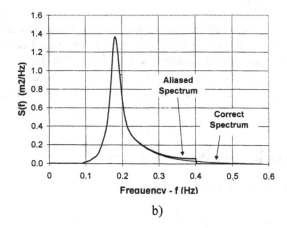

b)

Figure 3.9 Aliasing of a Wave Spectrum by High Frequency Components

From the area under the wave spectrum, assuming the wave height distribution to be Rayleigh, the various wave heights of Table 3.1 may be estimated as in Eq. 3.17. To distinguish between significant wave height derived from time domain analysis and its counterpart derived from frequency analysis, the latter is called the Characteristic Wave Height or Zero Moment Wave Height.

$$H_{ch} = H_{mo} = 4\,\sigma_f \qquad\qquad (3.41)$$

and for Fig 3.5, $H_{mo} = (4)(0.28) = 1.12$ m.

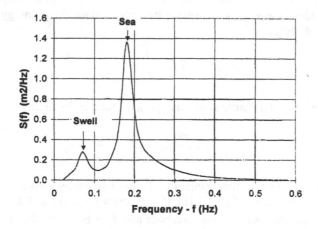

Figure 3.10 Wave Spectrum with Sea and Swell.

The representation of the wave energy distribution with frequency is an improvement over the time-domain analysis methods discussed earlier. With this information we can study resonant systems such as the response of drilling rigs, ships' moorings, etc. to wave action, since it is now known in which frequency bands the forcing energy is concentrated. It is also possible to separate sea and swell when both occur simultaneously (Fig. 3.10).

The moments of the wave spectrum also define spectral bandwidth

$$\varepsilon^2 = 1 - \left\{ \frac{m_2^2}{m_o m_4} \right\} \tag{3.42}$$

Cartwright and Longuet-Higgins (1956) show that for a narrow bandwidth ($\varepsilon \rightarrow 0$) all wave periods in a wave train are almost the same and the distribution of η is purely Rayleigh[3]. For $\varepsilon \rightarrow 1$, the distribution of η is random. This would obviously affect the wave height definitions used in Tables 3.1 and 3.2. For the record in Fig. 3.5, ε^2 was calculated to be 0.65.

Since there are many wave frequencies (or wave periods) represented in the spectrum it is usual to characterize the wave spectrum by its peak frequency f_p, the frequency at which the spectrum displays its largest variance (or energy). The peak period is then defined as

$$T_p = \frac{1}{f_p} \tag{3.43}$$

Other spectrum-based definitions of wave period found in the literature are

$$T_l = \frac{m_0}{m_l} ; \quad T_2 = \sqrt{\frac{m_0}{m_2}} \tag{3.44}$$

Theoretically T_2 is approximately equal to $\overline{T}$, as obtained by zero crossing analysis. For the spectra in Fig. 3.8, $f_p = 1.3$ Hz, if the narrow peak in Fig 3.8a is discounted. Thus $T_p = 7.6$ seconds and $T_1 = 6.4$ seconds and $T_2 = 6.1$ seconds. $\overline{T}$ was also 6.1 seconds in Example 3.2. Rye (1977) indicates that the moments for the spectrum are functions of the cutoff frequency (the highest frequency considered in the analysis) and thus ε, T_1 and T_2 should be viewed with caution.

Sometimes the angular frequency, ω, is used to define the frequencies in the wave spectrum. The total variance for the S(ω) spectrum is

$$\sigma_\omega^2 = \frac{1}{2\pi} \int_{\omega=0}^{\omega=\infty} S(\omega) d\omega \tag{3.45}$$

The results of the three methods of analysis for the waves of Fig. 3.5 are compared in Table 3.3.

---

3. The Rayleigh distribution is in fact based on the assumption that $\varepsilon \rightarrow 0$ (Section 3.2).

Table 3.3 Comparison of Analysis Methods

	Analysis		
	Initial	Zero Crossing	Frequency
σ (m)	σ =0.28	σ_z =0.24	σ_f =0.28
N		197	
T (sec)		$\overline{T}$ = 6.1	T_p = 7.6
			T_1 = 6,4
			T_2 = 6.1
H_s (m)	1.12	1.05	H_{mo} = 1.11
$\overline{H}$ (m)	0.70	0.68	0.70
H_{rms} (m)	0.79	0.76	0.79
$\overline{H}_{0.1}$ (m)	1.43	1.30	1.43
H_{max} (m)	1.92	1.56	1.92
ε^2			0.65

3.7 Uncertainties in Wave Measurements

At this point, we reflect upon how well we know the wave parameters: wave height, wave period and wave angle. First we define *uncertainty*. It quantifies the combination of errors, randomness and general lack of physical understanding. For most physical quantities, errors increase with the magnitude of the quantity. For example, the absolute error in measuring a wave height of 0.5 m will be less than the absolute error in measuring a wave height of 5 m. For this reason we normally use a relative error to define the accuracy of our quantities. The errors in a quantity such as wave height H are assumed to have a normal distribution with $\overline{H}$ as its mean value and σ_H as its standard deviation. The uncertainty in H is then defined as its coefficient of variation

$$\sigma'_H = \frac{\sigma_H}{\overline{H}} \tag{3.46}$$

More detail about uncertainties may be found in Thoft-Christensen and Baker (1982), Ang and Tang (1984), Madsen, Krenk and Lind (1986), Pilarczyk (1990), Burcharth (1992) and PIANC (1992).

From the definition of standard deviation, H is between $\overline{H}(1 \pm \sigma'_H)$ 68% of the time, there is a 95% probability that H is between $\overline{H}(1 \pm 2\sigma'_H)$, and virtually all values lie between $\overline{H}(1 \pm 3\sigma'_H)$.

Wave heights are based on measurements of instantaneous water levels, usually measured offshore, at frequent intervals (e.g. 10 Hz) over a recording period (e.g. 10 to 20 Minutes). Zero Crossing or Wave Spectrum Analysis is then used to reduce the instantaneous water level measurements to one single wave height value (H_s or H_{mo}) to represent the complete recording period. Along an exposed coast H_s=1m would be typical. Even for very carefully measured instantaneous water levels, using the latest equipment H_s=1m would contain an absolute error (standard deviation) $\sigma_{H,Measured}$=0.05 to 0.1 m (say 0.075 m). The uncertainty in a 1 m wave height would therefore be $\sigma'_{H,Measured}$=0.075. The errors in measuring smaller waves would be less and for larger waves they would be greater. Therefore an uncertainty $\sigma'_{H,Measured}$=0.075 would not be unreasonable for wave height measurements. For H=1m, there would be a 68% probability that 0.92<H<1.08 m, a 95% probability that 0.85<H<1.15 m and almost all values of H will lie between 0.78<H<1.22.

Such relatively accurate offshore wave height measurements are subjected to several conversions before they can be considered useful for subsequent computations. So far, the value of H_s=1m represents 10 to 20 minutes of record. For the 1 m wave height to represent a complete recording interval of 3 to 6 hours, it must be remembered that the environmental parameters such as wind speed and direction, water levels, etc. are not constant over the recording interval. This increases the uncertainty of the representative wave height values. The additional uncertainty depends on the variability of the conditions over the recording interval, but in most cases it would be reasonable to expect the uncertainty to double so that $\sigma'_{H,Interval}$=0.15.

Uncertainty of measured wave periods T is known to be greater than for H and reasonable estimates would be $\sigma'_{T,Measured}$=0.1 and $\sigma'_{T,Interval}$=0.2. Wave direction (α) is notoriously poorly measured. Even the best directional instrumentation has difficulty to produce wave directions within ± 3° for large, well-formed waves and may be as much as 10° wrong for smaller, more irregular waves. Estimates of wave direction by other means than directional measurement are much worse. Assuming the values of 3° and 10° to be maximum values of the errors in angle (assuming these values to be $3\sigma_\alpha$ removed from the mean), σ_α can be estimated as 1° in the first case and 3.3^0 in the second. Thus an average value of standard deviation is $\sigma_{\alpha,Measured}$=2°. This value of σ_α is independent of the incident wave angle and hence we cannot define σ_α'. However, in order to complete our subsequent discussion about uncertainties, we will relate σ_α to an incident wave angle of $10°$ with respect to the shoreline. In that case, $\sigma'_{\alpha,Measured}$ becomes 0.2. Many times the incident wave angle on a sandy shore is much smaller than $10°$, which would result in much higher uncertainty values. When the wave angle with the shoreline approaches $0°$ (as is often the case), the uncertainty for wave angle approaches infinity and the whole discussion about uncertainty loses its meaning. For the longer interval and $\alpha=10°$, a reasonable estimate is $\sigma'_{\alpha,Interval}=4°$.

These uncertainty values are only general indications. They are heavily influenced by assumed average values for wave heights and periods, and particularly angles of breaking. The actual values are not as important, however, as the fact that they clearly indicate that the basic coastal data of wave heights, wave periods and incident wave angles contain large uncertainties. Since these wave quantities are basic to all coastal design calculations, the effects of these uncertainties will pervade all subsequent calculations. The awareness of uncertainties is basic to our understanding of the fundamental issues of coastal engineering and management. For example, it explains why we can use small amplitude wave theory successfully for most design calculations. The discussion and evaluation of uncertainties will be extended in later chapters.

3.8 Common Parametric Expressions for Wave Spectra

Since the measured spectra show considerable similarity (they basically consist of a peak and two curves decreasing toward $f=0$ and $f=\infty$), attempts have been made to formulate parametric expressions. Only the most common expressions will be presented here. Phillips (1958) postulated that for the "equilibrium range" (for $f > f_p$) the spectral shape $S(f)$ is proportional to f^{-5}. He quantified his results as

$$\Phi_P = \frac{\alpha_P}{2\pi^4} g^2 f^{-5} \tag{3.47}$$

where Φ_p denotes the Phillips Function and the "Phillips constant" is

$$\alpha_P = 0.0074 \tag{3.48}$$

Pierson and Moskowitz (1964), added a low frequency filter to extend the Phillips expression over the complete frequency range

$$S_{PM}(f) = \Phi_P \, \Phi_{PM} \tag{3.49}$$

where:

$$\Phi_{PM} = \exp\left[-\frac{5}{4}\left(\frac{f}{f_p}\right)^{-4} \right] \tag{3.50}$$

The Pierson-Moskowitz spectrum is therefore

$$S_{PM}(f) = \frac{\alpha_{PM}}{(2\pi)^4} \frac{g^2}{f^5} e^{-(\beta/f^4)} \tag{3.51}$$

Commonly used expressions for α and β are:

$$\alpha_{PM} = 0.0081 \quad ; \quad \beta = \frac{5}{4} f_p^4 \tag{3.52}$$

The quantity β was also related to wind speed U so that this spectrum can be used to hindcast waves from wind data (Ch. 5)

$$\beta = 0.74\left(\frac{g}{2\pi U}\right)^4 \tag{3.53}$$

where U is the wind speed. The Pierson-Moskowitz spectrum is valid for a fully developed sea condition. For developing seas the Jonswap Spectrum was proposed by Hasselman et al (1973). It is essentially an enhanced Pierson-Moskowitz spectrum as shown in Fig. 3.11.

A developing seas filter, Φ_J, can be assumed so that the Jonswap spectrum is

$$S_J(f) = \Phi_P \cdot \Phi_{PM} \cdot \Phi_J \tag{3.54}$$

where

$$\Phi_J = \gamma^{e^a} \tag{3.55}$$

and

$$a = \left[\frac{-(f - f_p)^2}{2\delta^2 f_p^2} \right] \tag{3.56}$$

Typical values of δ are

$$\delta = 0.07 \quad for \quad f \le f_p$$
$$\delta = 0.09 \quad for \quad f > f_p \tag{3.57}$$

The Jonswap expression is therefore

$$S_J(f) = \gamma^{e^a} S_{PM}(f) = \frac{\alpha_J g^2 \gamma^{e^a}}{(2\pi)^4 f^5} e^{-\frac{5}{4}\left(\frac{f}{f_p}\right)^{-4}} \tag{3.58}$$

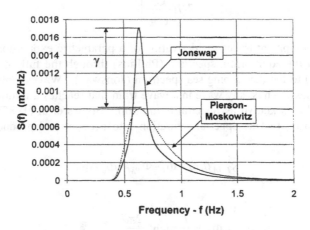

Fig 3.11 Jonswap and Pierson-Moskowitz Spectra

The coefficient α can be related to the wave generating conditions

$$\alpha_J = 0.076 \left(\frac{gF}{U^2} \right)^{-0.22} \tag{3.59}$$

where F is fetch length. Alternately, Mitsuyasu (1980) states

$$\alpha_M = 0.0817 \left(\frac{gF}{U_{10}^2} \right)^{-2/7} \tag{3.60}$$

and

$$f_p = 2.84 \left(\frac{gF}{U^2} \right)^{-0.33} \tag{3.61}$$

The peak enhancement factor

$$\gamma = \frac{S_J(f_p)}{S_{PM}(f_p)} \tag{3.62}$$

has an average value of 3.3 and typically lies between 1 and 7. Mitsuyasu (1980) postulates

$$\gamma = 7.0 \left(\frac{gF}{U_{10}^2} \right)^{-1\,7} \tag{3.63}$$

The above development traces the derivation of a parametric expression for a wave spectrum from the equilibrium spectrum (Phillips) through the fully developed sea spectrum (PM) to the developing sea spectrum (Jonswap). Waves, however have a limiting steepness. Thus, any wave in a wave train that reaches a limiting steepness will break. This is known as Spectral Saturation. Bouws et al (1985) modify the Jonswap spectrum to take spectral saturation into account and produce the TMA spectrum.

$$S_{TMA}(f,\,d) = \Phi_P \cdot \Phi_{PM} \cdot \Phi_J \cdot \Phi_d \tag{3.64}$$

where

$$\Phi_d = \frac{1}{2n} \tanh^2 \frac{2\pi d}{L} \tag{3.65}$$

In deep water, the value of Φ_d is one. In other words, the Jonswap spectrum through its own derivation takes into account the deep water wave steepness limitation and Eq. 3.63 modifies the Jonswap spectrum for wave breaking induced in shallow water.

3.8 Directional Wave Spectra

Until now η has been considered to be a function of time at a single location and we learned to calculate S(f) from such a time series. We also discussed some parametric expressions for such wave spectra. However, η is also a function of direction (of x and y). Measurement of wave direction involves correlating spectra for several synoptic, adjacent records of water levels, pressures and/or velocities. Some discussion may be found in Section 2.3.1, but detailed discussion is beyond the scope of this book. An example of a Directional Wave Spectrum (a function of both wave frequency and direction) is shown in Fig. 3.12 and a good description of directional wave spectra may be found in Goda (1985).

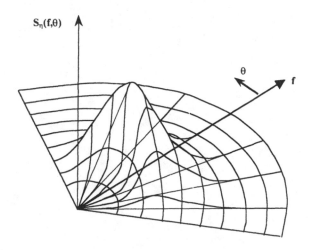

Fig. 3.12 Directional Wave Spectrum

To describe such a spectrum, the simplest approach is

$$S(f, \theta) = S(f) G(\theta) \tag{3.66}$$

G is called the Directional Spreading Function and θ is measured counter-clockwise from the wave direction. A necessary condition is obviously

$$\int_{\theta=-\pi}^{\theta=\pi} G(\theta)\, d\theta = 1 \tag{3.67}$$

Two common directional spreading functions used are the Cos-Squared function

$$G(\theta) = \frac{2}{\pi} \cos^2 \theta \quad \text{for } |\theta| < \frac{\pi}{2} \tag{3.68}$$

$$G(\theta) = 0 \text{ for all other values of } \theta$$

and the Cos-Power function (Mitsuyasu, 1980; Goda, 1985).

4. Long-Term Wave Analysis

4.1 Introduction

The analysis of long-term wave data provides a theoretical distribution of probability of occurrence of wave parameters over several years. Such analysis is most commonly carried out on long-term wave *height* data; a series of observed or hindcast wave heights spanning years or decades. Each wave height in the data series summarizes a short-term wave condition as discussed in Ch. 3, and thus represents waves existing over several hours. Normally, the parameter used to summarize the short-term wave height distribution is the significant wave height. Long-term distributions of wave periods and wave angles are usually considered to be a function of the long-term wave height distribution.

Long-term wave height analysis has two specific purposes: to organise the wave height data and to extrapolate the data set to extreme (high) values of wave heights occurring at low probabilities of exceedence. There are a number of ways in which this can be done. In this text, least squares regression analysis is used, simply because it is the most readily available and most universally understood statistical tool. This chapter will look at two basic methods of determining extreme wave height values – from *grouped data* obtained from a complete long-term data set, and from *ordered data* derived using a limited number of extreme values.

The principles will be presented using an example data set for wave heights on the East Coast of Lake Huron. This data set consists of 34.9 years of hourly significant wave heights obtained by wave hindcasting (Ch. 5). A one-month sample of this data set is plotted in Fig. 4.1. In Table 4.1 the wave heights for the whole data set are grouped into wave height bins of 0.25 m. The number of hours of occurrence of each wave height is shown in Column (3).

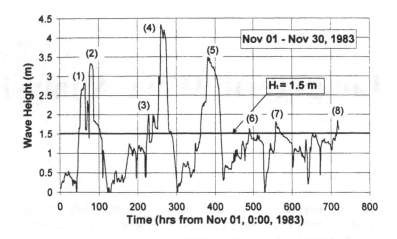

Figure 4.1 One Month of Waves on Lake Huron

4.2 Statistical Analysis of Grouped Wave Data

For the example wave data set, there are 282,306 values of hourly wave height. Theory of statistics requires that the individual data points used in a statistical analysis be statistically independent. From Fig. 4.1 it is clear that any hourly wave height depends very much on the wave heights of the previous hours and thus the theoretical condition of statistical independence is not met. To produce independent data points, we need to think of storms rather than individual hourly wave heights. The commonly used method to separate waves heights into "storms" is called Peak Over Threshold (POT) analysis. The basis for this method is demonstrated in Fig. 4.1. A Threshold Wave Height (H_t) of 1.5 m is arbitrarily introduced. Essentially this means a storm is defined as the time when the wave height exceeds 1.5 m. The only data points used in the POT analysis are the peaks (maximum wave heights) occurring during each storm. The month of record in Fig. 4.1 shows 8 occasions in which the wave height exceeded 1.5 m and the peaks of these 8 "storms" provide 8 data points. Using H_t = 1.5 m for the complete Lake Huron data set produces Column (4) of Table 4.1.

Table 4.1 Lake Huron Wave Heights - Hours of Occurrence

Bin Limits (m)		Threshold Wave Heights (m)			
Min	Max	0.0	1.5	3.0	4.0
(1)	(2)	(3)	(4)	(5)	(6)
0.000	0.250	52578			
0.251	0.500	60473			
0.501	0.750	50410			
0.751	1.000	35781			
1.001	1.250	25367			
1.251	1.500	18150			
1.501	1.750	12678	1019		
1.751	2.000	8894	549		
2.001	2.250	5961	382		
2.251	2.500	4143	254		
2.501	2.750	2826	174		
2.751	3.000	1869	113		
3.001	3.250	1238	81	121	
3.251	3.500	865	60	76	
3.501	3.750	483	40	44	
3.751	4.000	274	27	28	
4.001	4.250	171	19	19	24
4.251	4.500	87	10	10	10
4.501	4.750	29	4	4	4
4.751	5.000	8	2	2	2
5.001	5.250	11	1	1	1
5.251	5.500	4	2	2	2
5.501	5.750	4	0	0	0
5.751	6.000	2	1	1	1
	Total	282306	2738	308	44
	λ	8089	78.45	8.82	1.26

The data set represents maximum wave heights during storms where a storm was (rather arbitrarily) defined as when the wave height exceeds 1.5 m. It is not clear from Fig. 4.1 what H_t should be in order to make the storm peaks independent. Perhaps H_t=1.5 is not high enough. Theoretically this question can be answered by a correlation analysis. Columns (5) and (6) of Table 4.1 show the POT analysis for H_t=3.0 and 4.0 m respectively. All three of these data sets will be used in Section 4.5 and we will see that the final results of the Lake Huron data set are not very sensitive to the initially chosen values of H_t.

4.3 Transformation of Coordinate Axes

A probability that any wave height H' is less than a specified wave height H is defined as

$$P = P(H' < H) \tag{4.1}$$

Plotting P against wave height, results in the Cumulative Distribution Function (CDF). A Probability of Exceedence that H' is greater than a specified wave height H may also be defined as

$$Q = Q(H' > H) = 1 - P \tag{4.2}$$

Table 4.2 Analysis for Lake Huron Data with H_t=1.5 m

(1)	(2)	(3)	(4)	(5)	(6)	(7)	(8)	(9)
H	N	P	Q	z	ln H	G	W α=0.8	W α=1.3
1.75	1019	0.372	0.628	-0.326	0.560	0.012	0.384	0.555
2.00	549	0.573	0.427	0.183	0.693	0.584	0.816	0.883
2.25	382	0.712	0.288	0.560	0.811	1.081	1.316	1.184
2.50	254	0.805	0.195	0.859	0.916	1.528	1.848	1.459
2.75	174	0.869	0.131	1.119	1.012	1.959	2.421	1.723
3.00	113	0.910	0.090	1.339	1.099	2.359	2.996	1.964
3.25	81	0.939	0.061	1.550	1.179	2.772	3.627	2.210
3.50	60	0.961	0.039	1.766	1.253	3.232	4.366	2.477
3.75	40	0.976	0.024	1.976	1.322	3.713	5.176	2.750
4.00	27	0.986	0.014	2.190	1.386	4.244	6.105	3.044
4.25	19	0.993	0.007	2.442	1.447	4.916	7.326	3.406
4.50	10	0.996	0.004	2.683	1.504	5.611	8.638	3.769
4.75	4	0.998	0.002	2.849	1.558	6.122	9.632	4.031
5.00	2	0.99854	0.00146	2.976	1.609	6.528	10.436	4.234
5.25	1	0.99890	0.00110	3.063	1.658	6.816	11.014	4.377
5.50	2	0.99963	0.00037	3.378	1.705	7.915	13.276	4.910
5.75	0	0.99963	0.00037	3.378	1.749	7.915	13.276	4.910
6.00	1	1.00000	0.000					
Total	2738							

The data for $H_t=1.5$ m will now be analysed. The number of occurrences over the 34.9 years is set out in Column (4) of Table 4.1. These values must first be added and divided by the total number (2738) to yield P. For example, 1019 results are less than 1.75 m high. Therefore $P(H'<1.75)=1019/2738=0.372$. Similarly, (1019+549)=1568 results in Column (4) are less than 2.00 m. Therefore $P(H'<2.00) = 1568/2738=0.573$. This computation for all the values in Column (4) of Table 4.1 is given in Column (2) of Table 4.2. The results are plotted in Fig. 4.2. It is seen that the resulting CDF organizes the long-term wave height data, but is difficult to extrapolate. Since the most robust relationship for both interpolation and extrapolation is a straight line, a CDF such as in Fig. 4.2 needs to be transformed into a straight line by transforming the axes of the graphs. The equation for the transformed linear model will then be

$$Y = AX + B \qquad (4.3)$$

Here Y is the transformed probability axis, often called the reduced variate, and X is the transformed wave height axis. The coefficients A and B are the slope and intercept of the straight line relationship and they are determined by linear regression analysis.

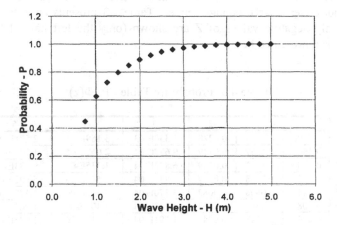

Figure 4.2 Cumulative Distribution Function for Data with $H_t=1.5$ m

4.3.1 Normal Probability Distribution

Although there is no reason to believe that the long-term probabilities of wave height are normally distributed, we will begin with this common distribution, because it forms the basis for the Log-Normal distribution of Section 4.3.2. The equation for the CDF, based on a normal distribution with sample mean $\overline{H}$ and sample standard deviation s_H is

$$P = \frac{1}{s_H \sqrt{2\pi}} \int_{-\infty}^{H} e^{-\frac{1}{2}\left(\frac{H-\overline{H}}{s_H}\right)^2} = \Phi\left(\frac{H-\overline{H}}{s_H}\right) = \Phi(Z) \qquad (4.4)$$

where H is wave height, $\overline{H}$ mean wave height, s_H standard deviation of wave height and Z the standard normal variate

$$Z = \left(\frac{H-\overline{H}}{s_H}\right) = \frac{1}{s_H} H - \frac{\overline{H}}{s_H} \qquad (4.5)$$

Standard Normal Probability Tables are used to relate P to Z. However, because we are interested in extreme values, the common versions of the normal probability tables do not cover a large enough range. Table 4.3 presents an extended range. Note that only negative values of Z are shown (only the left half of the normal distribution is represented here).

Table 4.3 Probability Table - $P=\Phi(z)$

Z	0	-1	-2	-3	-4	-5
0	0.5000	0.1587	0.0228	1.350E-03	3.169E-05	2.871E-07
-0.1	0.4602	0.1357	0.0179	9.677E-04	2.067E-05	1.701E-07
-0.2	0.4207	0.1151	0.0139	6.872E-04	1.335E-05	9.983E-08
-0.3	0.3821	0.0968	0.0107	4.835E-04	8.546E-06	5.802E-08
-0.4	0.3446	0.0808	0.0082	3.370E-04	5.417E-06	3.340E-08
-0.5	0.3085	0.0668	0.0062	2.327E-04	3.401E-06	1.904E-08
-0.6	0.2743	0.0548	0.0047	1.591E-04	2.115E-06	1.075E-08
-0.7	0.2420	0.0446	0.0035	1.078E-04	1.302E-06	6.008E-09
-0.8	0.2119	0.0359	0.0026	7.237E-05	7.944E-07	3.326E-09
-0.9	0.1841	0.0287	0.0019	4.812E-05	4.799E-07	1.824E-09

If Z is known, such a table will yield P=Φ(z) as defined by Eq. 4.4. For example, if Z=-3.4, then P=3.37x10^{-4}. Since the standard normal probability distribution is symmetrical, if Z=+3.4, P=1-3.37x10^{-4}=0.999663. The tables can also be used in reverse: Z can be obtained from P. We define that inverse operation symbolically as

$$Z = \Phi^{-1}(P) \tag{4.6}$$

If P=0.01, Z=-2.33; if P=0.99, z=+2.33. In addition to such tables, series solutions for P=Φ(Z) and Z=Φ^{-1}(P) may be found in Abramowitz and Stegun (1965) and most computer software provides these functions directly.

Equation 4.5 shows that Z is actually a linear function of H and therefore, the appropriate transformation of axes for a normally distributed CDF would be

$$Y = Z = \Phi^{-1}(P); \quad X = H \tag{4.7}$$

The slope and intercept of that straight line, according to Eq. 4.5, would be

$$A = \frac{1}{s_H}; \quad B = -\frac{\overline{H}}{s_H} \tag{4.8}$$

This transformation can test if a series of points is normally distributed. The test for normality is: Are the points on a straight line? The Y-Axis transformation of Eq. 4.7 is presented in Column (5) of Table 4.2. When this is plotted against H, it is seen in Fig. 4.3 that the points are much closer to a straight line than in Fig. 4.2. However, they do not form a straight line, which means that the points are not normally distributed.

4.3.2 Log-Normal Probability Distribution

The variables Z=Φ^{-1}(P) and ln H may be found in Columns (5) and (6) of Table 4.2. Using these values as Y and X produces almost a straight line (Fig. 4.4). The CDF of P vs ln H for this data set is therefore approximately log-normal, and

$$Y = Z = \Phi^{-1}(P) = \frac{\ln H - \overline{\ln H}}{s_{\ln H}} = \frac{1}{s_{\ln H}} \ln H - \frac{\overline{\ln H}}{s_{\ln H}} \tag{4.9}$$

or

$$Y = \Phi^{-1}(P); \quad X = \ln H; \quad A = \frac{1}{s_{\ln H}}; \quad B = -\frac{\overline{\ln H}}{s_{\ln H}} \tag{4.10}$$

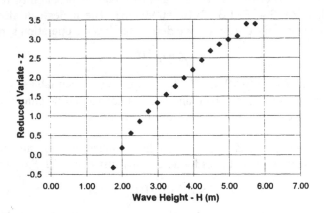

Figure 4.3 Normal Distribution for Data with H_t=1.5 m.

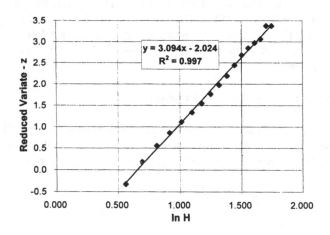

Figure 4.4 Log-Normal Distribution for Data with H_t=1.5 m.

The individual points do not lie exactly on a straight line, but their relationship may be approximated by the straight line. The equation of the straight line of best fit is obtained by linear regression analysis. For Fig. 4.4, A=3.09 and B=-2.02 while R^2=0.997. Hence, the values of the data are well represented by a log-normal distribution with mean and standard deviation

$$s_{\ln H} = \frac{1}{A} = \frac{1}{3.09} = 0.32; \quad \overline{\ln H} = \frac{B}{A} = -\left(\frac{-2.02}{3.09}\right) = 0.65 \qquad (4.11)$$

4.3.3 Gumbel Distribution

In addition to the log-normal probability distribution, it is possible to use distributions developed specifically for analysis of extreme values. These models were originally derived for a limited number of "ordered statistics" such as a set of maximum annual floods arranged in descending order. Extreme value analysis of the present wave data set will be discussed in Section 4.6. Here we will use two of the relationships with the grouped statistical data set of Table 4.1 and we will continue to use linear regression analysis to determine the line of best fit.

The Gumbel distribution is

$$P = \exp\left(-\exp\left(-\frac{H-\gamma}{\beta}\right)\right) \qquad (4.12)$$

This may be linearized by taking the logs of both sides

$$\ln P = -\exp\left(-\frac{H-\gamma}{\beta}\right) \qquad (4.13)$$

and taking logs again

$$-\ln\left(-\ln P\right) = \frac{H-\gamma}{\beta} \qquad (4.14)$$

or

$$-\ln\left(\ln\frac{1}{P}\right) = \frac{H-\gamma}{\beta} = \frac{1}{\beta}H - \frac{\gamma}{\beta} \qquad (4.15)$$

The reduced variate (Y), we will call G. The resulting transformation is

$$Y = -\ln\left(\ln\frac{1}{P}\right) = G; \quad X = H; \quad A = \frac{1}{\beta}; \quad B = -\frac{\gamma}{\beta} \qquad (4.16)$$

The Gumbel transformation of the data with H_t=1.5 is shown in Column (7) of Table 4.2 and in Fig. 4.5. It is seen that A=2.02 and B=-3.60. Therefore β= 0.50 and γ=1.80.

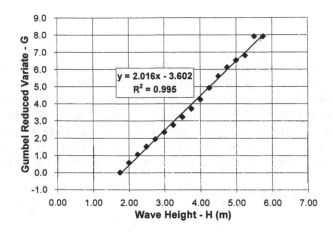

Figure 4.5 Gumbel Distribution for Data with H_t=1.5 m.

4.3.4 Weibull Distribution

The above distributions all have two parameters. A more versatile extreme value distribution is the three-parameter Weibull distribution

$$P = 1 - \exp\left(-\left\{\frac{H-\gamma}{\beta}\right\}^{\alpha}\right) \qquad (4.17)$$

which may also be expressed as

$$Q = \exp\left(-\left\{\frac{H-\gamma}{\beta}\right\}^{\alpha}\right) \qquad (4.18)$$

Linear transformation may be accomplished by taking the logs of both sides:

$$-\ln Q = \left(\frac{H-\gamma}{\beta}\right)^{\alpha}$$ (4.19)

which results in

$$\left(\ln \frac{1}{Q}\right)^{1/\alpha} = \frac{H-\gamma}{\beta}$$ (4.20)

Calling the reduced variate W, the transformation is[1]

$$Y = \left(\ln \frac{1}{Q}\right)^{1/\alpha} = W; \quad X = H; \quad A = \frac{1}{\beta}; \quad B = -\frac{\gamma}{\beta}$$ (4.21)

The Weibull distribution has three parameters (α, β and γ). Linear regression provides only two constants (A and B) and if we want to continue to use linear regression analysis, the determination of the third coefficient (α) will require some trial and error. Assuming different values of α will change the curvature of the points. Table 4.2, Column (8) and Fig. 4.6 show the Weibull analysis for the $H_t=1.5$ data set for $\alpha=0.8$. Table 4.2, Column (9) and Fig. 4.7 show the Weibull analysis for $\alpha=1.3$. Repeated regression analysis will determine what value of α provides the best straight line relationship. In this case $\alpha=0.8$ does not produce a straight line, while $\alpha=1.3$ produces a straight line with A=1.13 and B=-1.40. Thus for Figure 4.7, $\alpha=1.3$, $\beta= 0.88$ and $\gamma= 1.23$.

The parameter γ in the Weibull and Gumbel distributions has physical meaning. It is a lower limit of H (when $H=\gamma$, $Q=1$ or $P=0$). Thus γ is theoretically equal to the threshold value in a Peak over Threshold data set. This can be used as a check. Table 4.4 summarizes the above discussion.

---

1. Sometimes another Weibul transformation is used by taking the logs of Eq. 4.18 a second time.

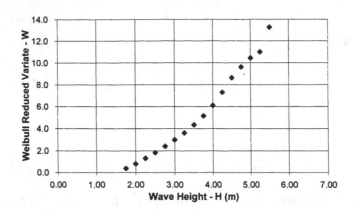

Figure 4.6 Weibull Distribution for Data with H_t=1.5 m. (α=0.8)

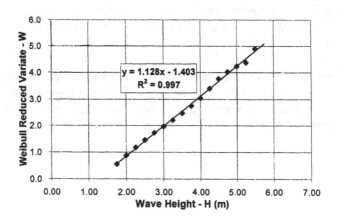

Figure 4.7 Weibull Distribution for Data with H_t=1.5 m. (α=1.3).

Table 4.4 Summary Of Distribution Models

Model	Equation	Y	X	A	B
Normal	$P = \Phi\left(\dfrac{H - \overline{H}}{s_H}\right)$	$\Phi^{-1}(P) = Z$	H	$\dfrac{1}{s_H}$	$-\dfrac{\overline{H}}{s_H}$
Log-Normal	$P = \Phi\left(\dfrac{\ln H - \overline{\ln H}}{s_{\ln H}}\right)$	$\Phi^{-1}(P) = Z$	ln H	$\dfrac{1}{s_{\ln H}}$	$-\dfrac{\overline{\ln H}}{s_{\ln H}}$
Gumbel	$P = \exp\left(-\exp\left(-\dfrac{H - \gamma}{\beta}\right)\right)$	$-\ln\left(\ln\dfrac{1}{P}\right)$	H	$\dfrac{1}{\beta}$	$-\dfrac{\gamma}{\beta}$
Weibull	$Q = \exp\left(-\left\{\dfrac{H - \gamma}{\beta}\right\}^{\alpha}\right)$	$\left(\ln\dfrac{1}{Q}\right)^{1/\alpha}$	H	$\dfrac{1}{\beta}$	$-\dfrac{\gamma}{\beta}$

4.4 Extrapolation

The above long-term wave height analysis meets both criteria expressed in Section 4.1; it organizes the data, and the co-ordinate transformations develop linear relationships that can be interpolated or extrapolated with some confidence to smaller exceedence probabilities. The wave height H for a return period of T_R years may now be determined. From the data, the number of events per year on which the analysis is based (λ) can be calculated. The exceedence probability of one event in T_R yrs would be

$$Q = \frac{1}{\lambda T_R} \qquad (4.22)$$

which also means

$$P = \left(1 - \frac{1}{\lambda T_R}\right) \qquad (4.23)$$

The Normal Probability distribution did not fit the data very well. If it had, Eqs 4.5,

4.7 and 4.8 would have lead to

$$H_{T_R} = \overline{H} + s_H \Phi^{-1}(P) = \overline{H} + s_H \Phi^{-1}\left(1 - \frac{1}{\lambda T_R}\right) \qquad (4.24)$$

For the Log-Normal distribution, Eqs 4.9 and 4.10 yield

$$\ln H_{T_R} = \overline{\ln H} + s_{\ln H}\,\Phi^{-1}(P) = \overline{\ln H} + s_{\ln H}\,\Phi^{-1}\left(1 - \frac{1}{\lambda T_R}\right) \qquad (4.25)$$

or

$$H_{T_R} = e^{\left(\overline{\ln H} + s_{\ln H}\Phi^{-1}\left\{1 - \frac{1}{\lambda T_R}\right\}\right)} \qquad (4.26)$$

For the Gumbel distribution, Eqs. 4.15 and 4.16 give

$$H_{T_R} = \gamma - \beta \ln\left(\ln\frac{1}{P}\right) = \gamma - \beta \ln\left(\ln\left\{\frac{\lambda T_R}{\lambda T_R - 1}\right\}\right) \qquad (4.27)$$

For the Weibull distribution, Eqs. 4.20 and 4.21 produce

$$H_{T_R} = \gamma + \beta\left(\ln\frac{1}{Q}\right)^{1/\alpha} = \gamma + \beta\left(\ln\{\lambda T_R\}\right)^{1/\alpha} \qquad (4.28)$$

4.5 Sensitivity to Distribution and Threshold Wave Height

Table 4.5 presents one interpolation and three extrapolations of wave heights for return periods T_R =20, 50, 100 and 200 years, for the three different distributions (Weibull, Gumbel and Log-Normal models) and for three different values of threshold wave height (H_t=1.5, 3.0 and 4.0 m). Neither the choice of model nor the value of H_t appears to make much difference for this data set. All yield quite similar results, even for T_R=200 yrs, where for the nine calculated values $\overline{H}$=6.38 and the uncertainty $(s_H / \overline{H})$=0.03=3%. It appears that H_t=1.5 m has successfully produced uncorrelated "storm" data. Similar calculations for H_t=0, produced quite different results, indicating that separation of the data into storms is a necessary step before statistical analysis.

Table 4.5[⊗] Wave Height Predictions (m) from Grouped Data

a) Weibull

					Return Period (Yrs)			
H_t	λ	α	β	γ	20	50	100	200
1.5	78.45	1.30	0.88	1.23	5.32	5.70	5.94	6.27
3.0	8.83	1.00	0.46	3.02	5.40	5.82	6.14	6.46
4.0	1.26	1.10	0.58	3.72	5.40	5.83	6.15	6.47

b) Gumbel

				Return Period (Yrs)			
H_t	λ	β	γ	20	50	100	200
1.5	78.45	0.50	1.80	5.48	5.94	6.28	6.63
3.0	8.83	0.43	3.13	5.35	5.75	6.05	6.34
4.0	1.26	0.46	3.94	5.42	5.84	6.16	6.48

c) Log-Normal

				Return Period (Yrs)			
H_t	λ	$\overline{\ln H}$	s	20	50	100	200
1.5	78.45	0.65	0.32	5.37	5.83	6.17	6.53
3.0	8.83	1.18	0.19	5.27	5.58	5.81	6.04
4.0	1.26	1.40	0.16	5.37	5.72	5.96	6.20

4.6 Extreme Value Analysis From Ordered Data

The above statistical methods derive robust estimates of extreme wave heights, based on the statistics for the whole 34.9 year data set. However, in many instances, only a few major events are known and it is necessary to base the analysis on such a limited number of extreme events. For purposes of demonstration and comparison, the 44 highest storm events occurring over the duration of the Lake Huron wave record were extracted (Table 4.6). This corresponds directly to the complete data set with H_t=4.0 m. These data were ranked in decreasing order and extreme value analysis was applied. The Weibull, Gumbel and Log-Normal distributions are all candidate distributions for extreme value analysis of ordered data. Extreme value analysis for waves is discussed in detail in Mathiesen et al (1994), Goda et al (1993) and Goda (1992).

Since only extreme values of H and their ranking are known, the ranking must somehow be converted into a plotting position, representing probability of exceedence. We use the expression:

$$Q = \frac{i - c_1}{N + c_2} \qquad (4.29)$$

where i is the ranking of the data point and N is the total number of points.

Table 4.6 Weibull Analysis of Ordered Set of Extreme Values ($\alpha = 0.8$)

i	H	Q	W	i	H	Q	W
1	5.95	0.010	6.675	23	4.22	0.505	0.621
2	5.38	0.033	4.642	24	4.21	0.527	0.572
3	5.26	0.055	3.775	25	4.20	0.550	0.526
4	5.03	0.078	3.227	26	4.20	0.572	0.482
5	4.82	0.100	2.832	27	4.17	0.595	0.441
6	4.75	0.123	2.524	28	4.17	0.617	0.402
7	4.71	0.145	2.274	29	4.16	0.640	0.365
8	4.68	0.168	2.064	30	4.16	0.662	0.330
9	4.63	0.190	1.884	31	4.14	0.685	0.297
10	4.54	0.213	1.727	32	4.14	0.707	0.266
11	4.49	0.235	1.588	33	4.13	0.730	0.236
12	4.43	0.258	1.463	34	4.09	0.752	0.208
13	4.40	0.280	1.351	35	4.09	0.775	0.182
14	4.38	0.303	1.250	36	4.08	0.797	0.156
15	4.36	0.325	1.157	37	4.07	0.820	0.133
16	4.35	0.348	1.071	38	4.07	0.842	0.111
17	4.34	0.370	0.993	39	4.06	0.865	0.090
18	4.33	0.393	0.920	40	4.05	0.887	0.071
19	4.29	0.415	0.852	41	4.04	0.910	0.053
20	4.25	0.437	0.788	42	4.04	0.932	0.036
21	4.24	0.460	0.729	43	4.03	0.954	0.022
22	4.23	0.482	0.673	44	4.01	0.977	0.009

The simplest estimate of plotting position assumes $c_1=0$ and $c_2=1$, but Table 4.7 presents coefficients for a so-called unbiased plotting position for each distribution. Table 4.6 shows the example set of ordered data for Lake Huron and their Weibull plotting position. Note that since α influences both the plotting position and the curvature of the Weibull graph, some trial and error is necessary.

Table 4.7 Constants for Unbiased Plotting Position

Distribution	c_1	c_2
Normal	0.375	0.375
Log-Normal	0.25	0.125
Gumbel	0.44	0.12
Weibull	$0.20+0.27/\alpha$	$0.20+0.23/\alpha$

The line of best fit for these points can be determined using the method of moments, the method of maximum likelihood or the least squares analysis used earlier. The literature does not indicate a preference and hence the least squares (linear regression analysis) method will be used again because it is generally available and well-known. The relevant equations are the same as those used earlier. The Weibull analysis with $\alpha=0.8$ brings the points successfully into a straight line (Fig. 4.8). The prediction results are shown in Table 4.8.

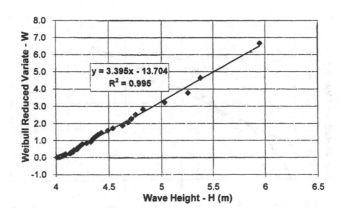

Figure 4.8 Weibull Distribution for Ordered Data Set ($\alpha=0.8$).

Table 4.8 Wave Height Predictions (m) from Ordered Extreme Values

a) Weibull

N	λ	α	β	γ	Return Period (Yrs)			
					20	50	100	200
44	1.26	0.80	0.29	3.97	5.22	5.68	6.05	6.43

b) Gumbel

N	λ	β	γ	Return Period (Yrs)			
				20	50	100	200
44	1.26	0.45	3.87	5.31	5.73	6.04	6.36

c) Log-Normal

N	λ	$\overline{\ln H}$	s	Return Period (Yrs)			
				20	50	100	200
44	1.26	1.36	0.19	5.44	5.86	6.16	6.45

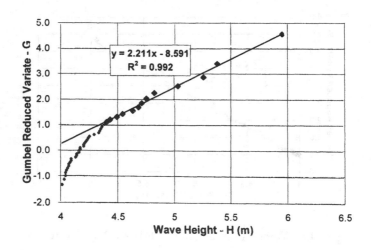

Figure 4.9 Gumbel Distribution for Ordered Data Set

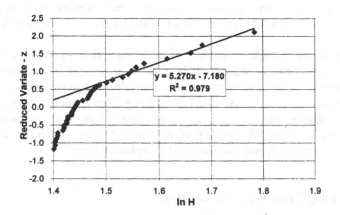

Figure 4.10 Log-Normal Distribution for Ordered Data Set

The Gumbel and Log Normal analysis do not result in linear data points (Figs. 4.9 and 4.10). The points with lower wave heights could be considered less important, and using only the highest 14 data points yields reasonably straight lines in Figs. 4.9 and 4.10. Predicted wave heights are again given in Table 4.8. The results in Table 4.8 are quite similar to those of Table 4.5. The Weibull extreme value analysis yields the most consistent results of the three, since its three parameters provide the best fit of a straight line through all the points. Note that in the analysis, we used the actual 44 highest events. Many times extreme value analysis is performed on data such as the annual maximum events. This would produce worse results, since some second highest annual events may be more important than some of the lower annual extreme events.

4.7 Conclusions About Wave Heights

The Lake Huron example shows that there are three distributions that can be used successfully to organize and extrapolate wave height data. These are the Weibull, Gumbel and Log-Normal distributions. The Weibull distribution is preferred because it contains an extra parameter and therefore it is more likely to produce a good fit to a straight line. This conclusion is also reached by Mathiesen et al (1994).

Since extrapolation to higher wave heights and longer return periods is a basic and very important part of any design, it is best to use all the available data in as many ways as possible (as was done in the present example) to gain confidence in the final results. However, the above discussion of the Lake Huron data set shows that any of the three distributions can be expected to give good results. Extreme value analysis, in which only a reasonable number of the highest wave conditions are known, involves much less work and can also yield acceptable values.

4.8 Other Long-Term Wave Distributions

Other long-term distributions, such as for wave period and wave direction could be derived the same way as the long-term wave height distribution. However, wave period and angle are much more difficult to measure and predict than wave height. Since wave height is normally the most important wave parameter in design, it is common to calculate the wave height distribution and then relate the others to wave height using the joint distributions. An example of a joint wave period-wave height distribution is given in Table 4.9.

Table 4.9 Joint Distribution of Wave Heights and Periods – Lake Huron
(Number of occurrences over 10 years)

H (m)	Wave Period (sec)								
	0.5	1.5	2.5	3.5	4.5	5.5	6.5	7.5	8.5
0.25	4854	16890	6582	629	197	55	21	2	0
0.75	0	120	15553	10838	720	211	63	6	0
1.25	0	0	7	6086	7263	346	83	8	1
1.75	0	0	0	16	3197	3195	107	7	1
2.25	0	0	0	0	3	1894	807	5	2
2.75	0	0	0	0	0	2	1341	4	1
3.25	0	0	0	0	0	0	47	432	2
3.75	0	0	0	0	0	0	0	56	17
4.25	0	0	0	0	0	0	0	0	38

This joint distribution may be simplified by relating wave period to wave height via the combinations of greatest frequency. (For example, in Table 4.9, interpolation gives T=5.8 seconds corresponding to H=2.25 m). Figure 4.11 shows that the appropriate equation to relate wave period to wave height is

$$T = c_3 H^{c_4} \qquad (4.30)$$

and Table 4.10 summarizes five such relationships. Because conditions at various locations differ, there is no general relationship. A similar analysis can be used to relate wave angle to wave height.

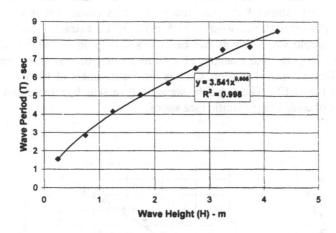

Figure 4.11 Wave Period –Wave Height Relationship for Lake Huron Data.

Table 4.10
Sample Wave Relationships

Location	c_3	c_4
Lake Huron	3.54	0.60
Lake Ontario	4.45	0.45
North Sea	3.94	0.38
Dubai	4.04	0.47
Israel	6.96	0.28

Wave persistence traditionally refers to duration of conditions in which wave heights are above or below a certain value. Persistence of high wave conditions is important to determining downtime of equipment and times during which construction can take place. For example, most pipeline dredges cannot work in wave heights greater than

1 m. Thus durations of wave action greater than 1 m will mean downtime on the dredging program. Persistence of calms is important when there is much wave action. Durations of calms are analysed to determine when to execute certain parts of the project construction that are sensitive to wave action. For example, moving or placing drilling platforms can only be done during such a window of calm wave action.

Persistence statistics may be extracted from the wave data via the POT analysis. Figure 4.1 clearly shows how persistence of calms (times when H<1.5 m) and persistence of storms (times when H>1.5 m) may be extracted from wave data. Different threshold wave heights can be analysed to produce persistence data for various wave heights. An example of a Weibull plot of wave persistence at the Olympic sailing site near Kingston, Canada for the summers of 1975 and 1976 may be found in Fig. 4.12. These graphs were used to schedule races and determine probabilities of completing specific race series.

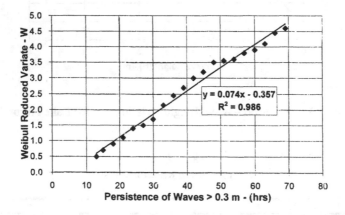

Figure 4.12 Wave Persistence During Sailing Season at Kingston

5. Wave Generation

5. 1 Wave Generation

When a gentle breeze blows over water, the turbulent eddies in the wind field will periodically touch down on the water, causing local disturbances of the water surface. Small ripples will form, but only where the eddies touch down, since the wind speed must be in excess of 0.23 m/s to overcome the surface tension in the water. Theory (Phillips, 1957 and Miles, 1957) shows wind energy is transferred to waves most efficiently when they both travel at the same speed. But the wind speed is normally greater than the wave speed. For this reason, the generated waves will form at an angle to the wind direction so that the propagational speed of the wave in the wind direction approaches the wind speed. At first, because the wave speed is very small, the angle between the wind and the wavelets will be large, forming "cats paws" on the water where the puffs of wind strike. Eventually, the generated wave crests will form a more regular pattern of crossing waves, as shown in Fig. 2.4. At any particular location this will yield short-crested, irregular waves. Even for large waves, when we step back far enough (for example, when we fly high above an ocean) crossing wave crests are clearly identifiable.

Once the initial wavelets have been formed and the wind continues to blow, energy is transferred from the wind to the waves mainly by two mechanisms. A simplistic picture is given in Fig. 5.1. Sheltering causes the wind speed downwind of the wave to be smaller or sometimes the reverse of the wind speed on the upwind side. The resulting shear on the water surface will tend to move water toward the wave crest from both sides. The form of the wave also causes the wind velocity to increase over the crest and to decrease in the trough. According to Bernoulli's principle, this means the pressure is lower over the crest and higher in the trough. Relative to the mean pressure, the effective pressure over the crest is negative and in the trough it is positive,

103

which will tend to lift the crest and push the trough further down.

Most of the wind energy is transferred to high frequency waves. Thus, the wind causes small waves to form on top of existing waves, rather than increasing the size of the larger waves directly by shear and pressure differences. This pool of high frequency wave energy is then transferred to lower frequencies by the interaction of the high frequency movement with the adjacent slower moving water particles. The process is similar to locally generated sea becoming swell as discussed in Ch. 3.

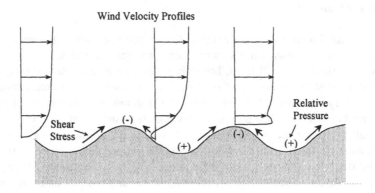

Figure 5.1 Wind Energy Transfer

Chapter 3 shows that wave height and period are closely related to wind speed. It should therefore be possible to estimate wave conditions from known wind conditions. In fact, it should be possible to reconstruct a wave climate at a site from historical, measured wind records. Such a computation is known as *wave hindcasting*. Wave *forecasting* is also possible by using forecast wind conditions. Since the procedures are identical we will cover both by the term wave hindcasting.

5. 2 Simple Wave Hindcasting

5.2.1 Introduction to Parametric Methods

The art of defining waves from wind fields came into prominence during World War II. When the Allied forces wanted to land troops on the beaches of Continental Europe, it was critical that weather forecasts could be translated into expected wave conditions. The result of this international war effort was written up by the scientists involved (Sverdrup and Munk, 1947). The method was later extended by Bretschneider (e.g., Bretschneider, 1958) to form the empirical method, now known as the SMB Method. The method is described fully in CERC (1977) and in earlier versions of that publication. In CERC (1984) this method was replaced by the Jonswap Method, based on research on wave spectra in growing seas by Hasselmann et al (1973).

The Jonswap and SMB methods are called parametric methods because they use wind parameters to produce wave parameters, rather than develop a detailed description of the physics of the processes. Although, these methods produce estimates of wave height and period only, they can be extended to provide estimates of the parametric wave spectra discussed in Section 3.7.

Parametric wave hindcasting determines wave height and period (H and T) from fetch (F), storm duration (t) and depth of water in the generating area (d). Fetch is the distance the wind blows over the water to generate the waves. For a lake or a bay, fetch is readily determined as the distance into the wind direction from a point of interest to the nearest shore. For irregular shorelines, CERC (1984) suggests that the distance to shore be averaged over 12° on either side of the mean wind direction. For hindcasts on large lakes or the open ocean, fetch is related to the sizes and tracks of the weather systems. Wave forecasting or hindcasting along an open shore is therefore more difficult. Fetches are large and defined by curvatures of the isobars describing the weather systems (CERC, 1984) and hence difficult to define accurately. Fortunately, the wave parameters are not very sensitive to absolute errors in fetch length for these large fetches. As a limit for storms of normal size, changes in wind direction make it unlikely that fetch would be greater than 500 km.

If F, t and d are all infinite, the result is a *fully developed sea* (Ch. 2). The waves are fully developed so that any added wind energy is balanced by wave energy dissipation rate resulting from internal friction and turbulence. In that case, the resulting wave conditions are a function of wind speed only, as described by the Beaufort Scale in Ch.2. When F, t or d are limited, the waves will be smaller.

5.2.2 Wind

Wind speed varies with distance above the water and the standard height used in wave hindcasting is 10 m. For wind records taken at a different height above the water, a logarithmic velocity profile is assumed so that

$$\frac{U_{10}}{U_z} = \left(\frac{10}{z}\right)^{1/7}$$

(5.1)

where z is the anemometer height. Wind speed is normally quoted as hourly average wind speed. Such hourly wind speeds can be introduced into an hourly wave hindcast (a hindcast that produces waves heights at hourly intervals as discussed in Section 5.3.1). To estimate the wave condition that has built up over a duration t, (without calculating the hourly waves from hourly wind speeds), the concept of effective wind speed is used. One such relationship may be found in CERC (1984). For durations less than 10 hours

$$\frac{U_t}{U_1} = 1 - 0.15 \log t$$

(5.2)

where U_t is the effective wind speed over time t and U_1 is the maximum of the hourly average wind speeds over time t.

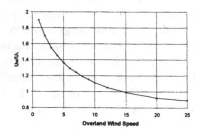 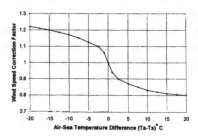

Figure 5.2 Wind Speed Corrections
(after Resio and Vincent, 1977)

The wind speed in a wave hindcast computation must be wind speed over the water. Normally we only know wind speeds from nearby airports, and we must take into account that winds over water are usually greater than winds over land because of the smaller friction over the water. Resio and Vincent (1977) and CERC (1984) provide some coefficients (Fig 5.2). It is seen that the difference between the wind over the water (U_w) and the wind speed over land (U_L) is greatest for small wind velocities. Figure 5.2 also proposes a correction factor for the air-sea temperature differences. CERC (1984) introduces an adjustment, based on the wind stress over the water surface, but comparisons of hindcast and measured waves generally show that the use of this correction is not recommended.

Wind direction can also be quite different over the water than over land (just ask any sailor). An airport wind direction could be up to 30° different from wind direction over water and again this effect is greatest for smaller wind velocities. Since wave direction is usually assumed to be the same as the wind direction, this assumption can be a source of substantial errors in wave direction, which in turn causes large errors in derived quantities, such as alongshore sediment transport rate (Ch. 12). Donelan (1980) and Donelan et al (1985) further show that the largest waves do not come from the wind direction, but from a combination of wind direction and the direction of the longest fetch. On the Great Lakes, therefore, the wave direction is always biased toward the long axis of the lake. Even more pronounced is the wind funneling that takes place along narrow bays, lakes, fjords and rivers. In that case, the wind tends to blow either up or down the bay or river, almost regardless of the wind direction elsewhere. Note that wind and wave directions are defined as the directions from where they come. For example, a wave direction of 90° (from North) means that waves come *from* the East.

5.2.3 Jonswap Parameters

The Jonswap method of wave hindcasting uses the following dimensionless expressions.

$$F^* = \frac{gF}{U^2}, \quad H^*_{mo} = \frac{gH_{mo}}{U^2}, \quad T^*_p = \frac{gT_p}{U}, \quad t^* = \frac{gt}{U}, \quad d^* = \frac{gd}{U^2} \qquad (5.3)$$

The Jonswap relationships are

$$H^*_{mo} = 0.0016\,(F^*)^{1/2} \qquad (5.4)$$

$$T^*_p = 0.286\,(F^*)^{1/3} \qquad (5.5)$$

and

$$t^* = 68.8\,(F^*)^{2/3} \tag{5.6}$$

Waves generated in deep water can be fetch limited, duration limited or fully developed sea. On a small body of water, the waves would be limited by a short fetch and H_{mo} and T_p can be calculated directly from Eqs. 5.4 and 5.5. On a larger body of water, the same equations apply, but wind duration may limit the size of the waves. Eq. 5.6 is then used to calculate an *effective fetch* (the fetch needed to produce the same wave height if the duration had been infinite)

$$F_{eff}^* = \left(\frac{t^*}{68.8}\right)^{3\,2} \tag{5.7}$$

When $F^* < F^*_{eff}$, the waves are fetch limited and Eqs. 5.4 and 5.5 are used. When $F^*_{eff} < F^*$, the waves are duration limited and Eqs. 5.4 and 5.5 are used with F^*_{eff} substituted for F^*. Thus the smaller value of F^* and F^*_{eff} is used. Finally, a fully developed sea, for a large body of water and a large duration, is calculated using the following upper limits

$$H_{mo}^* = 0.243 \quad ; \quad T_p^* = 8.13 \quad ; \quad t^* = 71,500 \tag{5.8}$$

The procedure of computing H_{mo} and T_p by Jonswap has been published as a nomogram in CERC (1984), which is shown here Fig. 5.3. The Jonswap method may be streamlined for computer calculation, as in Fig. 5.4 and the program **WAVGEN®**.

Equations 5.4 and 5.5 show that wave period is closely related to wave height, once fetch length has been defined

$$T_p^* = 20.9\,(H_{mo}^*)^{2/3} \quad or \quad T_p = 9.8\,U^{-1/3}\,H_{mo}^{2/3} \tag{5.9}$$

Thus Eqs. 5.4 and 5.9 could be used instead of Eqs. 5.4 and 5.5.

Because of the large uncertainties in the values of U, F and t and wave direction, and because the hindcasting relationships themselves are uncertain, all hindcasting results must always be regarded as approximate. It is essential that wave hindcasts are calibrated against observed values. This is discussed further in Section 5.4.

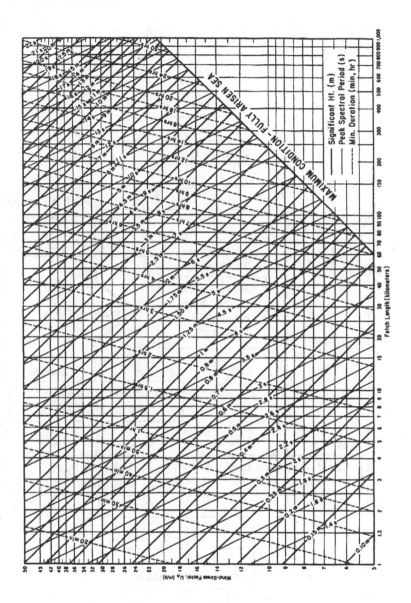

Figure 5.3 Wind Hindcasting Nomogram (after CERC, 1984)

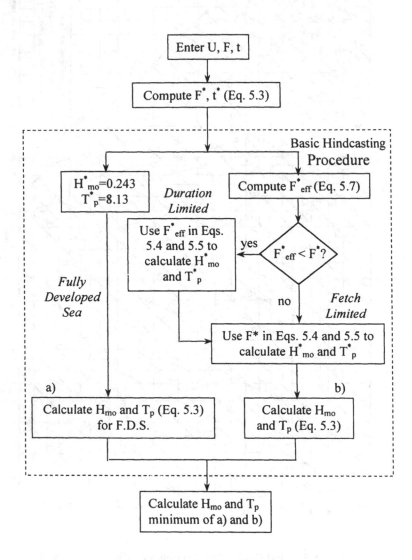

Figure 5.4 Basic Jonswap Hindcasting

Example 5.1   Simple Jonswap Wave Hindcast

Let us use the Jonswap method to calculate the wave conditions resulting from an effective wind speed U=20 m/s blowing for 6 hrs (t=21,600 sec) over a fetch of 100 km (F=100,000 m). According to Eq. 5.3, F^* = 2451.5 and Eqs. 5.4 and 5.5 yield the fetch limited values H^*_{mo}=0.079 and T^*_p=3.86 which in turn (using Eq. 5.3) produce H_{mo}=3.2 m and T_p=7.9 sec. These numbers may be confirmed from Fig. 5.3 at the intersection of U=20 m/s and F=100 km. Equation 5.7 yields F^*_{eff}=1909 (or F_{eff}=77.9·10^3 m). Since F^*_{eff}<F^* the waves are duration limited. Substituting F^*_{eff} into Eqs. 5.4 and 5.5 yields H^*_{mo}=0.070, T^*_p=3.55, and Eq. 5.3 gives H_{mo}=2.9 m and T_p=7.2sec. These results may also be found with Fig. 5.3 at the intersection of U=20 m/s and t=6 hrs, which occurs at F=78km. A quick check is required to see if the condition for a fully developed sea is exceeded. Comparison of the calculated values with Eq. 5.8 shows that both H^*_{mo} and T^*_p are considerably less than the upper limits for a fully developed sea, hence the correct answer is for the fetch-limited condition: H_{mo}=2.9 m and T_p=7.2sec.

5.2.4   Maximum Wave Conditions

For many designs and feasibility studies, it is important to identify maximum wave conditions. The above method can yield an estimate of maximum wave conditions, if effective wind speed (Eq. 5.2) is used several times for different values of t. But it is better to combine known storm segments or actual hourly wind speeds for the growing portion of the storm. To do this, the computation of Fig. 5.4 is repeated for each storm segment (or each hour), as illustrated in Fig. 5.5.

For the first storm segment, use F and t_1 (where the subscript refers to the first time segment) to determine $(H_{mo})_1$ and $(T_p)_1$ at the end of this first segment. The next storm segment will add to the wave energy generated during this first storm segment. Since wave energy is closely related to wave height, we first calculate a virtual storm duration – how long it would have taken for $(H_{mo})_1$ to be generated by the wind speed of the second storm segment U_2. We will call this (t_2'). To do that, we compute $(H^*_{mo,2})$ using $H_{mo,1}$ and U_2

$$H^*_{mo,2} = \frac{g\,H_{mo,1}}{U_2^2} \tag{5.10}$$

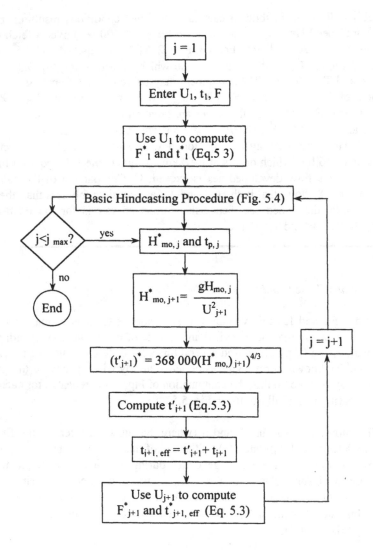

Figure 5.5 Jonswap Hindcasting for a Series of Input Conditions

Then Eqs. 5.4 and 5.6 are combined to yield

$$(t_2')^* = 368,000 \, (H^*_{mo,2})^{4/3} \tag{5.11}$$

From this, t_2' may be computed, using Eq. 5.3. We then add t_2' and t_2 to derive an effective duration $t_{2,\text{eff}}$ of the first two hours at the second wind speed and from this proceed to calculate $(H_{mo})_2$ and $(T_p)_2$. This is repeated for each storm segment until the maximum wave conditions or the maximum number of wind segments are reached. Implicit in such a computation is the assumption that wind direction remains constant. The method works well during the growing part of the storm, when generation of wave energy is far greater than wave energy dissipation.

5.2.5 Finite Water Depth

If the depth in the generating area is limited, friction with the bottom will result in smaller waves. CERC (1984) discusses wave generation in finite depth developed by Bretschneider (1958) from Bretschneider and Reid (1953). The expressions were further developed by Young and Verhagen (1996)

$$H^* = 0.24 \left\{ \tanh\left[0.49(d^*)^{0.75}\right] \tanh\left[\frac{0.0031(F^*)^{0.57}}{\tanh[0.49(d^*)^{.75}]}\right] \right\}^{0.87} \tag{5.12}$$

$$T^* = 7.54 \left\{ \tanh\left[0.33(d^*)\right] \tanh\left[\frac{.00052(F^*)^{0.73}}{\tanh[0.33(d^*)]}\right] \right\}^{0.37} \tag{5.13}$$

$$t^* = 537(T^*)^{7/3} \tag{5.14}$$

5.3 Hindcast Models

For many applications, simplistic hindcast methods are good enough for first estimates especially of maximum conditions. However, at other times, we need a long-term hindcast wave climate, at hindcast intervals of 1, 3 or 6 hours.

5.3.1 Parametric Models

Parametric, long-term wave hindcasting models are based on the concept of Fig. 5.5. They must, however, also account for wave decay and changes in wind direction. To demonstrate how this is done, imagine that we have correctly hindcast the wave climate at the end of a certain hour. The next hour will have its own wind speed and direction. If there is a large change in wind direction (>45°), we assume that the waves continuing in the old wave direction will stop growing and begin to decay. Waves in the new wind direction will begin to grow. The total wave energy (there may be other decaying wave trains from earlier wind direction changes) is combined to yield H_{mo} for the next hour. The composite wave period T_p and wave direction may be calculated by weighting the various contributing wave periods and directions according to the wave energy they contribute. If the change in wind direction is small (< 45°), we split up the wave energy into a portion that continues in the previous wave direction, $E (1-\cos\beta)$ and another portion, $E \cos\beta$ that accompanies the new wind direction. Here β is the difference between the new wind and the waves. Waves in the old direction begin decay and waves in the new direction begin to grow.

When the wind speed drops or changes direction, the existing wave energy decays, but at what rate? There are no clear recipes. Because of the short hindcast interval, we assume a simple linear decay rate for wave energy, represented a decay coefficient (the waves loose a certain fraction of their energy every hour). The correct value of such a coefficient can only be obtained by calibration against measured data and repeating the computation for different values of the decay coefficient can test its sensitivity.

An example plot from **HIND**, a model based on the above assumptions, is given in Fig. 5.6. It shows 25 days of calculated and observed waves at Grand Bend on Lake Huron. The waves were hindcast using the default coefficients for **HIND**, which are the values in Eqs. 5.3 to 5.11 and a decay coefficient of 0.3. In general, and in spite of a rapidly changing wind field, the hindcast wave heights, periods and angles are quite good. Some details such as storm peaks and decay after the peaks are not correct, indicating the necessity for calibration and further verification.

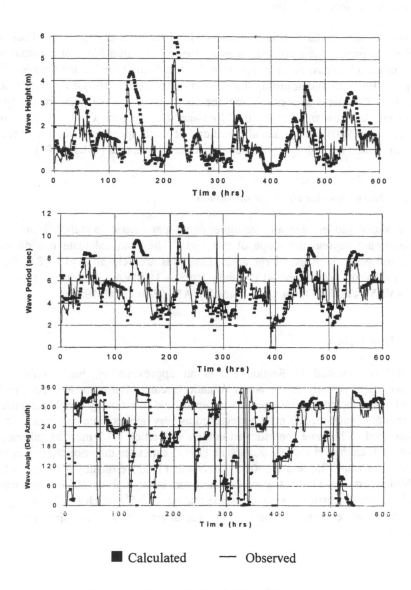

■ Calculated   ── Observed

Figure 5.6 Example Hourly Wave Hindcast

5.3.2 Wave Spectra Models

The discussion so far has concentrated on hindcasting H_{mo} , T_p and θ. These can be related to parametric wave spectra as discussed in Section 3.7. For example, we can formulate a hindcast Jonswap spectrum by substituting the hindcast value of T_p (or f_p) into Eq. 3.56 and adjusting the value of α to produce the correct hindcast value of H_{mo}. Chapter 3 also shows how the coefficients and the peak frequency of the Pierson Moskowitz (1964), the Jonswap (Hasselman et al, 1973) and the Mitzuyasu(1980) spectra are all functions of wind speed and that the spectrum is depth limited through spectral saturation, e.g., Bouws et al (1985, 1987).

5.3.3 More Complex Hindcasting Models

The above models are all one-dimensional parametric models. More complex models are beyond the scope of this text. They can calculate two dimensional spectral wave fields over large areas. Examples are Schwab et al (1984), Clancy et al (1986), the **WAM** model (Wamdi, 1988), **HISWA** (Holthuysen et al., 1989), **WAVEWATCH** (Tolman, 1991) and **SWAN**, (Booij et al, 1996).

5.4 Uncertainty

The basic method of Section 5.2 is an approximation, based only on some observations, mainly in the North Atlantic Ocean. There are also no clearly best methods to incorporate changes in wind speed and direction by simple parameters. As a result, all such models must be regarded as very approximate and must be carefully calibrated. The more sophisticated models of Section 5.3.3 must be calibrated also. For such calibrated models, Kamphuis (1999) estimates the uncertainties in hindcast waves as σ'_H=0.25, and σ'_T=0.3. Burcharth (1992) estimates $0.1 < \sigma'_H < 0.2$. The absolute error in hindcast wave direction can be as high as $30°$ in deep water. In subsequent discussions, we will assume that $\sigma_{\alpha, \text{Hind}}=8°$, which results in $\sigma'_{\alpha, \text{Hind}}=0.8$ for a $10°$ wave angle and causes very large uncertainty in wave direction in shallow water.

6. Tides and Water Levels

6.1 Introduction

Although coastal design is normally considered to be a function of wave conditions, it is primarily a function of water levels. It is water levels that control both flooding and wave exposure. Imagine a simple structure close to shore that is subject to waves. When the water level rises, the structure will be exposed to larger waves because the water depth determines where waves break and loose most of their energy (Ch. 7). This results in increased forces on the structure and overtopping of water that will damage the structure and areas behind it. Conversely, when the water level drops, the same structure may not be exposed to waves at all. Thus most damage to structures occurs when the water levels are high.

Similarly, high water levels cause retreat of sandy shores, even if they are backed by substantial dunes. The higher water levels allow larger waves to come closer into shore. These waves will erode the dunes and upper beach and deposit the sand offshore. If the water level rise is temporary, most of this loss will be regained at the next low water (Ch 11). Permanent water level rise, however, will result in permanent loss of sand (Ch 12). Shorelines consisting of bluffs or cliffs of erodable material, such as glacial till or soft rock are continuously eroded by wave action. (Ch 11). High water levels, however, will allow larger waves to attack the bluffs directly, causing a temporary rapid rate of shoreline recession.

There are several types of water level fluctuations and they can be classified according to their return period as:

117

- Short Term
  - Tides
  - Storm Surge and Barometric Surge
  - Seiche
- Seasonal
- Long Term
  - Climatic Fluctuations
  - Eustatic (Sea) Level Rise
  - Isostatic (Land) Emergence and Subsidence
  - Climate Change

6.2 Tides

Astronomic tides are often the defining water motion in coastal areas. They cause the water levels to rise and fall and cause large-scale currents patterns, sometimes with large velocities. Tides directly affect coastal morphology, navigation, fisheries, habitat and recreational activity. Because of their relative importance they are discussed extensively in this chapter.

The tides are the result of a combination of forces acting on individual water particles. These are:
- gravitational attraction of the earth,
- centrifugal force generated by the rotation of the earth - moon combination,
- gravitational attraction of the moon,
- gravitational attraction of the sun.

6.2.1 Equilibrium Tide (Moon)

Let us first neglect the force of the sun and assume that the whole earth is covered with water. The resultant force on the water particles is a small horizontal force. It moves the water particle A in Fig. 6.1 toward the moon and particle B away from the moon, resulting in two bulges of high water, (Defant, 1961; Ippen, 1966; Marchuk and Kagan, 1984; Neumann and Pierson, 1966). As we turn with the earth's angular velocity, ω_E, around the earth's axis at C_E in the direction of the arrow, we turn through this deformed sphere of water and experience two high water levels and two low water levels per day. The resulting tidal period would be 12 hrs, however, the moon-earth system also rotates around C_{ME} with velocity ω_{ME} in the same direction as the earth's rotation. The bulges follow the position of the moon and hence the tidal period is 12.42 hrs.

The tide in Fig. 6.1 is called *equilibrium tide* since it results from the assumption that the tidal forces act on the water for a long time so that equilibrium is achieved between the tide generating force and gravity (the slope of the water surface).

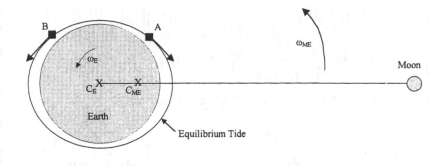

Figure 6.1 Equilibrium Tide

6.2.2 Equilibrium Tide (Sun and Moon)

The sun's gravity forms a second, smaller set of bulges toward the sun and away from the sun. Since our day is measured with respect to the sun, the period of the tide generated by the sun is 12 hrs.

Both these equilibrium tides occur at the same time and they will add up when the moon and sun are aligned (at new moon and full moon). At those times, the tides are higher than average. At quarter moon, the forces of the sun and moon are 90° out of phase and the equilibrium tides subtract from each other and at such a time, the tides will be lower than average. The higher tides are called *spring* tides and the lower ones *neap* tides. Fig. 6.2a demonstrates this. The phases of the moon are shown at the bottom of the figure and it is seen that, except for some phase lag, the maximum tides (spring tides) in Fig. 6.2a correspond to new and full moon, while the neap tides correspond to the quarter moon.

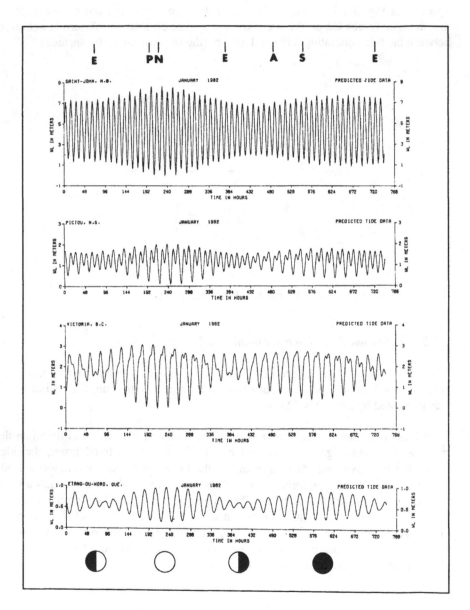

Figure 6.2 Tide Recordings (after Forrester, 1983)

6.2.3 Daily Inequality

Figure 6.1 was drawn looking down on the earth's axis. Since the equilibrium tide is three dimensional in shape (it forms a distorted sphere), the picture is the same when the earth is viewed from the side, as shown in Fig. 6.3. An observer, C, travelling along a constant latitude would experience two tides of equal height per day. However, the moon or sun is seldom in the plane of the equator. When the moon or sun has a North or South Declination with respect to the equator, as shown in Fig. 6.4, one bulge of the equilibrium tide will lie above the equator and one below the equator. An observer moving along a constant latitude would now experience two tides per day of unequal height. This is called *daily inequality*. The daily inequality is most pronounced when the moon or sun is furthest North or South of the equator. There is no daily inequality at the equator and it increases with latitude. Lunar daily inequality is demonstrated in Fig. 6.2b. The letters E, N and S at the top of Fig. 6.2 denote when the moon is in the plane of the equator, at the maximum North declination and maximum South declination. It is seen that the largest daily inequalities indeed correspond to N and S, except again for some phase lag. The daily inequality cycle generated by the moon's forces repeats itself every lunar month (29.3 days). For the tide generated by the sun, the daily inequality cycle has a period of a year and is greatest shortly after mid-summer and mid-winter, causing higher tides in early January and early July.

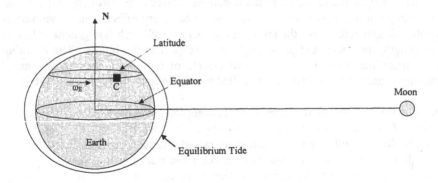

Figure 6.3 Equilibrium Tide (2)

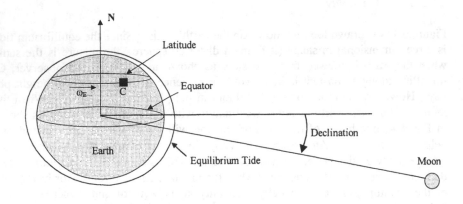

Figure 6.4 Daily Inequality

6.2.4 Other Effects

So far we have explained the characteristics of tides based on four influences, the gravitational attraction of the sun and moon, and the declination of the sun and moon. There are many other, secondary effects. For example, we have assumed that the sun and the moon travel in circular orbits relative to the earth. These orbits are actually elliptical and therefore the distances between the earth and the sun and moon change in a periodic fashion. All these secondary effects can be viewed as separate tide generators (like the moon in Fig. 6.1). Each such tide generator has its own strength, frequency and phase angle with respect to the others. The resulting tide is, therefore, a complex addition of effects of the moon, the sun and many secondary causes. Each component is called a *tidal constituent* (Dronkers, 1964).

Until now we have assumed that the earth is completely covered with water and that the same forces act everywhere continuously. It was seen that the tide moves relatively slowly, while the earth turns more rapidly through the tide. In reality, the earth's large land masses will not turn through the tide, but will move the water masses along with them, disrupting our simple picture. The only place where an equilibrium tide can possibly develop is in the Southern Hemisphere, where the earth is circled by one uninterrupted band of water. The equilibrium tide formed there progresses into the various oceans. It takes time to travel along those oceans and hence the actual tidal constituent (water level fluctuation) lags behind its related theoretical tidal constituent (from equilibrium theory), causing high water to occur after the moon crosses the local meridian, causing spring tide some time after full (or new) moon, etc.

The earth's geography not only confines the water and moves it along with the surface of the earth, but it also causes certain tidal constituents to resonate locally in the various oceans, seas, bays and estuaries. Thus some constituents are magnified in certain locations, while others simply disappear, making the tide at each location unique. One aspect that is often magnified by the land mass is the daily inequality, increasing the difference between the larger and smaller daily tides so that the small tides become virtually non-existent. The *semi-diurnal* (twice per day) tides then become *diurnal* (once per day). An example of this is shown in Fig. 6.2c. An extreme example in which geography and high latitude combine to produce a completely diurnal tide may be found in Fig. 6.2d.

6.2.5 Tide Analysis and Prediction

Tide Analysis consists of separating a measured tide into as many of its constituents as can be identified from the length of record available. The tide is assumed to be represented by the harmonic summation

$$\eta_T(t) = \sum_{i=1}^{I} a_i \cos(\omega_i t + \alpha_i) \qquad (6.1)$$

where $\eta_T(t)$ is the tidal water level at time t, a_i and α_i are the amplitudes and phase angles of the tidal constituents and ω_i are their angular frequencies (the angular frequencies of the tide generators that cause the constituents). For example, the semi-diurnal lunar constituent, usually identified as M2, has a period of 12.42 hours and therefore $\omega_{M2} = 2\pi/(3600 \times 12.42) = 1.405 \times 10^{-4} \text{ sec}^{-1}$.

Tide analysis, therefore, consists of determining values of a_i and α_i. The most important constituents have already been discussed; they are the semi-diurnal constituents generated by the moon (M2) and sun (S2) with periods of 12.42 and 12 hours, and the daily inequality (or diurnal) constituents with periods of 24.48 hrs and 24 hours. There are many other tidal constituents to be considered such as the ellipticity of the earth and lunar orbits, and local amplification of any constituent or combination of constituents. Each constituent may have superharmonics. One year's record will comfortably provide the amplitudes and phase angles of 60 such tide constituents. One important tidal constituent has a period of 18.6 years. It cannot be calculated from a reasonable record length and is therefore introduced by formulas. Factors are computed that adjust a_i and α_i as function of time relative to this 18.6 year cycle.

For practical purposes, we can think of tides as sinusoidal water level changes with the M2 period of 12.42 hrs, modified by a fortnightly cycle of spring and neap tides, and by daily inequality that varies on a 29.3 day cycle and an annual cycle.

Once the relevant constituents have been calculated through tide analysis, Eq. 6.1 may be used in *tide prediction* to calculate water levels in the future. Thus, the tides for all major ports around the world are predicted and published so that navigators know at least when high and low water will occur, etc. But for many construction projects, local tidal information will not be available and tides need to be measured and analysed specifically for a project. In that case, it is usual to collect rather short tidal records. For record lengths of a month or so, tide analysis can only yield the lunar and solar, semi-diurnal tides, daily inequality, and a few other constituents that can readily be separated. But that is sufficient for approximate predictions. Spectrum analysis can also be used to predict tides (Godin, 1972) and neural network computation has been shown to be effective (Tsai and Tsong, 1999).

6.2.6 Tidal Currents

To raise and lower water levels requires substantial currents. Because the wave lengths are long (order 100 km) compared to the depths (order m), the velocity of propagation of such currents is the shallow water long wave velocity (Ch. 2)

$$C = \sqrt{gd} \tag{6.2}$$

where C is the velocity of propagation, g the gravitational acceleration and d the water depth. The length of such a wave is

$$L = CT \tag{6.3}$$

where T is the wave period, which is 12.42 hrs for the important lunar semi-diurnal constituent.

In an ocean with a depth of 4 km, C=200 m/s and L=9000 km for T=12.42 hrs. In shallower water, both L and C decrease so that for d=10 m, C=10 m/s and L=450 km and for d = 3 m, C=5.5 m/s and L=245 km. Thus in shallow water, the wave length of the tide is still 200 to 500 km long. Such large systems are influenced by the earth's rotation (much like the large weather systems) and hence tides do not propagate in a straight line, but rotate. In fact, the tides are rotating, resonating fluid flow systems that consist of at least 60 major constituents and must fit into an almost limitless variety of inter-connected coastal shapes such as oceans, seas, bays and estuaries. More detail may be found in texts such as Neumann and Pierson (1966), Ippen (1966), Ross (1977), Marchuk and Kagan (1984), Carter (1988) and Open University (U.K), (1989), and in more technical references such as Defant (1961), Murthy (1984) and Reid(1990).

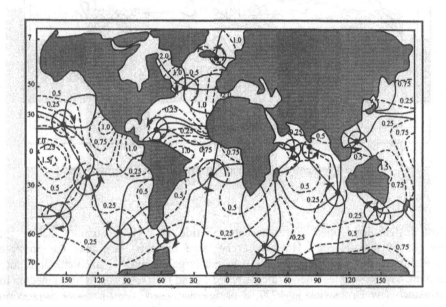

Figure 6.5 Semi-Diurnal Tide in Atlantic Ocean

Figure 6.5 shows the important M2 tidal constituent in the Atlantic Ocean. Forrester (1983) gives a detailed description of tides in Canadian waters and Fig. 6.6 shows the M2 constituent for the Gulf of St Lawrence. Both figures show co-range lines (dashed lines) where the tidal range (2 x amplitude) is the same, and co-tidal lines (solid lines) where the tide has the same phase. The tide clearly rotates as shown by the arrows. The centers of rotation are called *amphidromic points*; there is no vertical tidal fluctuation at those locations.

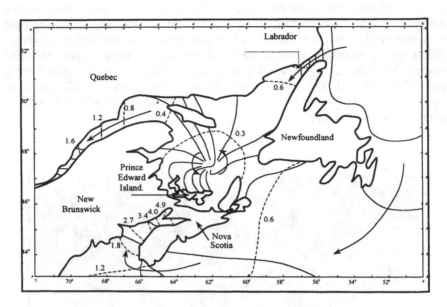

Figure 6.6 Semi-Diurnal Tide in the Gulf of St. Lawrence
(after Forrester, 1983)

Most design along open shores accepts the tide as a given quantity that is not changed by the project.   However, very large projects must take the possible interaction of the proposed changes with the large complex tidal circulations into account.   For example, when Prince Edward Island (Fig 6.6) was to be connected to mainland New Brunswick in 1997, a bridge was used.   A causeway connection would have induced major changes in the large-scale tidal patterns and would have impacted currents and fisheries for hundreds of kms.

In the deep, open ocean, the fluid velocity (tidal current or horizontal tide) is in phase with the tidal water level fluctuations (vertical tide).   At high water there is a maximum current velocity in the direction of tide propagation.   This is similar to progressive short waves discussed in Ch. 2, in which the horizontal component of orbital motion and the velocity of propagation are in the same direction at the moment of high water (Fig. 2.6).   When the tide approaches land, however, the phase relationship between horizontal and vertical tide changes.   In the case of a tidal inlet and bay as in Fig. 6.7, the water level fluctuations in the bay are driven by the tidal water level in the sea. Rising water levels in the sea cause a current to flow into the bay, raising its water level. This inflow of water is called *flood* and the outflow current during the other half of the tidal cycle is called *ebb*. For a small bay and a large entrance, there is no phase lag between the vertical tide in the bay and in the sea. At the time of high water in the sea, the maximum water level is also reached in the bay and will begin to lower. Thus, at that moment of high water in both the sea and the bay the flood current through the inlet becomes zero. This is called high water slack tide. Similarly a low water slack tide occurs at low water. Currents flowing through the inlet are maximum at the time of mean water. If we call the in-flowing (flood) current positive then the current leads the water levels by 90°. This is demonstrated in Fig 6.8 where Curve B (the current) leads the vertical tide (Curve A) by 90°. This is characteristic of complete wave reflection and may be compared with Table 2.4, where the upstream (flood) velocity u is 90° out of phase with the water level change η.

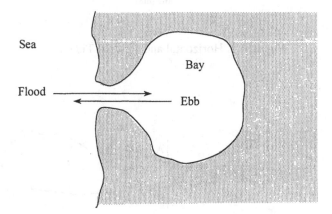

If the inlet is narrow or the bay is long, the maximum water level in the bay will occur later than in the sea, which means that flow will continue to enter the bay for some time after high water in the sea. In this case, the horizontal tide (current through the inlet) will lead the vertical tide (in the sea) by *less* than 90°, and the tidal wave is partly progressive, partly reflecting. This is demonstrated by Curve C in Fig. 6.8.

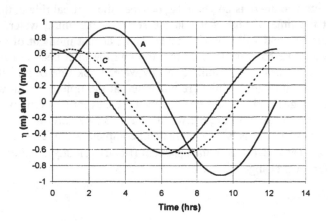

Figure 6.8 Horizontal and Vertical Tide

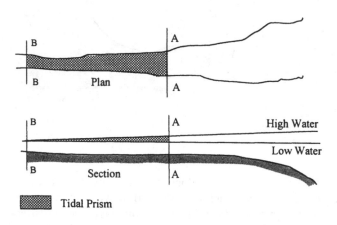

Figure 6.9 Tidal Prism in an Estuary

In an estuary (Fig. 6.9) the tide levels upstream of Section AA are the result of water flowing past AA. Because the distances along the estuary are substantial and may be of the order of the tidal wave length (200-500 km), the tide will take some time to travel upstream. Therefore the maximum water levels anywhere upstream of AA will occur later than at AA. Once again, because flood (inflow) continues after high water at AA, the horizontal tide at AA will lead the vertical tide by less than 90°. The differences between the high and low water levels everywhere upstream of AA, multiplied by the surface area above AA define the volume of water that must flow past AA every half tide cycle. This is called the *tidal prism* above AA and can be used to compute average current velocities at AA. At Section BB, the upstream limit of the estuary (head of tide), the tidal prism becomes zero. The phase difference between the current and the water levels will be 0° at the deep, wide seaward limits of the estuary and tend toward 90°, just downstream of BB.

Figure 6.10 shows an example of tides along the estuary of the St. Lawrence River. It is also seen that the tide shoals (increases in amplitude) as it moves upstream. Above Quebec City, however, the tidal amplitude decreases when the water becomes shallow and friction reduces the tidal motion. Figure 6.10 also shows that the tide becomes asymmetrical as it progresses upstream; the duration of the rising tide becomes shorter than the falling tide. If friction would not reduce the tide height, the wave would become a tidal bore, essentially a breaking tidal wave found in some relatively deep and short estuaries.

Figure 6.11 shows the horizontal and vertical tide at Portneuf. The asymmetry in the vertical tide is also reflected in the horizontal tide; the duration of the ebb flow is greater the flood. Because the head of tide is still 150 km further upstream, the tide at this location (400 km in the estuary) is still almost progressive in that the currents lead the water levels by about 20°. Clearly, the horizontal tide is closely related to the vertical tide and hence, tide analysis and prediction methods discussed in Section 6.2.5 can also be applied to tidal currents. The St. Lawrence is a very long estuary, used here to demonstrate the basic principles. Most estuaries are much shorter, will be more reflecting and behave much more like Fig 6.7.

6.2.7 Stratification and Density Currents

An estuary is defined as a tidal area where a river meets the sea. It has salt water on its downstream limit (sea) and fresh water on the upstream limit (river). The salt sea water normally has a salinity in the vicinity of 35 parts per thousand (ppt) and a density of 1035 kg/m^3. The fresh water has a density of 1000 kg/m^3.

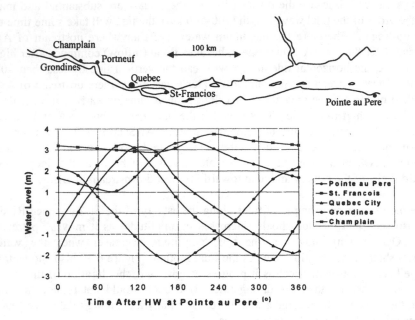

Figure 6.10 Vertical Tides along the St. Lawrence

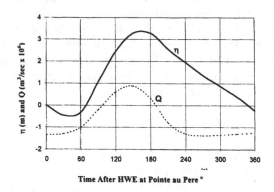

Figure 6.11 Horizontal and Vertical Tides at Portneuf

How the transition of salt to fresh water takes place depends on the amount of mixing in the estuary. In a well-mixed estuary (an estuary with much turbulence), salt and fresh water are thoroughly mixed at any location. Salinity simply varies along the estuary from 35 ppt in the sea to zero ppt in the river and at any specific location, salinity and density will vary with the tide as shown for Rotterdam in Fig. 6.12.

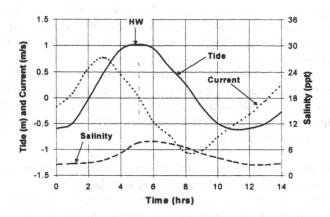

Figure 6.12 Salinities at Rotterdam

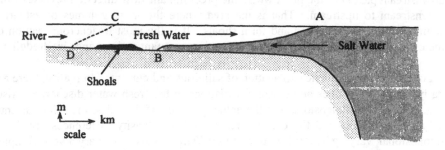

Figure 6.13 Stratified Estuary

If there is little mixing in the estuary, the lighter fresh water will lie over the heavier salt water, resulting in a stratified estuary. A *salt wedge* (AB) will form as shown in Fig. 6.13. Note that Fig 6.13 is highly distorted, since the estuary is many kilometers long, while the estuary depth is in meters. At any time and at any location, salinity will be a function of depth and there will be a sudden change in salinity and density at the interface of the salt water and the overlying fresh water. The sharpness of the interface is a function of the amount of mixing.

The salt wedge moves up and down the estuary with the tide level. Flood and the rising tide will move the salt wedge from AB to CD. Ebb and the falling tide will return the wedge to near AB. The fresh water river flow will flow out over the salt water and the incoming tide will predominantly flow in along the bottom, below the fresh water. Along the bottom the currents are downstream above (upstream of) position D and upstream below (downstream of) position B. Between D and B the flow direction changes 180° and there will be a location in the estuary where predominantly downstream flow changes to predominantly upstream flow. This is a crucial point in any stratified estuary.

Since an estuary is the downstream limit of a river, all the sediment carried by the river ends up in the estuary, usually as fine silt and clay. The clay is suspended in the flow and the clay particles carry a negative electrical charge that keeps them in suspension. When these particles meet the saline water, this charge is neutralized. The clay flocculates and settles out of the water, forming abundant layers of very loose mud on the bottom. This mud has a density of less than 1300 kg/m^3 and behaves essentially as a viscous fluid. The bottom currents move this material downstream past D to the point where the predominant flow direction reverses from downstream to upstream. That is the area where the large volumes of estuarine sediments will be deposited and form shoals. It is the most treacherous section of the estuary for navigation and usually it requires constant maintenance by dredging.

Figure 6.13 is a simplistic explanation of salinities and currents. In reality, there are daily variations in tides and seasonal variations in the fresh water discharge. Also, Fig. 6.13 is two-dimensional but the actual patterns of the tides, the tidal currents, salinities, densities and the currents resulting from density differences are three-dimensional, varying also across the estuary. This makes an estuary a very complex system that requires special care in design. Contrary to design in the open sea, much of the construction in an estuary will affect the tides. Dredging to improve navigation in the treacherous shallow water areas must be done with *care*. All the dredge spoil must be removed from the estuary, otherwise the converging currents will simply return it to the same location. Dredging the shoals increases the salinities further upstream and may affect marine habitat. For example, oysters can only live within a very narrow range of salinities and thus dredging sediment deposits may inadvertently kill oyster beds upstream. Similarly, water intake and sewer outfalls will be affected by the changes in salinity. Filling in low-lying land adjacent to the estuary not only destroys valuable, productive habitat, but it also decreases the tidal prism. This will in turn decrease tidal flow, encouraging sedimentation, and it will change salinities upstream, with its attendant consequences to the environment.

One other major design consideration in estuaries is that all basins (harbours, marinas, cooling water reservoirs) adjacent to an estuary will receive suspended sediment with each incoming flood and through density currents. This sediment settles into the basin and cannot be removed by the ebb currents. Thus such basins function as one-way sediment pumps, often resulting in very large maintenance costs.

6.2.8 Tidal Computation

The tides and tidal flows in an estuary are complex and require computational models to calculate water levels, flows, salinities, and densities. A detailed discussion of such models is beyond the scope of this text and may be found in many technical papers, as well as in Abbott (1979), Abbott and Basco (1989), Cunge et al. (1980), Dronkers (1964) and Murthy (1984). Such models use the equations of continuity and motion. The most sophisticated formulation uses three-dimensional versions of these equations, but most often a two-dimensional (horizontal) formulation (2-DH) is used[1]. In 2-DH models, it is assumed that all variables are constant over the depth of water. For stratified estuaries this assumption is obviously not valid, and it is customary to use several 2-DH models stacked on top of each other to represent layers in the flow.

For estuaries with a regular geometry, sometimes the equations can be simplified to give a one-dimensional (1-D) computation, which uses averaged values over the whole estuary cross-section. Such a 1-D model has severe limitations, but if the available input data are insufficient to calibrate a 2-DH model properly, as is often the case, then more sophisticated models will not yield better results than the simple 1-D models. Finally, for inlets as in Fig. 6.7, there exist very simple computational methods to compute water levels and flows.

---

1. Further discussion of this terminology may be found in Ch 13.

6.3 Storm Surge

The water level fluctuation of greatest concern in design is storm surge, which is an increase in water level resulting from shear stress by onshore wind over the water surface (Fig. 6.14). This temporary water level increase occurs at the same time as major wave action and it is the cause of most of the world's disastrous flooding and coastal damage. Parts of Bangladesh are flooded regularly by storm surge generated by passing cyclones, resulting in the loss of thousands of lives. In a 1990 cyclone, the water levels rose by 5-10 m and it was estimated that more than 100,000 lives were lost. The shorelines along the southern borders of the North Sea were flooded in 1953 because storm surge caused dike breaches. Property damage was very extensive and 1835 lives were lost in the Netherlands. The threat of severe storm surge from Hurricane Floyd in 1999 caused the evacuation of 3 Million people along the East coast of the United States and Canada. It resulted in 50 deaths in the United States, Bahamas and other Atlantic Islands.

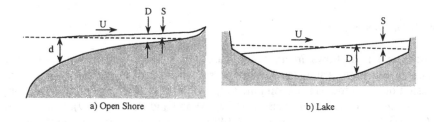

a) Open Shore   b) Lake

Figure 6.14 Definition Sketch for Storm Surge

During storm surge, the water level at a downwind shore will be raised until gravity (the slope of the water surface) counteracts the shear stress from the wind. Computations of storm surge are carried out using the same depth-averaged two dimensional equations of motion and continuity that are used for tidal computations. In this case wind-generated shear stress is the main driving force. For simple problems, the equations can be reduced to a one-dimensional computation

$$\frac{dS}{dx} = \frac{\zeta \, (U \cos \phi)^2}{g \, D} \tag{6.4}$$

where S is the storm surge (the setup of the water level by the wind), x is the distance over which the storm surge is calculated, ζ is a constant (=$3.2 \cdot 10^{-6}$), U is the wind speed, ϕ is the angle between the wind direction and the x-axis and D is the new depth of water (=d+S). Equation 6.4 shows that storm surge is greatest in shallow water; that is why Bangladesh on the delta of the Ganges, Brahmaputra and Meghna rivers and the Netherlands on the delta of the Rhine, Meuse and Scheldt are very susceptible to storm surge.

Example 6.1 One-Dimensional Surge Calculation[⊗]

Equation 6.4 may be solved numerically. The simplest numerical integration (Euler) starts in deep water with an initial condition S = 0 and moves toward shore. The distance to shore is divided into sections of length Δx for which depth is assumed to be constant. A value of ΔS is calculated for the first Δx and D=d+S may be calculated for the end of this first section. This value of D is then used to compute S for the second section and so on until the calculation reaches shore.

The following table presents S for a 10 km long offshore profile, divided into 6 sections for which the depth is assumed to be constant. For U = 20 m/sec and ϕ = 0°, the storm surge at the shore is shown to be 0.29 m.

Section	1	2	3	4	5	6
Δx (km)	3	2	2	1	1	1
d (m)	15	10	5	2.8	1.9	1.4
D (m)	15.0	10.03	5.05	2.90	2.05	1.61
ΔS (m)	0.026	0.026	0.052	0.045	.064	.081
S (m)	0.03	0.05	0.10	0.15	0.21	0.29

Equation 6.4 assumes steady conditions; the wind blows forever in one direction. Thus, it computes maximum surge, a value that can be used in feasibility studies and conservative desk design.

On an enclosed body of water such as a lake, the wind stress obviously results in a negative storm surge at the upwind shore as shown in Fig. 6.14b. An example of a measured storm surge on Lake Erie is presented in Fig. 6.15. A maximum water level difference in excess of 3 m existed between Bar Point near the West (upwind) shore and Port Colborne near the downwind shore, 300 km to the East. This makes the average water slope due to storm surge (dS/dx) about 1×10^{-5}. If we assume the whole lake to have an average depth of 25 m, then according to Eq. 6.2 the wind speed needed to generate this storm surge is about 27 m/s (or 55 knots), which is a severe storm for the area. The storm surge in Fig. 6.15 is therefore quite large for Lake Erie.

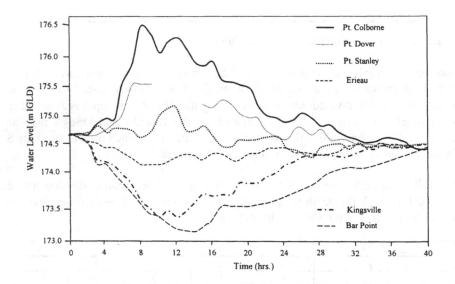

Figure 6.15 Measured Storm Surge on Lake Erie, Dec 1-3, 1985
(after Moulton and Cuthbert, 1987)

6.4 Barometric Surge

Since strong winds are the result of large pressure fluctuations, a barometric surge will accompany storm surge. Suppose there is a difference in barometric pressure Δp between the sea and the shore in Fig. 6.14a, or between the upwind and downwind shore in Fig. 6.14b, then an additional water level rise will be generated

$$\Delta h = \frac{\Delta p}{\rho g} \qquad (6.5)$$

where ρ is the density of water. Equation 6.5 results in a water level rise of about 0.1 m for each kPa of pressure difference. A major depression can easily generate a pressure difference of 5 kPa, resulting in a potential barometric surge of 0.5 m.

6.5 Seiche

When the wind that formed a storm surge stops blowing, the water level will begin to oscillate back and forth (seiche). The oscillations will continue for some time because friction forces are quite small. The wave length of the fundamental mode of the oscillation (a standing wave) for a closed basin (Fig. 6.16) is twice the effective basin length (B_e). In general, the wave length is $2B_e/(1+n_h)$ for the n_h harmonic. For an open ended basin (open coast), the fundamental wave length is 4 times the effective length of the shelf (B_e) over which the storm surge was initially set up. In general, for the n_h harmonic it is $4B_e/(1+2n_h)$.

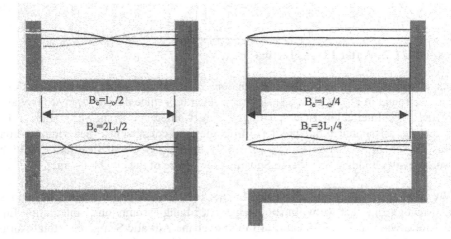

Figure 6.16   Seiche Wave Lengths

The period of oscillation (T=L/C) for a closed basin may be calculated as

$$T_m = \frac{2B_e}{(1+n_h)\sqrt{gd}}$$

(6.6)

and for an open ended basin,

$$T_m = \frac{4B_e}{(1+2n_h)\sqrt{gd}}$$ (6.7)

For the Lake Erie example in Fig. 6.15, the fundamental period of oscillation along Lake Erie (with an average depth of 25 m and an effective length of about 300 km) would be about 10.6 hours. The fundamental period for the seiche across the lake (about 55 km) is about 2.0 hours. The currents needed to displace the large volumes of water can be considerable. For Fig. 6.15, currents would be as high as 0.25 m/s in the lake itself. The oscillations may cause severe currents and water level changes in bays and rivers that connect to such a seiching water body. The currents can break ships and pleasure craft from their moorings. In the case of the Napanee River, which enters Lake Ontario (a tideless sea), the "tides" resulting from seiche on the lake were present so often that they were counted on by the sailing vessels to negotiate the river.

Example 6.2 Water Level Fluctuation at Venice

An interesting example of the combination of short term water level fluctuations may be found in Fig. 6.17, where the water level in Venice during the first few days of the 1992 International Coastal Engineering Conference is shown. Figure 6.17 shows the astronomical tide predicted from the tidal constituents. Superimposed on this we see an oscillation of about 23 to 25 hour period, which caused very high water levels (flooding San Marco Square in the centre of the city by 0.4 m).

We will now calculate what we might expect at Venice using the above equations and recognizing that we will make many simplifying assumptions. Integrating Eq. 6.2 for a wind speed of 25 knots (13 m/s) over the Adriatic Sea gives a total storm surge of 0.58 m. Inside the lagoon in which Venice is situated, there is a further storm surge of 0.05 m. The water level rise due to barometric pressure is calculated with Eq. 6.3 as 0.07 m. The total water level rise was therefore 0.58 + 0.05 + 0.07 = 0.70 m. This is close to the difference between the actual water level and the predicted astronomic tide in Fig. 6.17 for the two highest water level peaks.

The storm surge would not have created so much difficulty, if its peak had not coincided with the high water from the astronomic tides. This surge coincided with several high waters in a row, since the surge period was equal to about twice the basic tidal period. Let us see if we can calculate the surge period. The calculation depends very much on the assumed average depth of water over the portion of the Adriatic Sea involved in the oscillation. A reasonable estimate of average depth is 150 m and the effective length is 800 km. For these characteristics the period for the fundamental mode according to Eq. 6.7 is 23 hours.

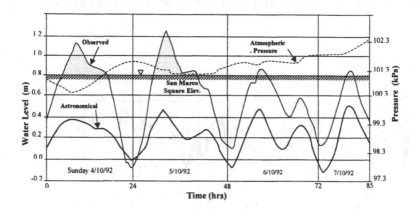

Figure 6.17 High Water at Venice

From these relatively crude computations, it is clear that the 'Aqua Alta' (high water) at Venice is a combination of storm surge, barometric surge and seiche. The simple equations permit a basic understanding of the complex problem. More elegant solutions are needed to solve actual design problems. Flooding problems in Venice are being studied with sophisticated numerical models, using the two-dimensional, depth-averaged equations of motion and continuity. To give better results than the above approximations, such models require extensive field measurements for calibration.

In passing, note that the barometric pressure was lowest when the first high water level occurred. Thus it appears that barometric pressure drop gave rise to high winds that caused a storm surge, which was enhanced by the barometric surge also resulting from the pressure drop.

6.6 Seasonal Fluctuations

Seasonal water level fluctuations do not occur along the open ocean, but they do occur on lakes and in the upper reaches of estuaries. Extreme fluctuations occur in power and water supply reservoirs. Normally, seasonal fluctuations are taken into account in design as a matter of course and hence they are not of much concern, even along the Great Lakes. Figure 6.18 shows examples for Lakes Michigan-Huron and Ontario. The seasonal fluctuations are about 0.5 and 1 m respectively.

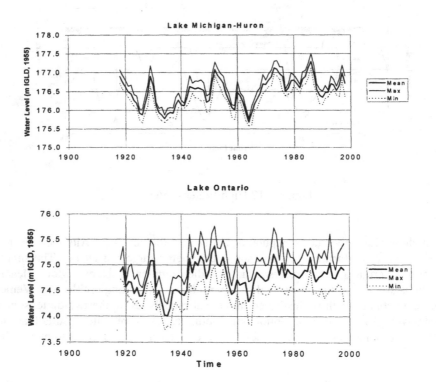

Figure 6.18   Monthly Water Levels on Lakes Michigan-Huron and Ontario
(after Monthly Water Level Bulletin, Environment Canada)

6.7 Long-Term Water Level Changes

6.7.1 Climatic Fluctuations

Water level fluctuations that are the response to long term climatic change such as extended wet and dry periods are a cause for concern. The danger of such longer term water level fluctuations is that everyone *forgets* in a few years how high (or low) the water can actually be. Therefore, a few years after high water, development begins once again to encroach on the shore, exposing new properties to extreme stress during the next high water. Similarly, once a low water has not occurred for a few years, docks and marinas will be built in areas of shallower water, so that when the next low water occurs, the water is too shallow for these facilities to be useful. One recurring theme resulting from such long-term fluctuations is that government funds are requested to help out the "unfortunate property owners". Fortunately there is a trend toward better coastal management with a longer "memory" that accounts for extreme events and monitors the activity in the shore zone closely.

Figure 6.18 shows a 1.3 m rise in mean water levels on Lake Michigan-Huron between 1934 and 1952, followed by a 1.4 m drop from 1952 to 1964. This is a much larger fluctuation than the annually expected 0.5 m fluctuation. In general, the total water level fluctuation along the Great Lakes (adding the annual and long-term change) is of the order of 2 m. Periods of major shore zone damage can be directly related to periods of high water levels, such as 1929, 1952, 1973, 1986 and 1997 for Lake Michigan-Huron. These high water levels allow the large waves to come closer into shore for several months to several years. When such water levels combine with short-term storm surges, structures are destroyed and protective beaches disappear This exposes the shore, which mainly consists of glacial till bluffs, to direct wave action and severe erosion. Extreme low water levels (such as 1934 and 1964 on Lake Michigan-Huron) also cause problems. Wells run dry, there is insufficient water for navigation and power generation, and pleasure craft cannot enter or leave marinas.

6.7.2 Eustatic (Sea) Level Change

The term *eustatic* refers to a global change in ocean water levels, resulting from melting or freezing of the polar ice caps and thermal expansion of the water mass with temperature change. Detailed descriptions may be found in Carter (1988) and Bird (1984). The sea levels 25,000 years ago were 150 m below the present level. Between then and 3,000 years ago, water level rose at about 7 mm/yr to almost the present water level. The present average rate of eustatic rise is small and therefore difficult to measure. The best estimates are 1 to 1.5 mm/yr. This relatively small rate of rise, nevertheless, submerges the ocean shores and is at least partly responsible for the fact that most beaches around the world are eroding over the long term.

6.7.3 Isostatic (Land) Rebound and Subsidence

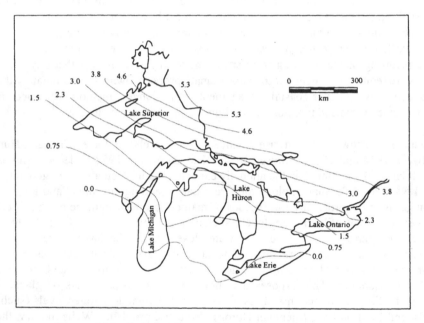

Figure 6.19   Relative Rates of Crustal Movement (mm/yr)
(after Clark and Persoage, 1970)

The common natural cause for isostatic (land) elevation change is a result of the adjustment of the earth's crust to the release of pressure exerted by the 1 to 2 km thick ice sheet that covered it during the last glaciation. Typically, the earth's crust was severely depressed by the ice and a rise (forebulge) was formed in the earth's crust ahead of the glaciers. When the ice retreated, the earth's surface rebounded (upward) where the glaciers had been and lowered where the forebulge had occurred. This process still takes place today, but at a much-reduced rate. Most areas in the higher latitudes experience isostatic rebound and areas at more intermediate latitudes experience some subsidence. Figure 6.19 shows the isostatic rebound over the Great Lakes and Fig. 6.20 shows a Northeast-Southwest line through the Northern United States, indicating both rebound and subsidence with a hinge line near Kingston (N.Y.).

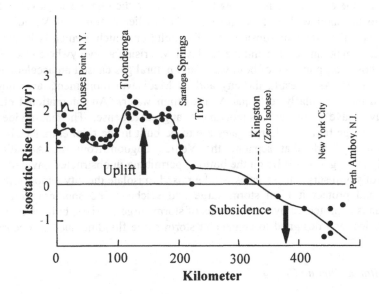

Figure 6.20 Isostatic Adjustment in Northern United States (mm/yr)
(after National Research Council, 1987)

In general, isostatic rebound decreases the impact of eustatic sea level rise, or even reverses it. For example, the measured rate of relative sea level rise (water level rise with respect to the land) at San Francisco is 1.3 mm/yr while at Juneau, Alaska the sea level drops at 13.8 mm/yr (National Research Council, 1987). On the Great Lakes, the effect of isostatic rebound is not quite so simple. All the land rises, but the relative rise of the land with respect to the water is controlled by the difference between the local rate of rebound and the rebound at the outlet of the lake. From Fig. 6.20 it may be seen that along Lake Michigan-Huron, the rate of rebound is 0 to

2.5 mm/yr. The outlet rises at about 0.5 mm/yr while most of the land rises at a greater rate and hence most of Lake Michigan-Huron has an emerging shore. Conversely, for Lake Ontario, the outlet rises at 2.5 mm/yr while the shore rises at 0.75 to 3.0 mm/yr, thus forming a submerging shore over most of the lake. Clearly the morphological development in these two lakes is totally different.

Although subsidence does occur naturally, often it is man-made. Pumping groundwater, petroleum and natural gas are common causes. Subsidence exacerbates the effects of eustatic sea level rise since the relative sea level rise with respect to the land will now be greater. The earlier example of Venice clearly demonstrates the effect of subsidence. The delta on which Venice is located was sinking at a small annual rate and the sea level was rising as everywhere else. In this century, however, pumping of both water and natural gas caused an accelerated rate of subsidence. As a result, the city and its Mediaeval monuments are subjected more and more regularly to 'Aqua Alta' or high water. An international effort is underway to save Venice and its monuments at great expense. The leading idea is to use storm surge barriers. Large gates are to be built in the tidal entrances between the offshore islands that separate the Venice Lagoon from the Adriatic Sea. Normally these gates will lie on the bottom, permitting unobstructed navigation, but at times of storm surge, these gates will be raised to isolate the city temporarily from the sea and protect it from storm surge and seiche. The southern part of the Netherlands is protected by such a series of storm surge barriers, built as part of the Delta Project and designed to counteract storm surge flooding such as occurred in 1953.

6.7.4 Global Climate Change

The final and potentially most dangerous water level change results from trends in global climate. In the discussion of eustatic sea level rise, we have already seen that global warming after the last glaciation has resulted in a sea level rise of 100 to 150 m through melting of the polar ice caps and thermal expansion of the water in the ocean. The present rate has slowed down to an estimated 1 to 1.5 mm/yr, but any additional warming would increase this rate of sea level rise.

Concern is centered around the production of the so-called greenhouse gases. These combustion products act as an insulating blanket over the earth, decreasing the net longwave radiation from the earth back into space and thus trapping the sun's heat to cause global warming. It is a controversial subject and indeed there is a contingent of respected scientists that disputes the whole idea. It is estimated (National Research Council, 1979) that a doubling of carbon dioxide (CO_2) would result in an average global temperature rise of 1.5 to 4.5 ^{0}C. At the poles the temperature rise is estimated to be two to three times the average. Monitoring stations such as Mauna Loa, Hawaii indicate an increase in CO_2 concentration from 315 to 340 parts per million (ppm) between 1958 and 1980 (National Research Council, 1983). Tree ring data show that from 1850 (prior to major industrialization) to 1950, there has been a 50 ppm increase in CO_2 concentration. Estimates of future concentrations vary greatly, but there is a 75% probability that by 2100, the pre-industrial CO_2 concentration will have doubled.

Global climate change models study how such an increase in greenhouse gases translates into temperature and water level rise. Such numerical models have produced several widely varying scenarios. Predicted rise in water level for the year 2025 varies from 0.1 to 0.2 m. For 2050, the estimates vary from 0.2 to 1.3 m and for 2100 the estimates are 0.5 to 2 m. The estimates for 2100 are made up of 0.25 to 0.8 m by thermal expansion of the water in the oceans, 0.1 to 0.3 m from the melting of Greenland and Alpine glaciers, and 0.1 to 1 m from Antarctic deglaciation. Breakup of the West Antarctic ice sheet resulting from these higher temperatures could contribute another 6 m of water level rise over 500 years. More details may be found in Wind (1987). Other studies about this potentially dangerous phenomenon and its major impacts are on-going.

The predictions of water level increases are by no means precise. There are many uncertainties in the estimates of production of greenhouse gases. Probably the most uncertain is what we are going to do about controlling emissions of gases. There is a great tension between appropriate environmental stewardship and impacts on national economies. The resolution of this is unpredictable and largely political. The methods to translate these uncertain atmospheric pollution figures first into global warming and then into water level rise also involves many assumptions.

The important question is: How do we prepare for global climate change as coastal engineers and managers? Helpful references are: Wind (1987) and National Research Council (1987). The following changes in physical processes may be identified:
- More severe storms (tornadoes and hurricanes) will occur more often.
- Storm surge will decrease a little because of the larger water depths, but will increase significantly because of the more severe storm activity.
- Offshore, the waves will be higher, because of more severe storms.
- Tides will not be significantly affected.
- Tidal prisms will increase, because the planform area of the bays and estuaries increases.

–   Breaking waves on shores and structures will be higher because larger depths
    all the way into shore will reduce bottom friction losses, as well as permit large
    breaking waves to come closer into shore.

These factors will result in the following responses:
–   Structures will be subjected to higher stress from the higher waves. Factors of
    safety will decrease.
–   Structure runup and overtopping will increase, adding to the risk of flooding
    and damage by overtopping. As an example, consider the Netherlands. The
    recently (1986) completed Delta Project, raised all dikes in response to the
    disastrous 1953 storm surge. It has been calculated that a 1 m sea level rise
    would reduce the present margin of safety by about 90% (Wind, 1987).
–   Sandy shorelines will retreat as shown in Ch 12.
–   Barrier islands will roll back more rapidly and marshes behind the barrier
    islands will disappear.
–   Deltas will not build out at the same rate; they may even retreat.
–   Bluffs and cliffs will retreat more rapidly.
–   Sediment transport rate will increase, possibly filling presently stable inlets and
    harbour entrance channels.
–   Salt water intrusion into groundwater table will increase.
–   Wetland areas will be inundated and disappear.

How can we prepare? Fortunately, the expected sea level rise is not beyond the
capability of present technology. Flood protection, shore protection and navigation
structures can be strengthened and raised to cope with the rise in water level. If all
else fails, a properly executed retreat can be planned in which buildings are moved
back from the shore or abandoned. There is no cause for alarm. Two important
aspects must be remembered. First, the magnitude of the problem is huge.
Population densities along the ocean shores are already high and rapidly increasing
(Ch. 10). Sustainable development of the coastal areas, in light of sea level rise will
be our most important task in the next century. Secondly, although flood defenses
such as dikes can be raised with presently available technology, the risk to the
people and properties behind those dikes increases.

The main casualties will be the already limited wetland areas. Their development
can keep up with the slowly rising sea levels and move inland, but they may have
problems adjusting to more rapid rise in water levels. Also, most of the properties
behind the wetlands are dedicated to man-made uses and it is unlikely that wetlands
will be allowed to intrude into this valuable real estate. Damage will also occur to
agricultural areas because of the additional salt water intrusion.

Sustainable shoreline development through maintaining the existing shorelines by retrofitting, retreat by moving infrastructure landward, and meeting the concerns for wetlands and agriculture, will require a complete restructuring of the present political and policy decision making processes. They were not designed to deal with slowly developing mega impacts over very large areas. Chapter 10 gives an example of how the Netherlands has developed a long-term national strategy involving protection and shoreline retreat.

For the Great Lakes, Cohen (1986, 1987) determined that for a doubling of CO_2, the temperature would rise on average from 3.1 to 4.8 °C. These higher temperatures would result in a higher evaporation rate and a 20% decrease in the net supply of water to the drainage basin. Thus, contrary to the maritime shores, global warming is expected to lower the levels of the Great Lakes. Allsopp and Cohen (1986) estimate that for a doubling of CO_2 the water levels in Lakes Superior, Michigan-Huron and Erie would drop by 25, 70 and 50 cm respectively. The water levels in Lake Ontario are controlled at the outlet. Such low water levels bring their own set of problems. Wave attack and damage to structures is no longer a concern, but the impacts on power generation and navigation depths will be large. Wetlands will benefit, however. They can follow the lowering of the water levels and migrate offshore, an area that is not occupied by man at the present time.

7. Wave Transformation and Breaking

7.1 Wave Transformation Equations

Wave transformation describes what happens to waves as they travel from deep into shallow water. It is defined by two vector equations: the *wave propagation equation*

$$\Delta \times \mathbf{k} = 0 \qquad (7.1)$$

and the *conservation of energy flux equation*

$$\Delta (E\,\mathbf{C_g}) = 0 \qquad (7.2)$$

The vector $\mathbf{k}$ is the wave number vector; its direction denotes the direction of wave propagation and its modulus is the wave number

$$k = \frac{2\pi}{L} \qquad (7.3)$$

According to small amplitude wave theory (Ch. 2), the wave length, L, is

$$L = \frac{gT^2}{2\pi} \tanh\,(kd) = L_o \tanh\,(kd) \qquad (7.4)$$

where T is the wave period, L_o is the deep water wave length and d is the depth of water. In Eq. 7.2, (EC_g) is known as the energy flux, E is the wave energy density

$$E = \frac{1}{8} \rho g H^2 \tag{7.5}$$

and $\mathbf{C_g}$ is the wave group velocity vector. The direction of $\mathbf{C_g}$ is the wave direction and its modulus is

$$C_g = nC \tag{7.6}$$

where C is the velocity of propagation of the individual waves,

$$C = \frac{L}{T} = \frac{gT}{2\pi} \tanh kd = C_o \tanh kd \tag{7.7}$$

and

$$n = \frac{1}{2}\left(1 + \frac{2kd}{\sinh 2kd}\right) \tag{7.8}$$

The simplest form of Eq. 7.2 occurs when it is written in the direction of wave propagation (s direction), when it becomes an ordinary differential equation

$$\frac{d}{ds}(EC_g) = 0 \tag{7.9}$$

Extensive developments of Eqs. 7.1 and 7.2 may be found in many individual technical papers and, for example, Liu (1990). General solutions of these equations are computationally difficult and intensive. The usual approach is via the Mild Slope Equation (Berkhoff, 1972) or the parabolic approximation to this equation (Radder, 1979). As computing power increases, the use such general solutions will obviously increase.

In most designs, simpler approaches are used and these will be described in the following sections. Wave transformation is concerned with the changes in H, L, C and α, the wave angle with the bottom contours; wave period T remains constant throughout the process. To derive the simpler solutions, wave transformation is separated into wave refraction and diffraction. Refraction is wave transformation as a result of changes in water depth. Diffraction is specifically not concerned with water depth and computes transformation resulting from other causes, such as obstructions. Discussions about wave refraction usually begin by calculating depth-related changes for waves that approach a shore perpendicularly. This is called wave shoaling.

7.2 Wave Shoaling

When waves approach perpendicular to a straight shoreline ($\alpha=0$), integration of Eq. 7.9 results in

$$EC_g = nCE = const \tag{7.10}$$

Using Eq. 7.5 we can write

$$nCH^2 = n_1 C_1 H_1^2 = n_2 C_2 H_2^2 = n_o C_o H_o^2 = const \tag{7.11}$$

from which we can relate the wave heights at any two water depths as

$$\frac{H_2}{H_1} = \sqrt{\frac{n_1}{n_2} \frac{C_1}{C_2}} \tag{7.12}$$

and H at any depth can be related to deep water wave height H_o as

$$\frac{H}{H_o} = \sqrt{\frac{n_o}{n} \frac{C_o}{C}} = \sqrt{\frac{1}{2} \frac{1}{n} \frac{1}{\tanh kd}} = \frac{H}{H_0'} = K_s \tag{7.13}$$

where H_o' is the deep water wave height for $\alpha=0$ and $K_s=H/H_o'$ is the *shoaling coefficient*. It is 1.0 in deep water, decreases with water depth to 0.91 and then rises to infinity as the water depth approaches zero. This is shown in Fig. 7.1.

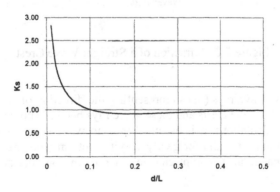

Figure 7.1 Shoaling Coefficient

7.3 Wave Refraction

7.3.1 The Equations

When waves approach the shore at an angle as in Fig. 7.2, wave refraction takes place in addition to wave shoaling. During refraction, the wave crests bend to align themselves with the bottom contours and the wave direction becomes more perpendicular to the shore. To understand the refraction process, imagine a long, straight wave crest approaching a shore at an angle as in Fig. 7.2. The depth of water increases from the inshore end of the wave crest to the offshore end. The wave speed of propagation, which is a function of depth, according to Eq. 7.7 also increases along the wave crest toward deeper water and this will cause the wave crest to bend and to become more parallel to the bottom contours. This process continues into shore, as demonstrated in Fig. 7.2.

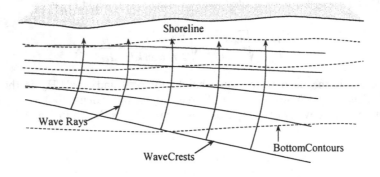

Figure 7.2 Refraction of a Straight Wave Crest

We can now draw wave rays (lines representing the direction of wave propagation) perpendicular to the wave crests and these wave rays bend as shown in Fig. 7.2. For wave refraction calculations, it is assumed that no wave energy crosses the wave rays. This assumption is valid for gently varying bottom contours. If obstructions to the wave propagation or abrupt changes in bottom contours are present, wave energy is transferred across the wave rays, which is wave diffraction (Section 7.5).

When the energy flux is conserved between the wave rays, then Eq. 7.9 yields

$$n\, C\, E\, b = const \tag{7.14}$$

where b is the distance between adjacent wave rays. Equation 7.5 may be substituted into Eq. 7.14 to yield

$$\frac{H_2}{H_1} = \sqrt{\frac{n_1 C_1}{n_2 C_2}} \sqrt{\frac{b_1}{b_2}} \tag{7.15}$$

to relate wave heights at any two specific locations. Wave height at any location can be related to deep water as

$$\frac{H}{H_o} = \sqrt{\frac{n_o}{n} \frac{C_o}{C}} \sqrt{\frac{b_o}{b}} = \sqrt{\frac{1}{2} \frac{1}{n} \frac{1}{\tanh kd}} \sqrt{\frac{b_o}{b}} = K_s K_r \tag{7.16}$$

where

$$K_r = \sqrt{\frac{b_o}{b}} \tag{7.17}$$

is called as the refraction coefficient.

7.3.2 Refraction Diagrams

We have not yet discussed how to draw a refraction diagram such as Fig. 7.2, other than that the wave rays are in the direction of wave propagation (or perpendicular to the wave crests). We will first discuss a graphical method called Huygen's method. On a contour map, we draw a straight wave crest in deep water, as in Fig. 7.2. We then calculate C at many locations along this wave crest and draw L=CT geometrically (with the arc of a circle) at each location. The tangent to these many arcs will be the next wave crest. This graphical construction is continued into shore. Finally, wave rays are drawn so that they are visually perpendicular to the wave crests. Equations 7.15 or 7.16 can then be applied using the changes in distance between the wave rays to compute wave heights. The graphical method is of the past. Today we use computers.

Since k in Eq. 7.1 is in the wave propagation direction, the shape of the wave rays may be computed by solving Eq. 7.1. Computer based methods are detailed in, for example, Ebersole (1985), de Vriend et al. (1993) and Liu (1990). More complex solutions compute wave refraction and diffraction simultaneously by solving the

Mild Slope Equation (Berkhoff, 1972) or the parabolic approximation to this equation (Radder, 1979). **REFDIF** distributed by University of Delaware is the best known example. Simpler programs have also been developed to calculate wave refraction only. The basic equations for these may be found in Dean and Dalrymple (1984, Ch 4). Many refraction programs are based on the work of Abernethy and Gilbert, (1975) and Brampton (1977).

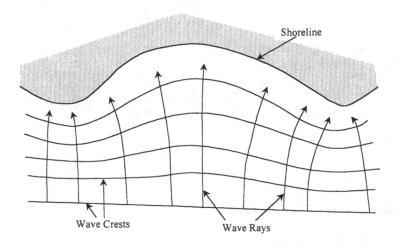

Figure 7.3 Wave Refraction Diagram

Figure 7.3 shows a refraction diagram with waves approaching a shoreline with more complex bathymetry such as a bay and headlands. Equations 7.15 and 7.16, and Fig. 7.3 indicate that waves will be higher at the headlands because of wave ray convergence and lower in the bays because of divergence of the wave rays. Note that the wave crests are only drawn for illustration. It is not necessary to draw the wave crests in order to define the refraction pattern; only the wave rays (and their spacing) are important. Projection of the wave rays from deep into shallow water, as shown in Figs. 7.2 and 7.3 is called forward tracking. We assume a wave direction in deep water and then move forward along the wave rays into shallower water. A complete refraction analysis consists of many such refraction diagrams (one for each possible wave direction - wave period combination of interest). It is also possible to compute wave rays by backward tracking in which a series of wave rays at different angles are projected to deep water from a single point of interest in shallow water. This solution only provides information at one inshore location, but each diagram includes many angles of interest and only one such diagram is needed for each wave period. Figure 7.4 shows both these types of refraction diagrams. In the upper figure, wave rays are projected seaward from the site of interest. In the lower figure, wave rays are projected into shallow water from locations and directions, identified as important in the upper figure. Clearly with such complex bottom topography, a refraction study must consist of many such partial refraction diagrams to define wave heights and directions at points of interest with sufficient accuracy and detail.

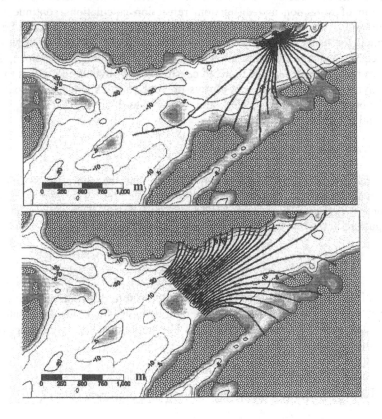

Figure 7.4 Refraction Diagrams for Kingston Harbour

Once the wave rays have been calculated, it is possible to use Eq. 7.2 or its simplifications, such as Eqs. 7.15 or 7.16 or the equations in Dean and Dalrymple (1984), to compute the wave height in any depth of water. The academic separation of refraction and diffraction may cause difficulty with refraction diagrams. When substantial irregularities occur in the bottom contours, the wave rays may cross, forming a caustic. In such a case, b→0 and, because we assume that no wave energy crosses the wave rays, wave height will become infinite according to Eqs. 7.15 or 7.16. This does not occur in nature. If diffraction had been taken into account simultaneously, wave energy would have leaked across the two converging wave rays and the caustic would not have formed.

In spite of the limitations of the refraction analysis, which is based on the artificial separation of refraction and diffraction, refraction calculations continue to be in widespread use because of their simplicity. Refraction computations can be readily performed for many possible scenarios and they can be applied to large areas at small cost. Since it remains attractive to use refraction calculations, specific remedies have been developed to avoid caustics. The obvious solution is to smooth the bathymetry, but that does not always prevent caustics when large-scale shoals are responsible. Hedges (1976), and Kirby and Dalrymple (1986) described additional remedies.

7.3.3 Snell's Law

For many practical problems we can assume that the shoreline and the depth contours are relatively straight and more-or-less parallel. The relevant definitions are shown in Fig. 7.5. Equation 7.1 is used to compute wave direction of propagation. It may be written in Cartesian Co-ordinates as

$$\frac{\partial}{\partial x}(k \sin \alpha) + \frac{\partial}{\partial y}(k \cos \alpha) = 0 \qquad (7.18)$$

where x is the cross-shore direction, y is the alongshore direction and α is the angle between a wave ray and the x-axis. For a straight shoreline and contours, the second term of Eq. 7.18 is zero and integration of the remaining ordinary differential equation yields

$$k \sin \alpha = const \qquad (7.19)$$

Since T remains constant throughout the refraction process, Eqs. 7.3, 7.4 and 7.7 yield

$$\frac{C}{\sin \alpha} = const \tag{7.20}$$

This is Snell's Law of wave refraction; it calculates wave angles as

$$\frac{\sin \alpha_2}{\sin \alpha_1} = \frac{C_2}{C_1} \quad and \quad \frac{\sin \alpha}{\sin \alpha_o} = \frac{C}{C_o} = \tanh \frac{2\pi d}{L} \tag{7.21}$$

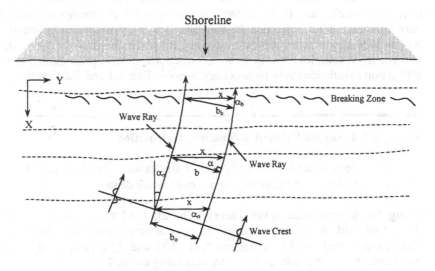

Figure 7.5 Wave Refraction Definitions

Since every wave ray in Fig. 7.5 refracts the same way, the distance parallel to the shore between the wave rays (y) remains constant and

$$\frac{b}{\cos \alpha} = const \tag{7.22}$$

which means

$$\sqrt{\frac{b_1}{b_2}} = \sqrt{\frac{\cos \alpha_1}{\cos \alpha_2}} \tag{7.23}$$

and the refraction coefficient may be approximated by

$$K_r = \sqrt{\frac{b_o}{b}} = \sqrt{\frac{\cos \alpha_o}{\cos \alpha}} \qquad (7.24)$$

7.3.4 Summary

In the simplest wave refraction calculation, the wave propagation equation (Eq. 7.1) is simplified to Snell's Law (Eq. 7.21) and the conservation of energy flux equation (Eq. 7.2) is reduced to the product of the shoaling and refraction coefficients (Eq. 7.15 or 7.16). Equation 7.13 defines the shoaling coefficient and Eq. 7.24 the refraction coefficient. If the bottom contours are not predominantly straight and parallel to each other, refraction diagram calculations, as described in Section 7.3.2, will be necessary from which local wave heights and angles may be deduced. If the bathymetry is complex, or if there are obstructions to wave propagation, refraction-diffraction calculations may be necessary to solve Eqs. 7.1 and 7.2 directly.

Example 7.1 Simple Refraction-Shoaling Calculation

A wave in deep water has the following characteristics: H_o=3.0 m, T=8.0 sec and α_o=30°. Calculate H and α in 10 m and 2 m of water depth.

Using Small Amplitude Wave Theory (Ch. 2): L_o=100 m. In 10 m of water, d/L_o=0.10 and from wave tables or a computer program, d/L=0.14, tanh $(2\pi d/L)$=0.71 and n=0.81. Equations 7.11, 7.21 and 7.24 yield K_s=0.93, α=20.9° and K_r=0.96, which results in H=2.7 m, according to Eq. 7.16.

In 2 m of water depth, similar computations yield: d/L_o=0.02, d/L=0.058, tanh $(2\pi d/L)$=0.35, n=0.96, K_s=1.23, α=10.0°, K_r=0.94 and H = 3.5 m.

The spreadsheet program **RSB**® computes refraction, shoaling and breaking of waves. Figure 7.6 only shows the "front end" of this program; the detailed calculations, which are similar to the wave table calculation in Fig. 2.12, are carried out elsewhere on the spreadsheet. The solid line in the graph shows the wave height development resulting from refraction and shoaling between 15 and 2 m. The answers for 10 and 2 m may be read in the appropriate columns.

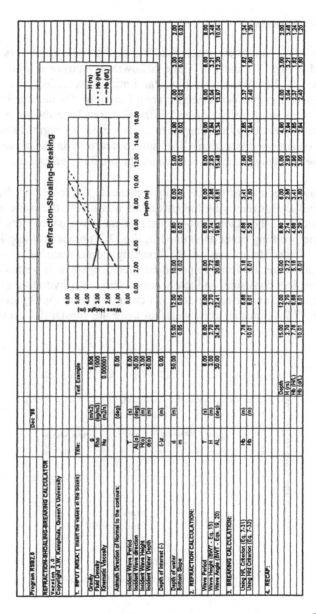

Figure 7.6 Refraction-Shoaling-Breaking Calculation with **RSB**[⊗]

7.4 Wave Breaking

Wave shoaling causes wave height to increase to infinity in very shallow water as indicated in Figs. 7.1 and 7.6. There is, however, a physical limit to the steepness of the waves, H/L. When this physical limit is exceeded, the wave breaks and dissipates its energy. Wave heights are a function of water depth, as shown in Fig. 7.7. Wave shoaling, refraction and diffraction transform the waves from deep water to the point where they break and then the wave height begins to decrease markedly, because of energy dissipation. The sudden decrease in the wave height is used to define the *breaking point* and determines the breaking parameters (H_b, d_b and x_b).

The breaking point in Fig. 7.7 really represents the location where the maximum wave breaks. But removing the maximum wave from the wave height distribution also reduces all the other short-term statistical wave parameters such as H_s. This makes the breaking point as defined in Fig. 7.7 unique and applicable to *all* wave heights. Other definitions of the breaking point, such as where white-capping begins, or where the front slope of the breaking wave is vertical, are much more subjective, less consistent and really only relevant for hydraulic model experiments. Because wave transformation does not change H very rapidly outside the breaking zone, Fig. 7.7 defines H_b quite well, but there are substantial uncertainties in x_b and hence d_b.

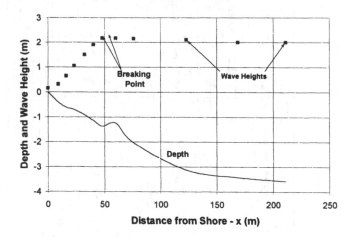

Figure 7.7 Determination of Wave Breaking Parameters

The breaking wave may have one of several shapes as it breaks (Fig. 7.8). The breaker type is a function of the beach slope m and the wave steepness H/L. These may be combined into a ratio, usually called the surf similarity parameter

$$\xi_b = \frac{m}{\sqrt{H_b / L_o}} \qquad (7.25)$$

Spilling breakers, according to Battjes (1974) occur when $\xi_b < 0.4$. They occur on flat beach slopes, for steep waves or both. Therefore when sea (which consists of steep waves) breaks on a flat sandy beach, the breakers are predominantly spilling breakers. Portions of the wave crest appear to break gently (spill). Several wave crests may be breaking simultaneously, giving the appearance of several rows of breaking waves throughout the breaking zone. Such beaches are often called dissipative beaches.

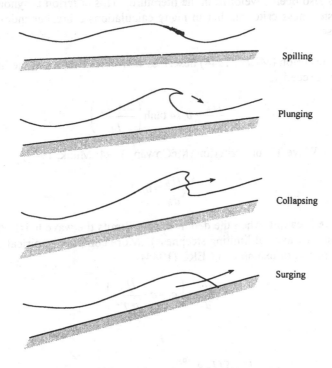

Figure 7.8 Breaker Types

Plunging breakers occur on steeper beaches and/or for flatter waves, when $0.4 < \xi_b$ < 2.0. As the name implies, the wave crest runs ahead of the main body of the wave and plunges forward violently. They are, for example, predominant when swell breaks on flat sandy beache . They are also the most common breaker type in hydraulic model studies, in which the beach steepness is often exaggerated. Collapsing breakers occur on steep beaches when $\xi_b > 2.0$. These waves, which are characterized by a wave front that more or less explodes forward, may be found where swell breaks on steep beaches made up of coarse material. Surging breakers occur on very steep beaches. The waves simply surge up and down the beach and there is very little or no breaking. Beaches with surging and collapsing breakers are often called reflective beaches.

Breaking criteria, defining where and how the waves break, are a function of the limiting wave steepness (H_b/L_b). A second limiting parameter, the breaker index (H_b/d_b) has also been developed in the literature. This criterion is synonymous with the wave steepness criterion, but in many calculations a breaker index criterion is easier to use.

The Miche criterion (Miche, 1944) describes wave breaking when the limiting wave steepness is exceeded

$$\frac{H_b}{L_b} = 0.14 \tanh\left(\frac{2\pi\, d_b}{L_b}\right) \tag{7.26}$$

The Solitary Wave Theory criterion (McCowan, 1894; Munk, 1949)

$$\frac{H_b}{d_b} = 0.78 \tag{7.27}$$

defines wave breaking when the depth of water limits the wave height (or when the depth produces waves of limiting steepness). More complex empirical criteria were also developed. For example in CERC (1984)

$$\frac{H_b}{d_b} = \left(c_1 - c_2\, \frac{H_b}{g\,T^2}\right) \tag{7.28}$$

where

$$c_1 = 43.75\,[1 - e^{-19m}]; \quad c_2 = \frac{1.56}{[1 + e^{-19.5m}]} \tag{7.29}$$

in which m is the beach slope. Goda (1970) developed

$$\frac{H_b}{d_b} = 0.17 \frac{L_o}{d_b} \left[1 - e^{-\left\{\frac{1.5\pi d_b}{L_o}\left(1 + 15m^{4/3}\right)\right\}} \right] \qquad (7.30)$$

The above breaking criteria were developed for *regular* waves – all waves have the same height and period as in small amplitude wave theory and in some hydraulic model tests. Kamphuis (1991a) proposed two criteria for the more practical case of *irregular* waves. The criteria are based on extensive model testing and use H_s as the definitive wave height at breaking.

$$H_{sb} = 0.095 \, e^{4.0m} \, L_{bp} \tanh\left(\frac{2\pi d_b}{L_{bp}}\right) \qquad (7.31)$$

is an extended version of the wave steepness criterion (Eq. 7.26) and

$$\frac{H_{sb}}{d_b} = 0.56 \, e^{3.5m} \qquad (7.32)$$

is an extended version of the depth limited criterion (Eq. 7.27). These expressions include the influence of beach slope m. Significant breaking wave height was determined because H_s was plotted against depth. The breaking wave length L_{bp} is based on d_b and T_p.

The breaking characteristics of waves may now be determined by combining wave refraction and shoaling calculations with one of the above wave breaking criteria.

Example 7.2 Refraction – Shoaling – Breaking Calculation

We extend the earlier example, $H_o = 3$ m, $T = 8$ sec and $\alpha_o = 30°$ to determine in what depth of water this wave breaks and what its breaking wave height, period and angle are. This is an iterative or trial and error calculation that will be demonstrated using the program **RSB**®, shown in Fig. 7.6. A simple trial and error analysis is presented here because it is easy to understand and most readily duplicated by hand calculator and wave tables. More sophisticated computations can solve the problem by iteration or root finding. In Section 2 of Fig. 7.6 the wave characteristics for shoaling and refraction are calculated for several water depths. Section 3 computes the maximum possible wave height according to Eqs. 7.31 and 7.32 for each depth. The wave height in Section 2 cannot exceed the possible maximum value calculated in Section 3. The height and angle of the breaking wave and the depth of water in which the wave breaks is determined when the shoaled and refracted wave height of Section 2 is just equal to the breaking criterion of Section 3. For this example with the beach slope m=0.02, the breaking wave height for Eq. 7.32 is found to be 2.9 m with a breaking angle of 15.3°, in a depth of water of 4.9 m. The two breaking criteria give slightly different answers.

7.5 Wave Diffraction

Wave diffraction is concerned with the transfer of wave energy across wave rays. This phenomenon was specifically separated from refraction. Refraction and diffraction of course take place simultaneously and therefore the above distinction is an academic separation of two closely related processes. The only correct solution is to compute refraction and diffraction together using computer solutions of Eqs. 7.1 and 7.2. It is possible, however, to define situations that are predominantly affected by refraction or by diffraction. Refraction is concerned with (gently) changing depth, causing the waves to shoal, and the wave crests and wave rays to bend. Wave diffraction is specifically concerned with zero depth change and solves for sudden changes in wave conditions such as obstructions that cause wave energy to be forced across the wave rays.

One classic example of pure wave diffraction is the obstruction to wave action by a breakwater as in Fig. 7.9. The breakwater separates a wave zone and a shadow zone and it is clear that the heavy line connecting them is a wave ray. In this case there will be definite transfer of wave energy across this wave ray. Wave crests will spill into the shadow zone and wave troughs will be filled with water from the shadow zone. Assuming that the depth is constant (i.e., no refraction), wave diffraction analysis calculates the wave energy that "leaks" into the shadow zone.

Such a calculation, even for the relatively simple layout in Fig. 7.9 is quite complicated (Dean and Dalrymple, 1984; Goda, 1985; Penney and Price, 1951). For preliminary calculations, however, it is often sufficient to use diffraction templates. One such template for the situation in Fig. 7.9 is presented in Fig. 7.10.

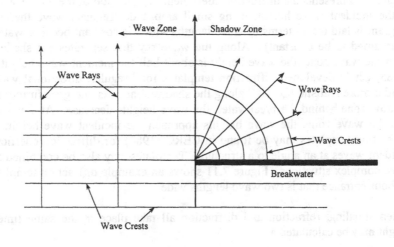

Figure 7.9 Pure Diffraction

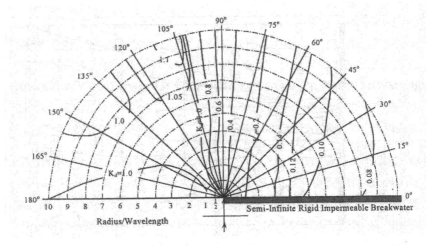

Figure 7.10 Diffraction at a Single Breakwater (after CERC, 1984)

Figure 7.10 presents the diffraction coefficient K_d, the ratio of the local wave height to the incident wave height, using small amplitude (regular) wave theory. The diagram is laid out in terms of wave length (which is constant because water depth is assumed to be constant). Along the wave ray that separates the shadow zone from the wave zone, the wave height is about half the incident wave height. Goda (1985, Ch 3) developed diffraction templates for irregular, directional waves that yield a more realistic $K_d = 0.7$ along the shadow line. As one goes further into the shadow zone behind the breakwater, the wave heights decrease. At some distance into the wave zone, the wave heights approach the incident wave height. Other diffraction templates may be found in CERC (1984) for diffraction coefficients of incident waves at an angle to a structure. Templates may also be combined to solve more complex situations. Figure 7.11 shows an example diffraction template for a harbour entrance that is two wave lengths wide.

When shoaling, refraction and diffraction all take place at the same time, wave height may be calculated as

$$\frac{H}{H_o} = K_s K_r K_d \qquad (7.33)$$

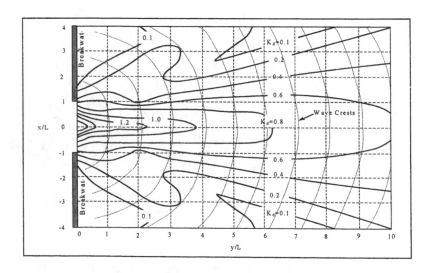

Figure 7.11 Diffraction at a Breakwater Gap

7.6 Uncertainty

In Ch. 3 the uncertainty in wave height measurements was shown to be $\sigma'_H=0.075$, $\sigma'_T=0.1$ and $\sigma'_\alpha=0.2$. In Ch 5 uncertainties in hindcast waves were estimated as $\sigma'_H=0.25$, $\sigma'_T=0.3$ and $\sigma'_\alpha=0.8$. Kamphuis (1999) shows that the additional uncertainties in the wave transformation and breaking formulations result in $\sigma'_H=0.45$, $\sigma'_T=0.3$ and $\sigma'_\alpha=1.0$ for breaking waves. Such large uncertainties are truly troublesome. The uncertainty in wave angle refers to a mean wave angle of $10°$. An uncertainty of 1.0 essentially means that in 16% of the cases, the wave angle with respect to the shore normal x-axis does not even have the correct sign, even though the assumed mean wave angle is $10°$. Any computations or models (Ch. 12 and 14) using input data with such uncertainties can only be approximate. The above estimates may represent upper limits. Nevertheless, it is clear that we must be aware of these limitations to our calculations and models and that all our results should be carefully calibrated. This will be discussed further in Ch. 12, 13 and 14.

8. Design of Structures

8.1 Introduction

The design process is either an open or hidden topic of discussion in all chapters of this book. Chapter 1 shows that design involves the synthesis of many concepts. This inevitably requires simplification and involves systems that are normally much larger than the direct area under consideration. Design also involves engineering time, the time of the order of a few hundred years.

Until recently, design was simply structural design; production of a structure that functioned. Today, the structural aspects are only one part of design. Physical, biological and sociological impacts are as important as structural stability and most designs will not be built, if they have not been thoroughly discussed with all stakeholders. Many times, non-technical stakeholders form part of the design team. All this makes design much more complex than a few decades ago. Today's design is certainly much more than putting together steel and concrete.

Modern design is concerned with at least the following aspects:
- physical (waves, tides, water levels, currents, strength of materials),
- structural (stability),
- sociological (esthetics, legal, planning),
- impacts
  - physical (sedimentation, erosion, salt spray, flooding),
  - environmental (water quality, visual impact),
  - biological (habitat, spawning, nesting).

Physical aspects of design are treated in Ch. 2 to 7. This chapter and Ch. 9 discuss structural stability. Environmental and biological impacts, and sociological aspects of design are found in Ch. 10, physical impacts of designs on the surroundings in Ch. 11, 12 and 15 and the use of models in coastal design is presented in Ch. 13 and 14.

8.2 Basics of Risk Analysis

8.2.1 Introduction

Design conditions even for major coastal projects are often vague and design parameters contain large uncertainties. Imposed forces, as well as the strengths and interactions of the various components are usually not clearly understood and the design process itself is ill defined. This is the background against which coastal design is made.

In the past, design was strictly based on *deterministic* expressions. The required sizes of a structure would be calculated from a formula derived from field observations or hydraulic model tests. For example, Eq. 9.33 defines the required armor unit mass for a breakwater of certain geometry to withstand a certain wave condition. It is derived from hydraulic model tests. The simple use of such an expression results in a single "correct" answer. A *factor of safety* against failure was usually included in such a design, to account for unknowns. The resulting equation is:

$$R = \Gamma S \qquad (8.1)$$

where R is the resistance (or strength) of the structure, S is the design load, usually related to a certain return period, T_R, and Γ is the factor of safety. Failure is assumed to be a step function. When $R > \Gamma S$, the structure stands (probability of failure, $P_F = 0$) and when $R < \Gamma S$ the structure fails ($P_F = 1$). More recently, *probabilistic* design methods have been introduced, in which the loads and the strengths of the structure are assumed to be statistical quantities and probability of failure is a continuous function through the design condition.

Deterministic design techniques are still used, however. First, we often do not know the statistical distributions for the various components of the loads and resistances. Second, design conditions are normally defined from model studies or limited field observations and such limited design input can only support deterministic design with a substantial factor of safety.

To determine probability distributions for the parameters would require many additional (and repeated) observations and/or model tests. At the same time, introducing even approximate or estimated parameters to describe the probability distributions gives additional insight about failure probabilities. That is why probabilistic design is introduced here and that is why only basic concepts of probabilistic design are treated.

8.2.2 Probability of Failure

Probabilistic design techniques are based on the *Limit State Equation*

$$G = R - S \qquad (8.2)$$

in which G is called the failure function. Equation 8.2 is a design equation and when G<0 the design condition fails. The quantities R and S and hence G are assumed to be functions of a number of variables:

$$R = f_r(R_1, R_2, R_3, \dots R_m) \qquad (8.3)$$

and

$$S = f_s(S_1, S_2, S_3, \dots S_n) \qquad (8.4)$$

where R_i and S_i denote various resistances and loads, respectively. For a coastal structure, such as a breakwater, R would be a function of strength of concrete, geotechnical properties of the supporting ground, shape of the structure, etc. and S would be a function of wave climate, water levels, ice conditions, etc. Each of the variables, R_i and S_i, is defined by a statistical distribution.

Probability of failure is defined as:

$$P_F = \mathrm{Pr}\,(G \le 0) \qquad (8.5)$$

Equation 8.5 is probability of failure of the design condition and it does not imply failure of the structure. Consider a rubble mound breakwater, as described in Ch. 9 (Fig. 9.8). It consists of many rocks. If we pose a design condition that no rocks be removed, then failure of the design condition means that a few rocks are removed. That does not constitute failure (collapse) of the complete structure. It is possible to pose several design conditions simultaneously, for example one condition relating to serviceability, another to major repair and rehabilitation, and another to structural collapse.

A design based on many probability distributions involves extensive computation. The complexity increases rapidly, when it is considered that Eqs. 8.2 to 8.5 should be applied to each member of a structure. Such complex design is beyond the scope of the present lecture. More details may be found in Ang and Tang (1984) and Pilarczyk (1990).

8.2.3 Levels Of Probabilistic Design

In the literature, three levels of probabilistic design are defined. Level III design involves the actual probability density functions of each R_i and S_i in Eqs. 8.3 and 8.4. Failure probability is determined by performing a large number of computations using many combinations of the possible values of the variables. This can be done using Monte Carlo methods. Level II design assumes that all probability density functions have normal distributions. This simplifies the problem, but it still involves many distributions. The usual design method is still Level I design. For that design level, a design equation is developed that contains only partial coefficients. The coefficients are derived from Level III or Level II calculations and take into account the effects of the probability distributions and a target P_F. Level I design is very similar to deterministic design (Eq. 8.1), except that safety factors are formally derived from probabilistic analysis.

8.3 Level II Demonstration

8.3.1 Equations

We will now demonstrate the effects of the probability distributions using the simplest Level II equation. It is based on Eq. 8.2 and assumes only one failure mode (one characteristic R and S value). We assume that one equation represents the whole problem and that the design equation is:

$$G = \frac{R_{ch}}{\gamma_r} - \gamma_s\, S_{ch} = 0 \tag{8.6}$$

where R_{ch} and S_{ch} are characteristic values of resistance and load, and γ_r and γ_s are partial safety coefficients pertaining to resistance and load. The characteristic values are assumed to be normally distributed with a mean value, μ, and a standard deviation, σ, where σ describes how well or how poorly we know the load or resistance. We define R_{ch} as:

$$R_{ch} = \mu_r + Z_r\,\sigma_r \tag{8.7}$$

where Z_r denotes the number of standard deviations that R_{ch} is removed from its mean. Usually, in conservative design, the (characteristic) resistance is taken to be smaller that its mean value and therefore Z_r is normally negative. If $Z_r=-1.64$, then R_{ch} represents the resistance (strength) that is exceeded 95% of the time by the structure (or member of the structure) to which Eq. 8.6 is applied. For 90, 98 and 99% exceedence, K would be -1.28, -2.05 and -2.33. Similarly, we define S_{ch} as:

$$S_{ch} = \mu_s + Z_s \sigma_s \qquad (8.8)$$

In conservative design, we normally use characteristic loads that are greater than the mean value. Hence Z_s is usually positive and $Z_s=1.28$, 1.64, 2.05 and 2.33 denotes loads that are exceeded 10, 5, 2 and 1% of the time.

In Eq. 8.6, γ_r is called the performance factor and γ_s is the load factor. They are the partial coefficients of safety, introduced to bring safety into the design, possibly as a response to uncertainties in R and S. Equation 8.6 can be rewritten as:

$$R_{ch} = (\gamma_r \gamma_s) S_{ch} = \Gamma S_{ch} \qquad (8.9)$$

If the product $(\gamma_r\gamma_s)$ is identified as the (global) factor of safety (Γ), then Eq. 8.9 resembles Eq. 8.1, except that R and S are based on statistical distributions.

8.3.2 Two Probability Distributions

First we will investigate some of the implications of the probability distributions. At the same time, the interaction of two distributions demonstrates Level II design with one failure mode, using Eqs. 8.6 or 8.9. Design consists of calculating the required strength of a structure to withstand the imposed loads. Combining Eqs. 8.7, 8.8 and 8.9 yields:

$$\mu_r = -Z_r \sigma_r + \Gamma (\mu_s + Z_s \sigma_s) \qquad (8.10)$$

and a design calculation determines μ_r.

Example 8.1 Simple Level II Design Calculation

We will determine the required strength of a structure for a characteristic load S_{ch} that has a mean value $\mu_s=80$ kN and is normally distributed with a standard deviation $\sigma_s=18$ kN. For the type of structure we are designing, R_{ch} is known to have σ_r=20 kN.

First, we shall choose Z_s=0, Γ=$\gamma_r\gamma_s$=1.0 and Z_r=-1.64, so that the mean load can be resisted by the structure 95% of the time. Equation 8.10 yields μ_r=113 kN. The normal probability distributions for R and S are shown in Fig. 8.1.

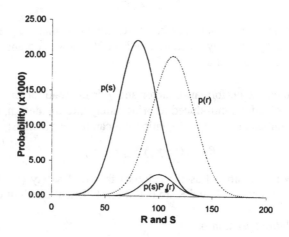

Figure 8.1 Two Probability Distributions with Z_r = - 1.64 and Γ = 1.0

Because the load and resistance distributions are independent of each other, a probability of failure for the design condition, P_F may be computed as:

$$P_F = \int_0^\infty \int_0^\infty p(s)\, p(r)\, dr\, ds = \int_0^\infty p(s)\, P_s(r)\, ds \qquad (8.11)$$

where p(s) is the probability density function of the load and $P_s(r)$ is the cumulative distribution function of the resistance r at any value of s (Fig. 8.2). The product {p(s) P(r)} for this example is shown in Fig. 8.1 and numerical integration yields P_F=0.112.

With the coefficient Γ, we can adjust the safety of the design. If we had used Γ=1.3, Eq. 8.10 would have yielded μ_r=137 kN. The distributions for this case are shown in Fig. 8.3. The distributions are now further apart and since there is less overlap, P_F for this second case is only 0.017. It may readily be seen that Γ and Z can be used interchangeably. For example, Γ=1.41 and Z_r=0 yields the same results as Γ=1.0 and Z_r=-1.64 in Fig. 8.1. Figure 8.3 would be the same if Γ=1.71 and Z_r=0.

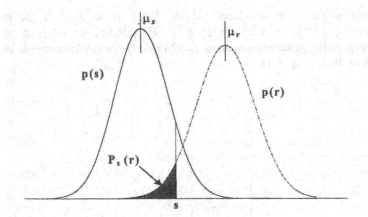

Figure 8.2 Probability of Failure Definitions

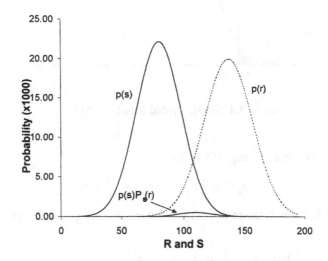

Figure 8.3 Two Probability Distribution with $Z_r = -1.64$ and $\Gamma = 1.3$

8.3.3 One Single Distribution

Two distributions were introduced in Sec. 8.3.2 to demonstrate the principles. However, it is difficult to evaluate Eq. 8.11. If p_r and p_s are independent, the two normal probability distributions can be combined into one, single normal probability distribution for G (Fig. 8.4).

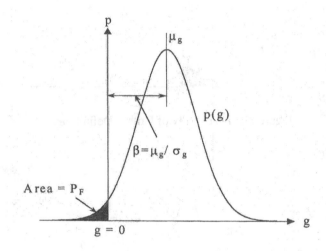

Figure 8.4 Single Probability Distribution

It can be shown (Ang and Tang, 1984) that:

$$\mu_g = \mu_r - \mu_s \quad ; \quad \sigma_g^2 = \sigma_r^2 + \sigma_s^2 \tag{8.12}$$

At failure, G=0 and thus the probability of failure (P_F) is equal to the shaded area in Fig. 8.4:

$$P_F = \Phi\left\{\frac{0-\mu_g}{\sigma_g}\right\} = \Phi\left\{-\frac{\mu_g}{\sigma_g}\right\} = 1 - \Phi\left\{\frac{\mu_g}{\sigma_g}\right\} = \Phi(-\beta) = 1 - \Phi(\beta) \tag{8.13}$$

where

$$\beta = \frac{\mu_g}{\sigma_g} = \frac{\mu_r - \mu_s}{\sqrt{\sigma_r^2 + \sigma_s^2}} \tag{8.14}$$

is the *reliability index* and Φ denotes the cumulative standard normal distribution function, which may be determined from published tables. Table 4.3 is such a table, but Φ may also be computed using expressions from Abramowitz and Stegun (1965), or by using pre-programmed software functions.

8.3.4 Example Calculations

The earlier discussion of simple Level II design is summarized in Cases 1 to 4 of Table 8.1. Cases 3 and 4 demonstrate that the safety based on the distributions of unknowns (σ) can be taken into account by adjusting the global factors of safety (Γ can represent Z).

Table 8.1[⊗] Calculation of Resistance and Probability of Failure

	Input Values						Calculated Values		
								Equations	
	Known			Specified			(8.10)	(8.14)	(8.13)
Case	μ_s	σ_s	σ_r	Z_s	Z_r	Γ	μ_r	β	P_F
1	80	18	20	0	-1.64	1.0	113	1.22	0.111
2	80	18	20	0	-1.64	1.3	137	2.11	0.017
3	80	18	20	0	0	1.41	113	1.22	0.111
4	80	18	20	0	0	1.71	137	2.11	0.017
3a	80	9	20	0	0	1.41	113	1.59	0.056
3b	80	9	20	0	0	1.34	105	1.22	0.111
3c	80	18	10	0	0	1.41	113	1.50	0.067
3d	80	18	10	0	0	1.34	107	1.22	0.111
5	80	18	20	0	0	1.0	80	0.00	0.5
6	80	18	20	0	0	1.43	114	1.27	0.10

Cases 3 a to d discuss the influence of the widths of the probability distributions on Case 3. Cases 3a and 3c show that a smaller σ, decreases P_F. Cases 3b and 3d show that if uncertainties are smaller, we can use a smaller value of Γ to obtain the same P_F. Thus Γ can represent both K and σ to obtain a specified value of P_F. This is the basis for the Level I design concept explained further in Section 8.6.

The fact that better knowledge (smaller σ) permits the use of smaller safety factors (Γ) to reach the same design safety (P_F) makes sense. But smaller Γ values also result in reduced cost, which is related to μ_r. This is an important concept. The values of σ_r and σ_s can only be decreased through careful field or laboratory research. Since field testing is expensive and usually not possible, we make use hydraulic and numerical model tests. Hence model testing is an integral part of the design process, as discussed in Ch. 13 and 14. Usually, the cost of such models is recovered from the resulting decrease in μ_r.

Case 5 in Table 8.1 demonstrates what happens when values are substituted directly into a design formula. Equation 8.9 is applied here with $K_r=K_s=0$ and $\Gamma=1$. Equation 8.14 yields $\beta=0$ for this case, regardless of the values of σ, which means that $P_F=0.5$. The consequence of recognizing that there are probability distributions is that probability of failure becomes a continuous function through the design condition and for simple substitution into a formula $P_F=0.5$, when the design condition is just met. Probability of failure is no longer a step function in which P_F changes suddenly from 0 to 1 when the design condition is reached.

In the examples in Table 8.1, σ, Z and Γ are used to calculate μ_r and in turn P_F. It is also possible to begin by specifying a target value of P_F and then calculate the reliability β and the required design value μ_r. The probability distributions are also most often expressed in terms of uncertainty ($\sigma'=\sigma/\mu$), rather than σ, and some reorganization of the equations permits calculation of μ_r and P_F from σ', Z and Γ or if we specify P_F, we can calculate β and μ_r.

8.4 Extension to More Complex Designs

The above discussion refers to two distributions and a linear limit state function (Eq. 8.2). Figure 8.5 shows Eq. 8.2 on an R vs. S set of co-ordinates as well as on normalized co-ordinates Z_r vs. Z_s, where:

$$Z_r = \frac{R - \mu_r}{\sigma_r}; \quad Z_s = \frac{S - \mu_s}{\sigma_s} \qquad (8.15)$$

It is seen that in both cases the failure surface is a straight line and it can readily be shown (Ang and Tang, 1984) that β represents the shortest distance from the origin to the (linear) failure surface for such a normalized plot. The intersection of the failure surface with the perpendicular is called the design point. If the failure surface is not linear and there are more than two distributions involved, it is usual to approximate the multi-dimensional hyperplane by a tangent line to the hyperplane at the design point, thus linearizing the analysis. This iterative technique is called First Order Reliability Method (FORM). A second order method (SORM) extends this analysis further. Details may be found in Pilarczyk (1990), Burcharth (1992), Thoft-Christensen and Baker (1982) and Madsen et al (1986). Such advanced methods (of Level II analysis) are used to determine the partial coefficients that are used with the simpler Level I methods.

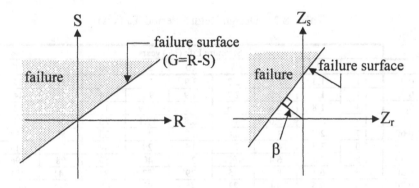

Figure 8.5 Definition of Failure Surface

8.5 Encounter Probability

To determine probability of failure for a design condition during the lifetime of a project (P_L), it is necessary to consider encounter probability of the loading condition. Encounter probability is related to return period by the binomial distribution:

$$T_R = \left(1 - (1 - P_E)^{1/N_L}\right)^{-1} \quad \text{or} \quad P_E = 1 - \left(1 - \frac{1}{T_R}\right)^{N_L} \qquad (8.16)$$

in which T_R is the return period in years, P_E is the encounter probability and N_L is the project design life in years. Example values may be found in Table 8.2.

Lifetime probability of failure (of the design condition) may now be computed from:

$$P_L = P_E\, P_F \qquad (8.17)$$

Consider a structure designed for N_L=50 yrs. If the design wave height is based on a 50-year return period of the design load, then P_E= 0.64 (Table 8.2). For the design condition in Case 1 of Table 8.1 (P_F=0.111), the lifetime probability of failure will be P_L=$P_E P_F$=(0.64)(0.111)=0.075. For Case 5 with T_R=50, P_L is (0.64)(0.5)=0.32. Note that many combinations of P_F and P_E (or T_R) can achieve the same value of P_L. If either P_E or P_F is not determined explicitly, it will be necessary to perform a number of computations to find the least cost combination that produces the desired value of P_L.

Table 8.2[⊗] Design Return Period T_R (yrs)

P_E	1	2	3	5	10	20	30	50	100
	\multicolumn{9}{c}{Structure Life (N_L) years}
0.95	1.1	1.3	1.6	2.2	3.9	7.2	11	17	34
0.90	1.1	1.5	1.9	2.7	4.9	9.2	14	22	44
0.80	1.3	1.8	2.4	3.6	6.7	12.9	19	32	63
0.70	1.4	2.2	3.0	4.7	8.8	17.1	25	42	84
0.60	1.7	2.7	3.8	6.0	11	22	33	55	110
0.50	2.0	3.4	4.8	7.7	15	29	44	73	145
0.40	2.5	4.4	6.4	10.3	20	40	59	98	196
0.30	3.3	6.1	8.9	15	29	57	85	141	281
0.20	5.0	9.5	14	23	45	90	135	225	449
0.10	10	19	29	48	95	190	285	475	950
0.05	20	39	59	98	195	390	585	975	1950
0.01	100	199	299	498	995	1990	2985	4975	9950

8.6 Level I Design

Most practical design at the present time is Level I design. It resembles deterministic design in that it applies Eq. 8.6. The γ values are formulated to take into account the uncertainties and are derived from experience with prototype structures and from model studies, using Level III or Level II analysis. The principle was demonstrated from Table 8.1. In Case 1 and 2, R_{ch} and S_{ch} are expressed as functions of their distributions. They were expressed in terms of only their mean values and Γ, in Cases 3 and 4. Cases 3a to 3d show that this Γ value is a function of the σ values for constant values of P_F. It is therefore possible to find an expression that uses single, defined characteristic values – the mean values of R and S in this case - and Γ (or γ) values to represent σ values for a target failure probability, P_F. Thus a Level I design expression is a pseudo-deterministic design formula that has a level of safety that is not arbitrarily chosen, but is based on available prototype and model information.

As an example, assume data have shown that $\sigma_s=18$ and $\sigma_r=20$ are representative of a certain type of structure and the loads to which it will be subjected. We want to develop a Level I design formula that uses the mean values of S and R as the characteristic values and represents a design condition with an inherent (target) $P_F=0.10$. Case 6 in Table 8.1 shows the design expression for to be:

$$G = \mu_r - 1.43\mu_s = 0 \tag{8.18}$$

Level I design now consists simply of applying Eq. 8.18 and adjusting μ_s according to how often we want the design condition to be encountered in nature (the encounter probability, P_E). If $N_L=50$ yrs and we use the expression with μ_s based on $T_R=N_L=50$ yrs, then we are designing for $P_L=P_EP_F=(0.1)(0.64)=0.064$. If we want P_L to be 0.001, then with $P_F=0.1$, $P_E=0.001/0.1=0.01$. For that we need to base μ_s on $T_R=5000$ yrs, according to Table 8.2 and $P_L=P_EP_F=(0.01)(0.10)=1.0 \times 10^{-3}$.

In standard structural design, the partial safety coefficients (γ) are provided in building codes, etc. They are based on statistical values derived from a large number of well-proven designs and tests. Such a large volume of accurate information about design values of σ and P_F is generally not available for coastal structures and hence detailed calibration of coefficients has generally not been possible. Pioneering work in that direction may be found in PIANC (1992), which reviews values of μ, σ' and γ for the stability of rubble mound breakwaters. They use Level II analysis of various model studies and field results and provide values of γ to be used in a Level I design formula. A subsequent PIANC analysis, reported in Burcharth and Sorensen (1998), provides similar γ values for vertical breakwaters.

8.7 Risk and Damage

Very important additional design concepts are risk and damage. If failure of the design condition occurs, how serious is it? The combination of lifetime failure probability and damage resulting from such failure determines *risk*.

This involves knowledge of how the structure fails. An example of how P_L, T_R and risk are related, derived from standard building codes, may be found in Table 8.3. Failure Type I is a slow, ductile-type failure where there is some residual strength to prevent collapse. Type II is more rapid, ductile failure with no residual strength and Type III is sudden brittle failure. It is seen that design probability of failure (inverse of risk) decreases as the type of failure becomes more sudden and as the expected damage becomes more serious.

Table 8.3 Design Probabilities of Failure

Damage	Failure Type		
	I	II	III
Less Serious	10^{-3}	10^{-4}	10^{-5}
Serious	10^{-4}	10^{-5}	10^{-6}
Very Serious	10^{-5}	10^{-6}	10^{-7}

A similar decrease in P_L must be included in the design of coastal structures to account for the type of failure and the consequential damage. A rubble mound structure consists of a large number of individual armor units (Ch. 9). If one or two armor units are moved from their original location, there is no serious damage and the risk resulting from such a failure is the cost of repair of the armor layer. As a result, the design value of P_L for a failure involving one or two armor stones can be quite high; $P_L=0.1$ for the probability of removal of a few armor units during the lifetime of the structure would be conservative. On the other hand, $P_L=0.1$ for a design condition of removal of 30% of the armor units would not be acceptable, since that would involve failure of the complete structure and possible serious damage to ships and docks. If a key structural member of an ocean drilling structure fails, the damage could be sudden collapse of the structure with serious loss of life. Thus, the risk associated with this type of sudden failure in combination with serious damage is high and P_L must be low (such as 10^{-3} to 10^{-4}). Similarly, to reduce the risk for a rubble mound perimeter protection of a nuclear power plant to acceptable levels, the return period for the zero damage design wave may be as high as 10^6 years.

Such long return periods are essentially only statistical concepts that do not have real physical meaning. Environmental conditions never remain steady for a million years.

The combination of damage and failure probability is risk. Social and economic considerations determine what risk is acceptable[1]. Acceptable risk, then determines design values of P_L and T_R.

8.8 The Design Wave

8.8.1 Wave Statistics

Wave heights, periods and angles clearly vary with time. Short-term wave statistics (Ch. 3) refers to the variation of wave height, period and wave angle over a short time span, such as within a single wave recording. Wave heights over such a short time span are usually represented by significant wave height, H_s, and further described by the Rayleigh distribution developed in Ch. 3.

When designing a facility, it is also necessary to know about the long-term wave statistics (Ch. 4). A long-term wave height distribution normally expresses probability of exceedence as a function of H_s (or H_{ch} or H_{mo}). We will use the symbol, H_s, to denote all three of these definitions. Extreme value distributions such as the Gumbel or Weibull distribution are fitted to the measured data to obtain wave heights for return periods greater than the record length by extrapolation. Since long series of measured wave data are scarce, long-term wave data often consist of wave hindcasts in which waves are derived from measured wind data (Ch. 5). With the uncertainties in wave measurements and hindcasting procedures, the uncertainties in a long-term wave climate are substantial.

8.8.2 Equivalence of Design Wave Height and Failure Probability

Since long-term wave heights are expressed as H_s it is convenient and consistent to use H_s as design wave height, and incorporate risk through P_L and T_R. In practice, additional design safety is sometimes introduced by using the short-term wave height definitions with a lower frequency of occurrence than the long-term significant wave height. In that case, long-term and short-term distributions are unnecessarily mixed. For example, for a structure for which failure would involve loss of life, the average of the highest 1% of the waves ($\overline{H}_{0.01} = 1.67 H_s$) may be used instead of H_s. That is simply equivalent to using H_s with a higher factor of

---

1. Social acceptability of risk is not straightforward. For example, killing 200 people every holiday weekend in automobile accidents seems to be acceptable, but killing 200 people on holiday weekends in air crashes or flooding is not!

safety or a lower probability of failure. Since forces on a structure are proportional to H^3, using a 1.67 factor to increase the design wave height involves an increase of $(1.67)^3 = 4.7$ in the factor of safety. But such a higher design wave value could be readily translated into either a higher value of T_R (lower P_L) while continuing to use significant wave height. For the Lake Huron data set, $\overline{H}_{0.01}$ for the offshore design wave height with a return period of 50 years is 1.67x5.7=9.5 m. This is the same as using H_s with a return period of about 30,000 yrs or an encounter probability (Eq. 8.16) of 0.0017. Using one wave height definition consistently that way gives a much better idea of the relative risks involved.

Since many structures are damaged specifically by the highest waves, it makes sense to use H_{max} consistently in those designs. In this book, we will not mix the two distributions, but we will use of either H_s or H_{max} for design wave height, depending on the cause of failure and on the historical approach used for that type of design. We will, for example, follow the historical approach and use H_s for design of rubble mound structures and H_{max} for vertical breakwaters. We define

$$H_{max} = K_{max} H_s \qquad (8.19)$$

From Eq. 3.12 we see that for a storm of 2000 waves, the Rayleigh distribution gives $K_{max}=2$. Goda (1985) suggests $K_{max}=1.8$. Research at Queen's University on stable breaking wave heights (breaking waves on a horizontal platform), shows that the Rayleigh distribution does not quite apply at breaking and that K_{max} for breaking waves approaches 1.5 (Rakha and Kamphuis, 1995).

8.8.3 Offshore Design Wave Height

Usually the available wave information pertains to offshore conditions in relatively deep water. To define a design wave, we must first determine lifetime failure probability. We saw that direct substitution into a formula with a return period $T_R=N_L$ was quite reasonable when calculating (zero damage) stability of rubble mound breakwaters; other structures may need long extrapolations if failure results in great damage.

From T_R, it is possible to calculate probability of exceedence using Eq. 4.23 and the wave height with that return period may be calculated using Eqs. 4.26 to 4.28. In Ch. 4, a 34.9-year wave hindcast data set for Lake Huron was discussed. For ordered extreme value data the wave heights were calculated in Table 4.8 up to $T_R=200$ yrs. That table is now extended in Table 8.4 to 10^6 yrs. The parameters used in Eq. 4.28 to calculate H_s in Table 4.8 are α=0.80, β=0.29, γ=3.97 and λ=1.26. The calculated waves for the larger return periods in Table 8.4 are very high and we need to check if such waves are physically possible. The highest measured hourly average wind speed over Lake Huron is 26 m/s and the maximum fetch is 400 km. Equation 5.6 shows that a 16 hr storm duration is needed to develop the maximum sea over this 400 km fetch. Equation 5.2 calculates the wind speed for 16 hours U_{16} to be 21 m/s. Introducing a measure of safety, to cover for the uncertainty in a simple wave hindcast, we assume U_{16}=23 m/s. The maximum wave height possible over the given fetch is therefore H_s=7.4 m. This limit is reflected in the second line of Table 8.4. Table 4.10 shows that for the same Lake Huron data set the wave period is related to the wave height as:

$$T_p = 3.54 H_s^{0.61} \qquad (8.20)$$

for wave heights smaller than 4.45 m. This expression was assumed to be also valid for larger wave heights and produced the wave periods shown in Table 8.4.

Table 8.4 Offshore Wave Conditions For Different Return Periods

	Return Period T_R (yrs)						
	1	50	100	10^3	10^4	10^5	10^6
H_s (extrapolated) - m	4.0	5.7	6.1	7.4	8.8	10.3	12.1
H_s (fully developed sea) - m	4.0	5.7	6.1	7.4	7.4	7.4	7.4
T_p - sec	8.3	10.2	10.6	12.0	12.0	12.0	12.0

8.8.4 Design Wave Height for Non-Breaking Waves

The design wave at a structure depends on the breaker location and therefore any design must first determine where the wave breaks.

Example 8.2 Design Wave Height in 12 m of Water

A rubble mound breakwater needs to be built on Lake Huron in 12 m of water. The foreshore slope is 1:50. We will use N_L=50 yrs and T_R=N_L. The 50-year significant wave height and peak period are 5.7 m and 10.2 seconds, according to Table 8.4. This wave approaches the shore with a deep-water angle of 30°. The breaking wave conditions may be computed using simple, first order shoaling and refraction calculations and the wave breaking criteria developed of Kamphuis (1991a).   The method is outlined in Fig. 7.6 and in the program **RSB**®. Figure 8.6 shows this calculation for the present example. It is seen that, considering an average of both breaking criteria d_b=9.2 m and at that breaking depth H_{sb}=5.4 m and α_b=16.3°.

A rubble mound structure in 12 m of water would therefore be offshore of the breaker and its design wave height would be:

$$(H_{des})_{non-breaking} = (H_s)_{at\ structure} \qquad (8.21)$$

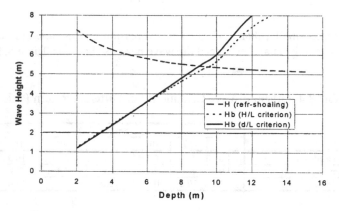

Figure 8.6 Shoaling Refraction Breaking Calculation

Table 8.5 Summary Table: Design Wave Height, Rubble Mound Structure

Foreshore Slope – m	0.02
Project Design Life – N_L (yrs)	50
Return Period –T_R (yrs)	50
Deep Water Significant Wave Height - H_s (m)	5.7
Peak Wave Period - T_p (sec)	10.1
Deep Water Wave Angle - α_o (degrees)	30
Breaking Wave Height - H_{sb} (m)	5.4
Breaking Depth - d_b (m)	9.2
Breaking Wave Angle - α_b (degrees)	16.3
Design Wave Height in 12 m - H_{des} (m)	5.2
Design Wave Angle in 12 m - α_{des} (degrees)	18.3

Figure 8.6 shows that the design wave height based on the 50 yr offshore significant wave height, shoaled into 12 m water depth is $H_s=5.2$ m. The example is summarized in Table 8.5.

For a vertical breakwater, we would use:

$$(H_{des})_{non-breaking} = (H_{max})_{at\ structure} = K_{max}(H_s)_{at\ structure} \qquad (8.22)$$

and with Goda's value of $K_{max}=1.8$, that would mean $H_{des}=9.7$ m.

8.8.5 Design Wave Height for Breaking Waves

A breakwater in 3 m water depth, on the other hand, needs to be designed for a breaking wave. Since the height of this breaking wave at the structure is very much influenced by the depth of water at the structure, this has traditionally been called *depth-limited design*. The concept of depth-limited design has been of vital importance in design of structures, since most structures are located in relatively shallow water, close to shore. The principle of depth-limited design may be expressed quite simply: If there is any possibility that a wave can break on the structure, the design wave for the structure must be the wave that breaks exactly on the structure. Any larger offshore waves will have broken further offshore and will have lost much of their energy (and wave height) by the time they reach the structure (Fig. 7.7 and Kamphuis, 1998). Conversely, any smaller waves, by definition, result in smaller forces on the structure.

If a structure is placed in a depth of water d_S, then the maximum possible significant height of the wave breaking directly on the structure could be obtained most simply by substituting d_s for d_b in Eq. 7.32.

$$H_{des} = (H_{sb})_{max} = 0.56\, d_s\, e^{3.5m} \qquad (8.23)$$

For $d_s = 3$ m, with m=0.02, $H_{des}=0.60d_s=1.8$ m. Such a wave would be generated whenever the maximum wave offshore exceeds about 1.8 m, which is much smaller than $H_s=5.7$ m, determined for $T_R=50$ yrs in Table 8.4. It will occur more frequently than once in 50 years. The first column of Table 8.4 indicates that it would occur much more often than once per year. In effect, this wave height is reached during any minor storm. Note that the wave height in Eq. 8.23 is a function of water depth at the structure and slope of the foreshore only. Neither the short-term nor the long-term offshore wave height distributions enter into the determination of the design wave height. The distributions only provide estimates of the frequency of occurrence of the design wave condition.

Kamphuis (1996, 1998) has shown that, contrary to the depth limited design philosophy, an increase in offshore incident wave height will increase the maximum significant breaking wave height at the structure and hence increase the damage. Therefore, the simplistic depth-limited design procedure used for so many years is not entirely correct. There are some secondary factors that influence the design wave. These are:

–   The incoming breaking waves cause an elevated water level at the structure (wave setup).
–   A long wave that accompanies the incident wave groups forms substantial long period water level fluctuations at the structure.

This produces a complicated interaction of incident short waves, incident long waves, reflected long waves and increase in water level at the structure. It is attractive, however, to continue to use a relatively simple methodology, based on depth-limited design concepts. All three of the above effects add to the depth of water in front of the structure, all are relatively small and all are more or less proportional to incoming wave height. Therefore Kamphuis (1998) modifies d_s in Eq. 8.23 by an amount proportional to the breaking significant wave height. The modified depth is defined as:

$$d_s' = d_s + C_H H_{sb} \qquad (8.24)$$

From many experiments, and using H_{sb} as representative of the incoming waves, C_H was determined to be 0.1 and hence a modified depth-limited expression is:

$$H_{des} = (H_{sb})_{max} = 0.56\, d_s'\, e^{3.5m} = 0.56\, (d_s + 0.1 H_{sb})\, e^{3.5m} \qquad (8.25)$$

Example 8.3 Design Wave Height in 3 m of Water

Extending the earlier Lake Huron example, we found in Example 8.2 that H_{sb}=5.4 m and d_b=9.2 m. A straightforward depth-limited design, using Eq. 8.23 yields H_{des}=$(H_{sb})_{max}$=1.80 m. Modified depth-limited design with Eq. 8.25, on the other hand produces H_{des}=$(H_{sb})_{max}$=2.1 m. Thus, taking into account the additional water level fluctuations near the breakwater caused an increase in $(H_{sb})_{max}$ of 17%. This may not appear to be significant, but considering that forces vary with H^3 this correction represents an increase in forces on the structure of 60%.

The effect of T_R on Eq. 8.25 is small. For example, for $T_R=1$ yr, $(H_s)_{offshore}=4.0$ m, $H_{sb}=3.8$ m and Eq. 8.25 yields $(H_{sb})_{max}=2.0$ m. This means that H_{des} for $T_R=50$ yrs is not much different from H_{des} for $T_R=1$ yr (or for $T_R=1$ month, or $T_R=1000$ yrs for that matter).

The frequent occurrence of the design wave, defined in Eq. 8.25, and the fact that damage on a rubble mound breakwater is cumulative, may prompt us to be cautious and ensure truly zero damage, each time the design wave is reached. Although traditional practice uses H_s in rubble mound breakwater design, we might consider using:

$$(H_{des})_{breaking,cum} = (H_b)_{max} = (0.56)K_{max}(d_s + 0.1H_{sb})e^{3.5m} \qquad (8.26)$$

where cum refers to cumulative damage. For this example, Eq. 8.26, with $K_{max}=1.5$ for the breaking zone, as determined by Rakha and Kamphuis (1995), gives $H_{des}=3.1$ m. Equation. 8.26 would only apply when damage is cumulative, such as in the armor unit stability calculations. For other parts of the design such as structure height, based on wave runup, overtopping, freeboard, etc., Eq. 8.25 is used.

8.8.6 Model Study

The uncertainties surrounding wave measurements, hindcasting, long-term distributions, the use of H_{max} and the definitions of K_{max} make the choice of design wave height difficult. We could use higher safety factors to account for these uncertainties, but that makes the resulting designs very conservative. One method to reduce the uncertainties, the factors of safety and the costs of the structure is to design the project using models. This alternative is discussed in Ch 13 and 14.

8.9 Water Levels

Water level is the other major design factor and a complete discussion, similar to the one for waves, could be presented if statistical data about water levels are available. However, for long-term water level data, particularly for extrapolations to higher return periods, long-term environmental changes such as subsidence, eustatic water level rise, isostatic rebound and global warming need to be taken into account, as discussed in Ch. 6. Two other important water level components to be considered in design are longer term fluctuations such as occur on the Great Lakes and short term fluctuations due to storm surge and seiche that may occur at long intervals. We usually do not have the data to produce statistical distributions for such water level fluctuations, and in any case, it is unlikely that they could be described by simple statistical distributions.

In the absence of long-term statistical data, it is general practice to design for a very high water level, since design wave heights are larger for deeper water and hence wave overtopping, forces and moments increase with water depth. The safety of structures should be computed with respect to this design water level, but sensitivities to adjacent higher and lower water levels should also be computed. Water levels and safety factors should be carefully monitored throughout the lifetime of the structure. In the Netherlands, safety of all structures is reviewed every 5 years, in light of wave action, structural deterioration and new findings about water levels. Because additional study normally results in greater mean values and/or scatter, and because risk becomes less acceptable with time, design water levels and risk factors are usually adjusted upward in time, requiring costly retro-fitting of facilities that now no longer meet the safety standards.

Certain portions of designs are sensitive to water levels other than the maximum. For example, the toe of a rubble mound breakwater, the berm of a composite structure and the depths of navigation channels are all sensitive to fluctuations in low water levels.

9. Breakwaters

9.1 Vertical Breakwaters

9.1.1 Introduction

There are several types of vertical breakwaters, and some examples may be found in Fig. 9.1. The main component of a vertical breakwater cross-section is normally a concrete *caisson*; a large, hollow concrete box that is floated to the site and filled with granular material to sink it to form a stable structure. The caissons are placed on a prepared pad of rock or *berm*. Sometimes the berm is a partial rubble mound structure, thus forming a *composite breakwater*. In some locations it has become practice to decrease wave reflection and prevent waves breaking directly onto the caisson by covering its seaward face with an armor protection.

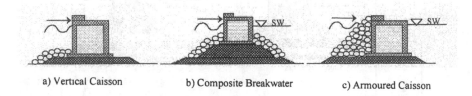

a) Vertical Caisson   b) Composite Breakwater   c) Armoured Caisson

Figure 9.1 Vertical Breakwaters

Typical *failure mechanisms* of vertical breakwaters are shown in Fig. 9.2 and structural stability is determined by the failure mechanism that results in least resistance of the breakwater. Only three common types of failure will be discussed

191

in this chapter by way of example: stability against sliding of the caisson over its base, overturning of the caisson around its landward corner, and stability of a sandy base. Other failure mechanisms are discussed in Burcharth and Sorensen (1998).

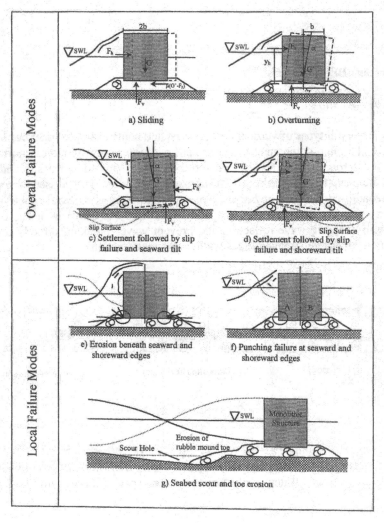

Figure 9.2 Failure Mechanisms

9.1.2 Forces for Non-Breaking Waves

Vertical breakwater caissons are monolithic gravity structures and the mass of the structure must resist the imposed forces. Because damage to a vertical breakwater is closely related to the maximum wave that reaches the structure, it has traditionally been assumed that the design wave

$$H_{des} = H_{max} = K_{max} H_s \qquad (9.1)$$

as presented in Ch 8 and Goda (1985).

A structure located in deeper water will be subjected to non-breaking waves, and the basic forces for non-breaking waves are shown in Fig. 9.3. On the seaward side the force is a combination of hydrostatic forces resulting from the still water depth, a wave-generated rise in water level, Δ_H, and the force from the non-breaking, standing wave against the seaward side. On the landward side, there is the hydrostatic force representing the still water depth. There are also a buoyancy and an uplift force under the caisson.

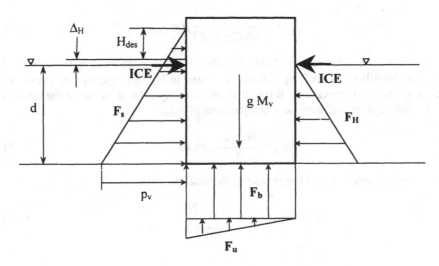

Figure 9.3 Forces for Non-Breaking Waves

In some areas, ice forces are important. These will be large and are often the determining factor for structure stability. Ice forces may be caused by pressure of a static ice sheet or by impact by floating ice. For a vertical structure, ice exerts a force at the water level that can be as high as the crushing strength of the ice (1.5 MPa) multiplied by the ice thickness[1]. Ice-generated force may occur both on the seaward and the landward side of the structure, but the design condition is for the worst case, when ice pushes only on one side of the structure. When there is ice, there are no waves (or only small waves) and we assume that ice forces and wave forces do not occur simultaneously.

The non-breaking wave force on the seaward side of the structure is assumed to result from an approximately *hydrostatic* pressure, generated by the highest water level reached by the design wave. More sophisticated methods, such as the methods of Sainflou and Miche-Rundgren may be found in CERC (1984). These methods are refinements of the hydrostatic approximation and yield similar results. Assuming perfect wave reflection off the breakwater face, the standing wave will have a height that is twice the incident wave height. Table 2.4 shows that the pressure of a standing wave of height H_{des} on the seaward side of the structure, at any depth d will be

$$p_{sw} = \frac{1}{\cosh\left(2\pi d / L\right)} \rho g H_{des} \qquad (9.2)$$

From higher order wave theory, Table 2.4 also shows that the mean wave level will be above the still water level by a distance related to H^2. Since it will take time for such a water level increase to form, it is not directly related to the instantaneous wave height and we will assume it is a function of H_{rms}.

$$\Delta_H = \frac{H_{rms}^2 k}{2} \coth kd \qquad (9.3)$$

Thus on the seaward side of the structure, the total pressure is

$$p = \rho g \left(d + \Delta_H + \frac{H_{des}}{\cosh(2\pi d / L)} \right) \qquad (9.4)$$

---

1. In theory, such large forces can be prevented by sloping the structure. However, that introduces other problems such as a decrease in caisson mass (and stability) and damage through ice ride-up and overtopping by the ice.

and on the landward side, the pressure is simply hydrostatic. If we take into account the usual configuration of a vertical breakwater, placed on a prepared berm foundation and having a freeboard f_v above still water, and we subtract the landward triangular hydrostatic pressure distribution from the seaward forces, we obtain Fig. 9.4.

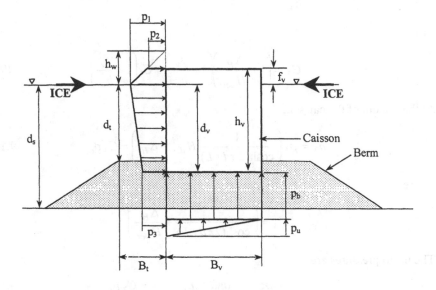

Figure 9.4 Wave Force Definitions

We define the following:

$$h_w = H_{des} + \Delta_H = C_w H_{des} \quad where \quad C_w = 1 + \frac{\Delta_H}{H_{des}} \qquad (9.5)$$

The pressure against the structure at still water level is

$$p_1 = \rho g(H_{des} + \Delta_H) = C_1 p_o \qquad (9.6)$$

where

$$p_o = \rho g H_{des} \quad and \quad C_1 = C_w = \left(1 + \frac{\Delta_H}{H_{des}}\right) \tag{9.7}$$

At the top of the structure

$$p_2 = \left(1 - \frac{f_v}{h_w}\right)p_1 = C_2 p_o \quad if \quad f_v < h_w \quad else \quad p_2 = 0 \tag{9.8}$$

and

$$C_2 = \left(1 + \frac{\Delta_H}{H_{des}}\right)\left(1 - \frac{f_v}{h_w}\right) = C_1\left(1 - \frac{f_v}{h_w}\right) \tag{9.9}$$

At the bottom of the caisson,

$$p_3 = \rho g\left(\frac{1}{\cosh(2\pi d / L)}H_{des} + \Delta_H\right) = C_3 p_o \tag{9.10}$$

where

$$C_3 = \left(\frac{1}{\cosh(2\pi d / L)} + \frac{\Delta_H}{H_{des}}\right) \tag{9.11}$$

The uplift pressures are

$$p_b = \rho g d_v \quad and \quad p_u = p_3 = C_3 p_o \tag{9.12}$$

The horizontal force per unit length of the breakwater is the horizontal force resulting from waves and water levels (F_w) or from the ice force

$$F_h = F_w = \frac{p_1 + p_3}{2}d_v + \frac{p_1 + p_2}{2}f_v \quad or \quad F_h = F_{ice} \tag{9.13}$$

The vertical force is the mass of the structure acting down (F_m), the uplift force resulting from buoyancy (F_b). If there is no ice, there is also the wave-generated uplift force (F_u)

$$F_v = F_m - F_b - F_u = gM_v - B_v\left(p_b + \frac{p_u}{2}\right) \tag{9.14}$$

The overturning moment around the landward bottom corner of the structure with waves is

$$\tilde{M}_o = \tilde{M}_w + \tilde{M}_u = p_3 \frac{(d_v)^2}{2} + (p_1 - p_3) \frac{(d_v)^2}{3}$$

$$+ p_2 f_v (d_v + \frac{f_v}{2}) + \frac{p_1 - p_2}{2} f_v (d_v + \frac{f_v}{3}) + p_u \frac{B_v^2}{3} \qquad (9.15)$$

For ice

$$\tilde{M}_o = \tilde{M}_{ice} = F_{ice} d_v \qquad (9.16)$$

and the restoring moment is

$$\tilde{M}_r = \tilde{M}_m - \tilde{M}_b = gM_v \frac{B_v}{2} - p_b \frac{B_v^2}{2} \qquad (9.17)$$

9.1.3 Forces for Breaking Waves

The more usual (and more serious) situation is when the breakwater is located in a depth of water where breaking waves will occur on the structure. In that case, in addition to ice, hydrostatic and uplift forces, the wave forces are increased. If the breaking is sudden, such as for plunging breakers, there will be a (large) *impact force* resulting from the direct wave impact (wave slamming) against the structure. Such impact forces are very high, but have a very short duration (Fig. 9.5). What does this mean for design of a structure with a large mass, such as a vertical breakwater?

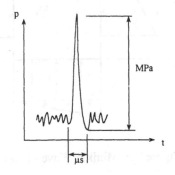

Figure 9.5 Impact Force History

One method that attempts to define forces due to direct impact of breaking waves is the Minikin method (CERC, 1984). The assumed forces are shown in Fig. 9.6. Impact from breaking waves is assumed to produce a parabolic pressure diagram centered on the still water level and its effect is added to the hydrostatic forces system. The dynamic pressure, force and moment about the bottom are approximated by

$$p_d = 100 \frac{d_i}{d_s L}(d_s + d_i) \rho\, g\, H_{des}$$

$$F_d = \frac{1}{3} p_d H_{des}\left(1 - 2\left\{\frac{h_w - f_v}{H_{des}}\right\}^2\right) \quad for \quad h_w > f_v$$

$$F_d = \frac{1}{3} p_d H_{des} \quad for \quad h_w \le f_v$$

$$\widetilde{M}_d = F_d d_v$$

(9. 18)

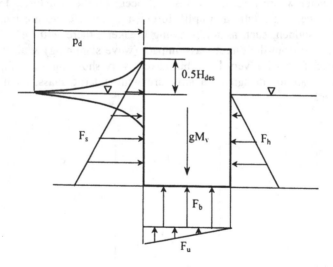

Figure 9.6 Minikin Wave Forces

The structure, resulting from this design will be large and costly and essentially, if the possibility of wave slamming exists, we must find another location or use a different design. Goda (1985) presents a checklist to assess the possibility of wave slamming.

The most often-used design method for breaking waves is the method by Goda (1985). It defines the higher forces due to breaking waves in terms of a pseudo-static design and uses Fig 9.4 with different definitions for the pressures. Such pseudo-static design accounts for higher forces due to waves that are in the process of breaking slowly, such as spilling breakers, but it does not represent direct impact by suddenly breaking waves, such as plunging breakers. Direct impact over a long section of structure may therefore cause damage to the Goda design. Care must be taken also to ensure that the berm under the vertical section does not cause the wave to slam against the structure. Takahashi et al (1994) have extended the Goda method to take this into account.

The coefficients for Goda's method (as extended by Takahashi), along with the coefficients for non-breaking waves and the Minikin method, as presented in CERC, 1984, are summarized in Table 9.1.

Table 9.1 Design Parameters for Vertical Breakwaters

	Non-Breaking	Goda	Minikin[*]
C_w	$1+\Delta_H/H_{des}$	$0.75(1+\cos\alpha)$	0.5
C_1	$1+\Delta_H/H_{des}$	$0.5(1+\cos\alpha)(\zeta_1+\zeta_4\cos^2\alpha)$	$0.5/\cosh(2\pi d/L)$
C_2	$C_1(1-f_v/h_w)$		
C_3	$1/\cosh(2\pi d/L)+\Delta_H/H_{des}$	ζ_3	C_1
C_u	C_3	$0.5(1+\cos\alpha)\zeta_1\zeta_3$	C_3
ζ_1		$0.6+0.5((4\pi d/L)/\sinh(4\pi d/L))^2$	
ζ_2		Smallest of: $((d_{5H}-d_s)/3d_{5H})(H_{des}/d_t)^2$ and $2d_t/H_{des}$	
ζ_3		$1-((h_v-f_v)/d_s)(1-(1/\cosh(2\pi d_s/L))$	
ζ_4		$Max\{\zeta_2,(\zeta_5\,\zeta_6)\}$	

(*) This is only the standing wave portion for the Minikin method. The dynamic wave (Eq. 9.18) needs to be added.

Here α is the angle of wave approach with respect to the breakwater and d_{5H} is the depth of water a distance of 5 wave heights in front of the structure. The remaining coefficients for Goda's method may be calculated as in Eq. 9.19 below.

$$\delta_{11} = 0.93\left(\frac{B_t}{L} - 0.12\right) + 0.36\left(\frac{d_s - d_t}{d_s} - 0.6\right)$$

$$\delta_1 = 15\delta_{11} \quad for \quad \delta_{11} > 0; \quad \delta_1 = 20\delta_{11} \quad for \quad \delta_{11} \leq 0$$

$$\delta_{22} = -0.36\left(\frac{B_t}{L} - 0.12\right) + 0.93\left(\frac{d_s - d_t}{d_s} - 0.6\right)$$

$$\delta_2 = 3\delta_{22} \quad for \quad \delta_{22} > 0; \quad \delta_2 = 4.9\delta_{22} \quad for \quad \delta_{22} \leq 0 \qquad (9.19)$$

$$\varsigma_5 = \frac{1}{\cosh \delta_1 \cosh^{1/2} \delta_2} \quad for \quad \delta_2 > 0; \quad \varsigma_5 = \frac{\cos \delta_2}{\cosh \delta_1} \quad for \quad \delta_2 \leq 0$$

$$\varsigma_6 = 2.0 \quad for \quad \frac{H_{des}}{1.8d_t} > 2; \quad \varsigma_6 = \frac{H_{des}}{1.8d_t} \quad for \quad \frac{H_{des}}{1.8d_t} \leq 2$$

9.1.4 Stability Design

Although a complete Level II risk analysis for the structure is preferred, the calculations in this chapter will design for stability against sliding, overturning and failure of a sandy base, using Level I design equations. We will use Eq. 8.6

$$G = \frac{R_{ch}}{\gamma_r} - \gamma_s S_{ch} = 0 \qquad (9.20)$$

the simplest limit state equation for one failure mode. We assume that the effects of σ_r and σ_s are included in γ_r and γ_s. For sliding, the equation may be written as

$$\frac{f_f \left(F_m - F_b - \gamma_w F_u\right)}{\gamma_s} - \gamma_w (F_w + F_d) = 0 \qquad (9.21)$$

where f_f is the friction coefficient between the structure and the sub-base, γ_s and γ_w are the partial coefficients for resistance against sliding and wave loading, and F_d is the dynamic force that only applies to the Minikin design. For ice

$$\frac{f_f \left(F_m - F_b\right)}{\gamma_s} - \gamma_{ice} (F_{ice}) = 0 \qquad (9.22)$$

where γ_{ice} is the partial coefficient for ice loading. Traditionally the global factor of safety for deterministic design against sliding $(\gamma_s\gamma_w)$ has been 1.2 to 1.5. These γ

values themselves reflect the fact that the forces are only known approximately but the large range really indicates our uncertainty about the uncertainty. Burcharth and Sorensen (1998) derive partial coefficients based on extensive Level II analysis of stability of existing vertical breakwaters and breakwater models. They used Goda's expression and took into account commonly occurring values of σ. They derive several γ coefficients for several target values of P_F. For $P_F=0.01$, the coefficients of Burcharth and Sorensen (1998) can be approximated as $\gamma_w=1.25$ and $\gamma_S=1.3$. The partial factor for ice force has not been researched that well and we will use 1.2. Typical friction coefficients may be found in CERC (1984); for concrete on rock or gravel $f_f=0.6$ and on sand $f_f=0.4$. Takahashi (1996) recommends $f_f=0.6$ for rock.

For overturning, the moments are calculated around the landward bottom corner of the structure, resulting in

$$\frac{(\widetilde{M}_m - \widetilde{M}_b)}{\gamma_O} - \gamma_w \, (\widetilde{M}_w + \widetilde{M}_d + \widetilde{M}_u) = 0 \qquad (9.23)$$

or for ice

$$\frac{(\widetilde{M}_m - \widetilde{M}_b)}{\gamma_O} - \gamma_{ice}\widetilde{M}_{ice} = 0 \qquad (9.24)$$

Traditionally, the overall factor of safety ($\gamma_O\gamma_w$) has been 1.2 to 1.5. Burcharth and Sorensen (1998) may be approximated by $\gamma_w=1.25$ and $\gamma_O=1.3$. The safety factors are a function of the quality of the available data, as was seen from Table 8.1. In particular they can be reduced when model studies have been performed and Burcharth and Sorensen (1998) introduce different coefficients for design based on model study and no model study.

9.1.5 Geotechnical Stability

Normally, vertical breakwaters are placed on sand or rock.

One method to analyze soil stability for sand is to calculate the stress on the "column" of soil below the structure and compare it with a critical value (Fig. 9.7)[2]. The stress is transmitted through the granular rock berm at an angle of approximately 45° and therefore the width of the soil column affected is

---

2. This is only introduced as an example. Other types of soil stability analyses, such as slip circle analysis are also required to ensure that the structure is stable, particularly for the structure placed on a cohesive bed.

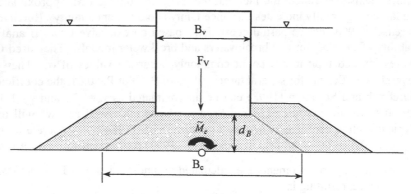

Figure 9.7 Soil Loading

$$B_c = B_v + 2d_B \qquad (9.25)$$

where d_B is the height of the berm below the caisson. The extreme stress at the harbor side of the soil column may be computed as

$$\sigma_c = \frac{F_v}{B_c} + \frac{\tilde{M}_c \frac{B_c}{2}}{I_c} = \frac{F_v}{B_c} + \frac{6\tilde{M}_c}{B_c^2} \qquad (9.26)$$

where F_v is the total vertical force, $\tilde{M}_c$ is the moment about the center of the top of the soil column and I_c is the moment of inertia of the soil column section. The allowable soil pressure on a sandy bottom for a column of width B_c may be defined as a function of the "blow count" N_b.

$$\sigma_b = (0.00016N_b^2 - 0.006)B_c \quad \text{(MPa)} \qquad (9.27)$$

Practice in Japan limits σ_b to 0.6 Mpa. For structures placed on a rock berm, the berm is simply considered as surcharge of height d_B, resulting in an additional allowable stress

$$\sigma_d = (0.026 \ N_b + 0.047)^2 \frac{d_B}{3} \quad (MPa) \qquad (9.28)$$

The maximum allowable stress in the sandy soil is therefore the sum of σ_b and σ_d. Because the soil is underwater, however, we use the underwater soil density, which is about half the density in air. Therefore

$$\sigma_u = \frac{1}{2}(\sigma_b + \sigma_d) \qquad (9.29)$$

Equations 9.27 to 9.29 have been derived from Peck et al (1974). Burcharth and Sorensen (1998) present discussions for other soil failure mechanisms.

9.1.6 Other Design Considerations

The above design calculation considers forces per unit length of the structure. To compute total forces on a structural unit, we multiply the calculated forces by the length of the unit. The resulting force from such a computation assumes that the whole structure is subjected to the calculated unit force. Particularly with short-crested waves and with waves arriving at an angle, this is seldom the case and the portion of the structure directly subjected to the design wave forces will be in part supported by the lateral connections to the remainder of the breakwater. Even though Goda's method takes wave angle into account with the term $(1+\cos \alpha)$, the above calculations could be quite conservative and model testing with directional wave spectra is advisable.

Wave transmission over the structure results in wave agitation behind the structure and possibly in damage to ships and facilities in the harbor. Experimental curves for regular waves, presented in Goda (1985), may be approximated by

$$\frac{H_T}{H} = 0.2\left(\frac{d_v}{d_s}\right)^2 - 0.4\left(\frac{d_v}{d_s}\right) + 0.58 - 0.32\frac{f_v}{H_{des}} \quad for \quad -1 < \frac{f_v}{H_{des}} \leq 1 \quad (9.30)$$

where d_v and f_v must be related to the mean wave level for the case of non-breaking waves. Goda states that Eq. 9.30 can be used for irregular waves by substituting H_s for H. A usual limit on transmitted wave height, H_T, for small craft harbors that is about 0.3 m. Often this will result in a breakwater that is visually too high and will be very costly. In general, breakwater height is determined by a combination of wave agitation, esthetic considerations and cost/benefit analysis for a higher structure.

When the wave angle of incidence to the structure is large, a *Mach Stem* wave may occur (Wiegel, 1964). The wave no longer breaks on the structure, but runs along it. The structure forms a converging wave ray with the incoming wave rays, resulting in a wave that increases in height as it runs along the structure. This can lead to large increases in hydrostatic pressure with distance along the structure, to dangerous flooding along the breakwater crest, and to scour of the berm when the Mach Stem wave breaks.

Since direct wave impact forces are very high, considerable effort has been expended to reduce the impact forces. Tanaka (1994) discusses some of these:
- – Placing wave-dissipating armor against the seaward face as in Fig. 9.1c,
- – Using a perforated seaward face that absorbs wave energy,
- – Curving the seaward face in the vertical plane,
- – Curving the seaward face in the horizontal plane,
- – Using an "elastic" seaward face or foundation,
- – Building a submerged breakwater seaward of the vertical breakwater.

9.2 Design Examples[3]

9.2.1 Vertical Breakwater in 12 m of Water with a Short Fetch

We will first design a vertical breakwater for a short fetch such as on a wide river. The incident waves are fetch limited with maximum conditions: $H_s=1.2$ m, $T_p=3.2$ sec and $\alpha=0°$. Using Eq. 9.1, with $K_{max}=1.8$ results in $H_{max}=2.2$ m. The breakwater will be located in 12 m of water and the foreshore slope m=0.1. The incident waves are clearly non-breaking waves and we will use Eqs. 9.2 to 9.17 to represent the non-breaking wave forces. Table 9.2 presents the calculations for a breakwater placed on a 3 m high berm with a freeboard of 1 m.

We use Level I design with Eqs. 9.20 to 9.29. Design of a vertical breakwater essentially consists of defining all the variables except caisson width, and then determining the correct caisson width so that it is safe against sliding and overturning and can be supported by the soil on which the structure rests. The other variables, such as water depth, berm height, freeboard, etc. are then changed and the calculations are repeated, until suitable solutions are found. Table 9.2 is therefore only one calculation of a number of parallel computations that must be made.

---

3. These examples were worked out using the spreadsheet program **VBWdes.xls**®

First we apply Eq. 9.20 for sliding, setting $R_{ch}=F_v=F_m-F_b-F_u$, $S_{ch}=F_w$ and $\Gamma_S=\gamma_S\gamma_w=1.3$. This is like using Eq. 8.1 or 8.9 and is similar to deterministic design. We find a required $B_v=3.3$ m for sliding. For overturning, we use Eq. 9.23 with $\Gamma_O=\gamma_O\gamma_w=1.3$ and find $B_v=4.8$ m. For both cases, the soil stress remained below the critical value, assuming N=20 and $\gamma_{soil}=1.2$. These are shown as Cases A and B in Table 9.2 and clearly overturning is the governing condition of these two, since it requires the wider structure.

Table 9.2 Vertical Breakwater on a Wide River, in 12 m on a River

Case	A	B	C	D	E	F
d_s (m)	12	12	12	12	12	12
d_v (m)	9	9	9	9	9	9
d_t (m)	9	9	9	9	9	9
d_5 (m)	12.6	12.6	12.6	12.6	12.6	12.6
Γ_S	1.3					
Γ_O		1.3				
Γ_{soil}	1.2	1.2	1.2	1.2	1.2	1.2
γ_S			1.3	1	1.5	1.3
γ_w			1.25	1.25	1.25	
γ_O			1.3	1	1.5	1.3
γ_{ice}						1.2
f_v (m)	1		1	1	1	1
B_v (m) - Hydrostatic	3.3	4.8	5.4	4.7	5.9	13
B_v (m) - Goda	3.5	4.5	5.1	4.5	5.6	13
Critical Condition	S	O	O	O	O	Soil

Critical condition - S = Sliding; O = Overturning; Soil = soil

Case C uses partial safety factors $\gamma_w=1.25$, $\gamma_S=1.3$, $\gamma_O=1.3$, which are similar to the coefficients determined by Burcharth and Sorensen for a target $P_F=0.01$ and we find $B_v=5.4$ m . We then test the sensitivity to the partial factors, using $\gamma_S=\gamma_O=1.0$ and 1.5 (Cases D and E) and finally assume an ice force resulting from a 0.3 m thick ice sheet, using $\gamma_{ice}=1.2$ (Case F). The particular γ values are not introduced here because they are "correct". They are simply values that are in common use.

All calculations were repeated introducing Goda's expressions for the forces (Table 9.1, Column 3), even though these were developed for breaking waves. It is seen that the two methods do not differ very much and that the hydrostatic method is more conservative. With the exception of Case F, the critical condition was always overturning, as might be expected for a tall, narrow caisson. If ice is possible, the ice forces are seen to control the design width of the structure and in that case the soil strength becomes the critical condition.

9.2.2 Vertical Breakwater in 12 m of Water on an Open Coast

We will now design a vertical breakwater in 12 m of water at the location in Lake Huron for which the wave climate was presented in Ch. 4 and design waves were derived in Ch. 8. The foreshore slope is m=0.02 and we assume N_L=50 years. We will also introduce some concepts in addition to the ones presented in Section 9.2.1. By way of example, we will design with a substantial margin of safety. We will use a Level I design expression with safety factors that approximate P_F=0.01 *and* we use P_E=0.1 or T_R=475 yr.

For T_R=475, the deepwater wave is H_s=6.9 m, T_p=11.5 sec (Section 8.8.3 and Table 8.4) and for that wave α_o=30°. In the actual design, we would take into account other water levels and their accompanying storm surge and wave conditions. For this example we will assume that 12 m of water represents a maximum water depth at the structure. That means we design for the largest waves. But such a water depth occurs only a few months every decade or so. The normal water level would be around 11.5 m and at times, it can be as low as 10 m (Fig. 6.18). For a complete analysis, we need at least three parallel designs to investigate the impacts of different water levels and we need to take into account the effects of storm surge and seiche. In tidal areas, we need to design at least for extreme high and low water and include seiche and storm surge. Only one design for one water level is presented here.

The breaking conditions for the design wave are d_b=11.2 m, H_{sb}=6.6 m and α_b=16.0°. Since d_b is the depth in which the largest wave breaks, we could in theory design for a non-breaking wave shoaled into 12 m of water: H_s=6.5 m, T_p=11.5 sec and α=16.5°. However, the maximum wave (Eq. 9.1) is so close to breaking that we would need to calculate for both breaking and non-breaking waves. Since highest forces will come from the breaking waves, we will only calculate the breaking wave forces using the Goda method in this example. It is seldom that a non-breaking condition can be used on an open coast. Usually, non-breaking wave design can only be used for breakwaters in deep water, subjected to fetch-limited waves.

The required freeboard to produce a 0.3 m high wave inside the harbor, as estimated from Eq. 9.30 is in excess of 8 m, which is unrealistic. We will use a design structure height of f_v=2 m. This could be a consensus level, based on an esthetic limit defined through a roundtable discussion with all stakeholders. This lower breakwater crest will still form an obstruction that is 4 m above water when the lake levels are low and will interfere substantially with the view of the horizon from shore. The transmitted wave height at the 12 m water level will be a dangerous 2.2 m, according to Eq. 9.30 and even higher if a storm surge arrives at that time. These two unwanted consequences of the consensus decision to incorporate 2 m of freeboard must be carefully communicated, along with the fact that Eq. 9.30 may contain quite large uncertainties. In this design example we will use a 6 m high berm with B_T=3 m and a 8 m high vertical caisson.

We follow the same procedure as in Section 9.2.1 and the results are summarized in Table 9.3. From Cases A and B, using Eq. 9.20, we see that sliding governs the overall design, requiring a caisson width of 29 m. When Eqs. 9.21 and 9.23 are used with γ_S=γ_O=1.3 (Case C) the required structure width becomes 41 m. This number is very sensitive to p_u, which, like all the other forces in Table 9.1 contains substantial uncertainties. Case D shows that increasing the freeboard to 3 m would result in a width of 37 m. Reducing the γ values to 1.1 results in a 35 m wide structure (Case E). Decreasing the berm height and using a taller caisson (Cases F and G) leads to different results again. The failure mechanism also changes from sliding to overturning for the narrower caisson in Case F and to soil failure in Case G (for N=20 and γ_{soil}=1.2). A 0.3 m ice sheet against the structure (Case H) requires B_v=9.5 m, which is less than the width required to resist the wave forces.

Table 9.3 Vertical Breakwater in 12 m on an Open Coast

Case	A	B	C	D	E	F	G	I
d_s (m)	12	12	12	12	12	12	12	12
d_v (m)	6	6	6	6	6	9	9	6
d_i (m)	6	6	6	6	6	9	9	6
d_S (m)	12.8	12.8	12.8	12.8	12.8	12.8	12.8	12.8
Γ_S	1.3							
Γ_O		1.3						
Γ_{soil}	1.2	1.2	1.2	1.2	1.2	1.2	1.2	1.2
γ_S			1.3	1.3	1.1	1.3	1.3	
γ_w			1.25	1.25	1.1	1.25	1.25	
γ_O			1.3	1.3	1.1	1.3	1.3	
γ_{ice}								1.2
f_v (m)	2	2	2	3	2	2	3	2
B_v (m)	29	15	41	37	35	33	30	9.5
Critical Condition	S	O	S	S	S	O	Soil	O

S = Sliding; O = Overturning, Soil = Soil strength

Table 9.3 provides a partial background to select a design. Other such tables for different water depths and more extensive sensitivity analysis with respect to the Γ and γ parameters would be needed for a more complete assessment, and it is clear that there are differences between the various design values. From Table 9.3, we would want to select Case C, since it incorporates commonly used coefficients. The large width of this structure reflects the low probability of failure incorporated in this design - $P_L=P_EP_F=(0.01)(0.1)=10^{-3}$ - in combination with the design wave of Eq. 9.1. We would certainly need to review these design conditions and study this design in a model, to reduce the uncertainties and the safety coefficients used. Above all, alternatives such rubble mound breakwaters should be investigated.

9.2.3 Vertical Breakwater in 3 m of Water

For a breakwater in 3 m of water, a limited depth calculation (Eq. 8.23) yields $H_{sb}=1.8$ and the modified limited depth equation (Eq. 8.25) yields $H_{sb}=2.2$ m. For the design wave we use Eq. 9.1 with $K_{max}=1.5$, as determined by Rakha and Kamphuis (1995) for breaking waves. This results in $H_{des}=(H_b)_{max}=3.3$ m. The breaking wave angle is calculated to be $\alpha_b=10°$. Note that if the 3 m water depth represents a high water, then the mean and low water levels would be 0.5 and 2 m lower on Lake Huron. These will result in totally different design conditions that must be investigated. We will only present the one design here for $d_s=3$ m. To produce a transmitted wave less than 0.3 m requires a freeboard of 0.61 m. We will use a 0.5 m freeboard, resulting in $H_T=0.33$ m for this design water level and we will use a 1.5 m high berm, resulting in a 2 m high caisson and $B_T=3$ m.

The results of the computations are presented in Table 9.4 for the same safety factors as used earlier. The forces were computed using both the Goda and Minikin methods. Clearly the Minikin design is much more conservative than the Goda design, because it attempts to incorporate wave slamming. It is clearly economical to avoid direct wave impact, in which case the results calculated by Goda's method govern. Sliding is the critical design condition, except in Cases A and B. Comparing Cases C with D (and G with H) indicates that a smaller width is needed if a larger freeboard is used. However, there is no cost saving, because both structures need almost the same amount of concrete. Similarly, Cases C, D, G and H show that there is no advantage in lowering the berm by 0.5 m. Finally to resist forces of a 0.3 m thick ice sheet a very wide structure is needed (Case I). From Table 9.4, we might again choose Case C as our design, because it uses the safety factors determined from Level II probability analysis by Burcharth and Sorensen (1998). Since the breaking wave condition occurs very often, P_E in this design is 1.0

and hence P_L=0.01. If ice forces can occur, we would certainly opt for a rubble mound breakwater design, but even for the wave-based designs, the caissons are very wide.

Table 9.4 Vertical Breakwater in 3 m of Water

Case	A	B	C	D	E	F	G	H	I
d_s (m)	3	3	3	3	3	3	3	3	3
d_v (m)	2	2	2	2	2	2	2.5	2.5	2.5
d_t (m)	2	2	2	2	2	2	2.5	2.5	2.5
d_5 (m)	3.2	3.2	3.2	3.2	3.2	3.2	3.2	3.2	3.2
Γ_S	1.3								
Γ_O		1.3							
Γ_{soil}	1.2	1.2	1.2	1.2	1.2	1.2	1.2	1.2	1.2
γ_s		1.3	1.3	1.0	1.5	1.3	1.3	1.3	1.3
γ_w			1.25	1.25	1,0	1.25	1.25	1.25	
γ_{ice}									1.2
γ_o			1.3	1.3	1.0	1.5	1.3	1.3	1.3
f_v (m)	0.5	0.5	0.5	1	0.5	0.5	0.5	1	0.5
B_v (m) Goda	6.8	6.8	8.2	7.3	6.9	9.5	8.2	7.4	23
B_v (m) Minikin	18	6.6	24	20	15	28	21	18	23
Critical Condition	S	Soil	S	S	S	S	S	S	S

9.2.4 Summary

These examples show that there is no single "correct" solution to a design problem. We have only used Level I design with commonly used coefficients. The safety factors we use and our assumptions with respect to probability of failure are still tentative. Our only strength is that we have computers that can easily calculate alternatives, thus providing a basis from which we can make more informed decisions. Much more prototype evaluation needs to be done (and is being done) to refine our estimates of uncertainty and determine appropriate safety factors. Even then , it is good practice to test the structure in a model study.

In the end, we continue to face uncertainty, and structural design in the coastal zone will continue to be a delicate balance between economic cost, structural safety and imposed environmental boundary conditions. It is no great trick to design a totally safe structure. It is very difficult to design a structure that is safe, economically viable and fits within socially acceptable norms.

9.3 Rubble Mound Breakwaters

9.3.1 Filter Characteristics

Rubble mound breakwaters are built up like filters. They consist of layers of stone as in Figs. 9.8 and 9.9. The center core of the breakwater is made up of quarry run rock of the most economically available size. The outside layer consists of large armor units, that can be either rock or specially designed concrete units. This primary armor layer is intended to be statically stable with respect to the environmental conditions imposed on it (the waves and currents do not move the armor stones under design conditions). It is usual to build the primary armor layer roughly two unit diameters thick and to place the units randomly, meaning that they are not especially fitted together. If the armor units were placed directly over the core, the finer core material would be removed by the waves through the openings of the armor layer. It is therefore necessary to construct the breakwater as a filter of three or four layers so that the material from any layer is not removed through the layer above it.

A typical example filter relationship to prevent removal of the lower material through the upper layer is

$$D_{15} \ (upper \ layer) < 5 \ D_{85} \ (lower \ layer) \tag{9.31}$$

where D is the nominal size and D_{85} means that the nominal size of 85% of the sample is less than D_{85}. For rock, the nominal armor unit diameter is defined as

$$D_a = D_{50} = \left(\frac{M_a}{\rho_a} \right)^{1/3} \tag{9.32}$$

where M_a is the armor unit mass and ρ_a is the armor density. Figures 9.8 and 9.9 give an initial estimate of the rock mass needed in the secondary layer ($M_a/10$) and a possible third layer ($M_a/200$). Final rock sizes obviously depend on the rock gradations that can be obtained from the quarry.

When a breakwater is built on erodible material, the toe filter is of particular interest. It is located where the largest stone (the primary armor) and the base on which the breakwater is built (often fine material such as sand) are adjacent to each other. To prevent removal of the base material through the armor, this toe filter also needs to be built up of several layers, but the layers must be compact so that the total depth of the filter remains small. The toe filter is crucial to the operation of the breakwater. If it fails, the base material will be removed and the lowest armor stones will drop down into the resulting cavity and endanger the stability of the whole primary armor layer. If the breakwater is located in shallow water under breaking waves, the toe filter will be exposed to extreme wave action. In breaking wave conditions therefore, this toe filter must be completely protected by the primary armor as shown in Fig. 9.9. It is also customary to use geotextiles in the toe filter and to dig down into the base material to make room for a toe filter of appropriate thickness.

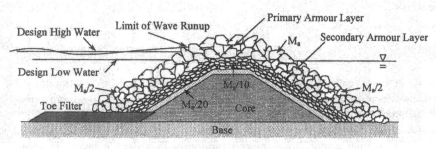

Figure 9.8 Rubble Mound Breakwater in Deep Water
(after CERC, 1984)

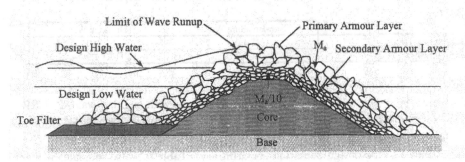

Figure 9.9 Rubble Mound Breakwater in Shallow Water
(after CERC, 1984)

9.3.2 Rock Armor

Stable rock armor mass has traditionally been calculated with the Hudson formula (CERC, 1984)

$$M_a = \frac{\rho_a H_{des}^3}{K_D \left(\frac{\rho_a}{\rho} - 1 \right)^3 \cot \theta} = \frac{\rho_a H_{des}^3}{K_D \Delta_a^3 \cot \theta} \tag{9.33}$$

where ρ_a is armor unit density, ρ is the fluid density, θ is the angle of the front slope of the structure with respect to horizontal and Δ_a is the relative underwater density of the armor

$$\Delta_a = \frac{\rho_a - \rho}{\rho} = \frac{\rho_a}{\rho} - 1 \tag{9.34}$$

K_D is an empirically determined damage coefficient. It is a function of all the variables involved in armor stability that are not included in Eq. 9.33, but primarily, it is a function of the type of armor, its shape, its location along the breakwater and the amount of damage considered to be acceptable. Typical published values of K_D for rough angular armor stone, placed randomly in a double layer are shown in Table 9.5. The term "Zero Damage" means that there is nominally no removal of the armor units from the face of the breakwater.

Table 9.5 Published Damage Coefficients Rock, Zero Damage

	Non-Breaking Waves	Breaking Waves
Structure Trunk	4.0	2.0
Structure Head	3.2	1.9

Equation 9.33 was based on hydraulic model tests with regular waves. CERC (1977) applies Eq. 9.33 to irregular waves by assuming that H_s can be used as H_{des}. In the later edition CERC (1984), the use of $H_{des} = \overline{H}_{0.1} = 1.27 H_s$ is advocated. These adjustments to the design condition are arbitrary. Table 9.5 also indicates that the tests showed a different K_D value for breaking waves. To be consistent, we assume that K_D is essentially a property of the armor stone and use $K_D = 4$ for armor stone on a breakwater trunk. We also use $H_{des} = H_s$ but may use H_{max} to account for

cumulative damage by breaking waves, as advocated in Ch. 8. The increased exposure for the head of the breakwater is then taken into account by a lower value of K_D, because the same stone will be less stable on the head of a breakwater than on its trunk. We use the 20% decrease in K_D shown in Table 9.5.

These arbitrary adjustments to the basic design variables in the literature are an indication of the uncertainties in Eq. 9.33. These uncertainties can again be taken into account by appropriately conservative values of the safety coefficients, which is a costly solution. The uncertainties and hence the final construction costs, particularly for large and costly projects are usually reduced through physical model studies (Ch. 13).

Equation 9.33 can be re-arranged as

$$N_s = \frac{H_{des}}{\Delta_a D_a} = (K_D \cot \theta)^{1/3} \tag{9.35}$$

where N_s is known as the stability number. Van der Meer (1987) derives expressions that include some additional characteristics of the incident wave climate. He uses H_s exclusively as the design wave height and is not concerned about cumulative damage for depth-limited design. For plunging breakers, Van der Meer derives

$$N_s = \frac{H_s}{\Delta D_a} = 6.2 \, P_b^{0.18} \left(\frac{S_a}{\sqrt{N_w}} \right)^{0.2} \xi_m^{-0.5} \tag{9.36}$$

For surging breakers

$$N_s = \frac{H_s}{\Delta D_a} = 1.0 \, P_b^{-0.13} \left(\frac{S_a}{\sqrt{N_w}} \right)^{0.2} \sqrt{\cot \theta} \, \xi_m^{P_b} \tag{9.37}$$

Here P_b represents an overall porosity of the breakwater. Van der Meer suggests that for an armor layer over an impermeable layer $P_b = 0.1$, for armor over a filter over a coarse core $P_b = 0.4$. For a structure built entirely out of armor stone $P_b = 0.6$. Armor damage, S_a is defined as

$$S_a = \frac{A_e}{D_a^2} \tag{9.38}$$

where A_e is the erosion area in the breakwater profile between the still water +/- one wave height. For zero damage we would need $D_a=\infty$. Van der Meer recommends using $S_a=2$ as equivalent to zero damage. For failure of the breakwater, normally defined as the point when the secondary armor layer becomes exposed, $S_a=15$.

The surf similarity parameter ξ_m, for the breakwater slope is related to the mean wave period.

$$\xi_m = \frac{\tan \theta}{\sqrt{s_m}} \qquad (9.39)$$

where s_m is the mean wave steepness.

$$s_m = \frac{H_s}{L_{o,m}} = \frac{2 \pi H_s}{g T_m^2} \qquad (9.40)$$

The transition from plunging to surging waves on the breakwater takes place at a critical value of ξ_m

$$\xi_{mc} = \left(6.2 \, P_b^{0.31} \, \sqrt{\tan \theta} \; \right)^{\frac{1}{(P_b - 0.5)}} \qquad (9.41)$$

according to Van der Meer. Essentially, the surging breaker expression is needed only for very flat waves.

Since Van der Meer uses H_s consistently for design wave, N_w is the only parameter to take into account that the design condition is reached many times over the design life of the structure. For example, if the design wave period is 9 sec and occurs 6 hours per year on average, then during a 50 year structure lifetime $N_w=50 \times 6 \times 3600/9=120,000$. However, this would result in a very small value of N_s and a very large value of D_a. Van der Meer (1993) and Pilarczyk and Zeidler (1996) recommend that a maximum value of $N_w=7500$ be used. Figure 9.10 shows the effect of the wave steepness using Eqs. 9.36 for cot $\theta=1.5$, $P_b=0.4$, $S_a=2$ with $N_w = 1000$ and 7500. It also compares these Van der Meer values with Eq. 9.33.

9.3.3 Concrete Armor

Armor units need not be rock. They can be manufactured out of concrete and a whole gallery of different units is available. Some are shown in Fig. 9.11. At first sight, such units would be very helpful, since it is possible to shape the units so they interlock better than rock armor and hence provide greater stability for the same mass. In simple terms, for the Hudson formula (Eq. 9.33), more interlocking will increase K_D and hence the required armor mass for a specific design condition will decrease. Sample published values of K_D for zero damage on a breakwater trunk are given in Table 9.6.

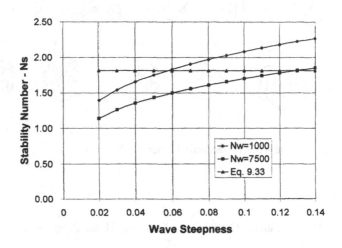

Figure 9.10 Comparison of Van der Meer with Eq. 9.33

Experience with such units has shown, however, that the units must still depend primarily on their mass for stability. Prototype units with relatively thin members (such as Dolos) can break under the stresses imposed upon them. Armor unit strength was initially not simulated in hydraulic model tests and results showed Dolos to be very stable ($K_D=32$) because of their interlocking. Conservative practice, based on field experience and additional model testing with Dolos that were scaled for strength, now recommends $K_D=16$ for Dolos.

Van der Meer presents a different expression than Eqs. 9.36 (or 9.37) for concrete units that he tested. For tests, which were limited to $\theta=1.5$ and for zero damage:

$$N_s = c_1 \, s_m^{c_2} \tag{9.42}$$

where c_1 and c_2 are constants that depend on the type of unit as shown in Table 9.7.

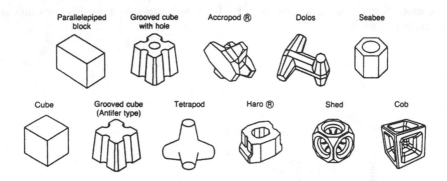

Figure 9.11 Sample Concrete Armor Units

Table 9.6 Damage Coefficients Concrete Units, Zero Damage

Armor Unit	K_D
Rock	4
Tetrapods	8
Tribars	10
Dolos	32 (16)
Modified Cubes	7.5

Table 9.7 Van der Meer's Coefficients

	c_1	c_2
Cubes	1	-0.1
Tetrapods	0.85	-0.2
Accropods	3.7	0

9.3.4 Armor Unit Density

Armor unit size D_a varies inversely with underwater relative armor density Δ_a in both the Hudson equation and the Van der Meer equations. If concrete is used, it is possible to increase ρ_a substantially through the use of heavy aggregate, such as blast furnace slag. This is an effective method to reduce the required armor unit mass. For example, a relatively small increase in concrete density from a normal concrete (ρ_{a1}=2200 kg/m³) to a heavier concrete (ρ_{a2}=2600 kg/m³) results in Δ_{a1}=1.2, Δ_{a2}=1.6 and therefore D_{a2}=0.75D_{a1} or M_{a2}=0.42M_{a1}, a reduction in armor mass of more than 50%.

9.3.5 Primary Armor Layer

On the seaward side, it is customary to extend the armor layer from the breakwater crest down to about 1.5 H_s below the lowest water level. Because the wave action is less at greater depth, smaller armor units can be placed below – 1.5 H_s. Figure 9.8 indicates a preliminary size (M_a/2). Primary armor is placed on the back of the structure down to the lowest water level, because overtopping waves will put severe down-slope stress on any armor units above water. If the structure is in shallow water, then primary armor covers the complete structure, including the toe filter.

The primary armor layer is usually placed in a double layer. Since the nominal armor unit size as defined in Eq. 9.32 is the size of a cube, a shape factor k_a is introduced to account for the shape of the unit as well as for its random placement. The armor layer thickness is therefore

$$r_a = n_a k_a D_a \qquad (9.43)$$

where n_a is usually 2. Typical values of k_a are given in Table 9.8.

The number of armor units required per unit length of the structure is

$$N_a = \frac{A_a \, n_a k_a \, (1-e)}{D_a^2} \qquad (9.44)$$

where A_a is the surface area (per unit length of the breakwater) to be covered by the armor units and e is the porosity of the armor layer. The values in Eq. 9.44 are approximate. They depend heavily on the rock that comes out of the quarry and the methods and care of placement. However, their values have a major influence on both the armor layer thickness and the number of units required (the cost of the armor layer). As a result, it is virtually impossible to estimate numbers of armor units accurately, and this can cause major differences between estimated and real costs of armor in a design.

Table 9.8 Shape Factor and Porosity

	k_a	E
Rock	1	0.37
Modified Cubes	1.1	0.47
Tetrapods	1.04	0.50
Tribars	1.02	0.54
Dolos	0.94	0.56

9.3.6 Breakwater Crest

The crest of a rock armor breakwater is usually made up of the same rock as the rest of the armor layer and it is normally about three stones wide. The crest of a breakwater with concrete units is usually a monolithic cap unit, which provides support for the armor units (Fig 9.12). This cap can carry traffic and infrastructure. Because the cap is impermeable, there is often concentrated damage at the interface between the cap and the concrete armor units. Since the uprush of the water cannot pass through the cap, it can only go up through the topmost primary armor units. The resulting high vertical fluid velocities will decrease the stability of the units near the cap so that they are easily displaced or broken.

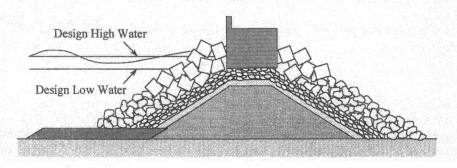

Figure 9.12 Artificial Armor Units with Concrete Cap

The first estimate of the design crest elevation of a rubble mound breakwater is the limit of runup of the largest waves, superimposed on the highest water level. Such a crest height would prevent all wave overtopping and as a result prevent any generation of waves behind the structure by overtopping waves. Such a crest elevation may be high enough that it forms an ugly visual barrier. Total cost of a rubble mound breakwater is also very sensitive to crest height. As a result, crest height is often reduced, allowing overtopping of the larger waves. The combination of safety in the harbor, negative esthetic impact and cost of the structure combine to determine the actual breakwater crest elevation.

Much work has been done to determine wave runup - the vertical distance above still water level reached by the waves. CERC (1984) contains sets of curves. A relatively simple estimate of runup (Van der Meer, 1993) is

$$\frac{R_{2\%}}{H_s} = 1.5\, r_f\, \xi_p \qquad for\; \xi_p < 2;$$

$$(9.45)$$

$$\frac{R_{2\%}}{H_s} = 3\, r_f \qquad for\; \xi_p \geq 2$$

where $R_{2\%}$ is the runup exceeded by 2% of the waves, r_f is a factor which takes into account friction, any horizontal berm sections in the front face, the angle of approach and whether the waves are short crested. The surf similarity parameter, ξ_p, is based on the peak period of the wave spectrum. For a simple rock breakwater and with waves coming normal to the front face, $r_f=0.5$. For Dolos $r_f = 0.45$ and for a smooth slope, $r_f = 1.0$. This factor r_f is reduced by incident wave angle. For the usual short crested waves r_f may be multiplied by a factor which reduces linearly with wave angle from 1 at 0° to 0.8 at 90° degrees.

A rubble mound breakwater will inevitably settle after its construction. If the base under the structure is solid (sand, gravel or rock), it is usual to add 0.3 m to the design crest elevation. For softer bases, the breakwater base is sometimes widened to decrease the stresses in the soil. Sometimes the soil directly below the structure is removed and replaced with granular material. If the breakwater is expected to settle substantially, accurate settlement calculations are necessary to determine the design crest elevation.

9.4 Design Examples

We will now design the armor layer for rubble mound breakwaters in 12 m and in 3 m of water at the location in Lake Huron for which the wave climate was presented in Ch. 4 and the design waves were derived in Ch. 8. We assume N_L=50 years and m=0.02. In this case we will use T_R=N_L, which means P_E=0.64.

We use the Hudson expression (Eqs. 9.33 and 9.35). These two equations are deterministic, but the equations as well as the design parameters contain uncertainties and therefore we apply probabilities and partial coefficients of safety as in Eq. 9.20. To separate loads and resistances, Eq. 9.33 may be re-written as

$$S_{ch} = H_{des}; \quad R_{ch} = \Delta_a \left(\frac{K_D \cot\theta}{\rho_a} \right)^{1/3} M_a^{1/3} = K_M M_a^{1/3} \quad (9.46)$$

where

$$K_M = \Delta_a \left(\frac{K_D \cot\theta}{\rho_a} \right)^{1/3} \quad (9.47)$$

Because Eq. 9.33 is not an exact representation of armor unit stability we introduce a third partial coefficient (γ_e) to describe the uncertainty in the equation itself. Adding this into Eq. 9.20 gives

$$G = \frac{1}{\gamma_e} \frac{R_{ch}}{\gamma_r} - \gamma_s S_{ch} = \frac{K_M M_a^{1/3}}{\gamma_e \gamma_r} - \gamma_s (H_{des}) = 0 \quad (9.48)$$

Three design approaches will be compared
 -   Deterministic design with Γ=1.0 (or simple formula substitution),
 -   Level I design with the coefficients derived by PIANC (1992),
 -   Level II design with one failure mechanism (Eq. 9.48).

9.4.1 Breakwater in 12 m of Water

In Ch. 8 we found that H_{des}=5.2 m for T_R=N_L=50 years, as shown in Tables 8.4 and 8.5. We will first demonstrate simple formula substitution with R_{ch} and S_{ch} defined as in Eq. 9.46 and Γ=1.0. The slopes for a rubble mound breakwater are normally as steep as possible (cotθ = 1.5 to 2.0), to minimize the amount of material required to construct the breakwater. We will use $\cot\theta=1.5$. The rock density is $\rho_s=2650$ kg/m^3 and since Lake Huron is fresh water, $\rho=1{,}000$ kg/m^3. For $K_D=4.0$, $K_M=0.217$ and Eq. 9.33 yields a required armor stone mass $M_a=13.8$ tonnes. Since most quarries cannot readily produce stones in excess of 10 tonnes consistently, it is most likely that concrete armor units will be used. Using Dolos armor units with $\rho_a=2400$ kg/m^3 and $K_D=16$ (Table 9.6) gives $K_M=0.302$ and Eq. 9.33 yields $M_a=5.1$ tonnes. This is shown as Case A in Table 9.9.

Table 9.9[⊗] Level I Design Calculations

Case	μ_s	$\gamma_e\gamma_r$	γ_s	μ_r	M_a	P_F	P_E	P_L
A	5.2	1.0	1.0	5.2	5.1	1.0	0.64	0.64
B	5.2	1.06	1.03	5.7	6.6	0.5	0.64	0.32
C	5.2	1.29	1.07	7.2	13.5	0.16	0.64	0.1

PIANC (1992) used Level II analysis to derive values of partial safety coefficients for Level I design. For $S_{ch}=H_s$ and $T_R=N_L$, γ is a function of target P_F, as in Table 9.10.

Table 9.10[⊗] Partial Coefficients PIANC (1992)

P_F	$\gamma_e\gamma_r$	γ_s
0.5	1.06	1.03
0.16	1.29	1.07
0.1	1.38	1.08
0.05	1.51	1.11
0.01	1.79	1.17
0.001	2.27	1.25

Level I design, using the PIANC coefficients for $P_F=0.5$ ($P_L=0.5\times0.64=0.32$) results in $M_a=6.6$ tonnes (Case B). We can reduce this high P_L by choosing a higher P_F for the design formula. To achieve $P_L=0.1$, $P_F=0.1/0.64=0.16$ and the partial coefficients are $\gamma_s\gamma_r=1.29$ and $\gamma_e=1.07$. The resulting Dolos armor mass is 13.5 tonnes (Case C).

To look at this problem as a simple Level II design with one failure mechanism, we assume that the uncertainty in wave height is $\sigma_H'=0.15$, the uncertainty in concrete armor mass is $\sigma_M'=0.05$ and $\sigma_{cot\theta}'=0.1$, $\sigma_{K_D}'=0.15$, $\sigma_\Delta'=0.05$ and $\sigma_\rho'=0.05$. Therefore

$$\sigma_r'=\sigma_\Delta'+\frac{\sigma_{cot\theta}'+\sigma_{K_D}'+\sigma_\rho'+\sigma_M'}{3}=0.17 \qquad (9.49)$$

We will first use Eq. 9.48, $Z_r=Z_s=0$, a global factor of safety $\Gamma=\gamma_s\gamma_r=1$ and $\gamma_e=1.05$. This results in Case D in Table 9.11. For Case E, we use $\Gamma=\gamma_s\gamma_r=1.1$ and in Case F $\Gamma=\gamma_s\gamma_r=1.21$, specifically adjusted to yield $P_L=0.1$.

Table 9.11[⊗] Level II Design Calculations ($\mu_s=5.2$, $P_E=0.64$)

Case	σ_s'	σ_r'	$\gamma_e\gamma_r$	γ_s	μ_r	M_a	β	P_F	P_L
D	0.15	0.17	1.0	1.05	5.5	5.9	0.214	0.42	0.27
E	0.15	0.17	1.1	1.05	6.0	7.9	0.627	0.27	0.17
F	0.15	0.17	1.21	1.05	6.6	10.4	1.013	0.16	0.10
G	0.30	0.17	1.1	1.05	6.0	7.9	0.432	0.33	0.21
H	0.075	0.17	1.1	1.05	6.0	7.9	0.737	0.23	0.15
J	0.15	0.33	1.1	1.05	6.0	7.9	0.369	0.36	0.23
K	0.15	0.09	1.1	1.05	6.0	7.9	0.865	0.19	0.12
L	0.295	0.17	1.29	1.07	7.2	13.5	1.01	0.16	0.10
M	0.15	0.25	1.29	1.07	7.2	13.5	1.01	0.16	0.10

Cases D, E and F show that increasing the γ values introduces additional safety (larger stone sizes and smaller P_F for the same values of T_R, μ_r and P_E). Cases G to K address sensitivities of P_F to uncertainties in R and S and show again that smaller uncertainties (smaller values of σ_r' and σ_s') decrease P_F and therefore P_L. As in Table 8.1 smaller values of γ can also be used to achieve the same P_F and a smaller stone size, once again indicating that it pays to reduce uncertainties as much as possible.

Tables 9.9 and 9.11 show a difference in the stone size, calculated with Level I PIANC expression, and the particular Level II formulation used to generate Table 9.11. To achieve the same value of $P_L=0.1$ needs a 13.5 tonne Dolos in Case C and a 10.4 tonne Dolos in Case F. The difference can be accounted for by assuming larger values of σ' in Table 9.11. Cases L and M show the results of such an adjustment. To obtain the same answer as in Table 9.9 requires doubling σ_s' or increasing σ_r' by 50%. PIANC is evidently more conservative than Case C.

9.4.2 Breakwater in 3 m of Water

For a breakwater in 3 m of water, Example 8.3 found that, using Eq. 8.25, $(H_{sb})_{max}=2.1$ m. We will first use this as H_{des}. Using deterministic methods and substituting directly into Eq. 9.46 results in $R=S=(H_b)_{max}=2.1$ m and, using rock, with $K_M=0.217$, the required armor stone mass may be calculated as $M_a=(R/K_M)^3=0.9$ tonnes (Table 9.12, Case N). Using the PIANC expression for $P_F=0.5$, $\mu_r=(1.06)(1.03)(2.1)=2.3$ m or $M_a=(2.3/0.217)^3=1.2$ tonnes (Case M). The encounter probability for this example is always $P_E=1$, because the breaking waves occur regularly. Thus we cannot reduce P_E to design a safer breakwater. The only alternative is to decrease P_F (to increase the γ values) in the Level I design formula. Using the PIANC coefficients for $P_L=P_F=0.1$ from Table 9.10 yields $\mu_r=(1.38)(1.08)(2.1)=3.1$ m or $M_a=(3.1/0.217)^3=3.0$ tonnes (Case O).

Table 9.12[⊗] Level II Design Calculations ($\mu_s=H_s=2.1$ m, $P_E=1.0$)

Case	σ_s'	σ_r'	$\gamma_c\gamma_r$	γ_s	μ_r	M_a	β	P_F	P_L
N			1.0	1.0	2.1	0.9			1.0
M			1.06	1.03	2.3	1.2		0.5	0.5
O			1.38	1.08	3.1	3.0		0.1	0.1
P	0.15	0.23	1.1	1.05	2.4	1.4	0.51	0.31	0.31
Q	0.15	0.23	1.38	1.08	3.1	3.0	1.31	0.095	0.095
R	0.17	0.23	1.38	1.08	3.1	3.0	1.28	0.10	0.10

Taking probability distributions into account in a Level II analysis, we can again assume the uncertainty in wave height to be $\sigma_H'=0.15$. For rock armor the uncertainty in armor stone mass would be larger than for concrete, such as $\sigma_M'=0.25$ (a nominal stone mass of 10 tonnes means that for 68% of the stones, $7.5<M_a<12.5$ tonnes). If σ' for $\cot\theta$, Δ and ρ_s are all equal to the earlier values, then Eq. 9.49 gives $\sigma_r'=0.23$. Case P presents the calculation for $\gamma_s\gamma_r=1.1$ and $\gamma_e=1.05$. That results in $\mu_r=2.4$ m and $M_a=1.4$ tonnes. Using the PIANC coefficients (Case Q), we find $M_a=3.0$ tonnes, with $P_L=0.095$. This time the PIANC assumptions match the estimates of σ' used in Cases P and Q quite closely as is seen in Case R, where σ_s' is adjusted to arrive at $P_L=0.10$.

Table 9.13 presents the same computations using H_{max} as design wave height. All results are similar, except that the required armor mass is much larger. It has in fact increased by $(H_{max}/H_s)^3=(3.1/2.1)^3=3.2$. The reasons for the two different design waves were given in Ch. 8. In Table 9.12, we use H_s because that has been customary. In Table 9.13, we use H_{max} to ensure absolutely zero damage, because damage is cumulative. There is not enough information to chose between the two approaches. Table 9.13 is probably conservative, but a breakwater designed according to Table 9.12 may sustain damage over the long term. This is once again an indication of uncertainties in our design. It points out the necessity for careful documentation of field experience and further model study in the future.

Table 9.13[⊗] Level II Design Calculations (μ_s=H_{max}=3.1 m, P_E=1.0)

Case	σ_s'	σ_r'	$\gamma_c\gamma_r$	γ_s	μ_r	M_a	β	P_F	P_L
N			1.0	1.0	3.1	2.9			1.0
M			1.06	1.03	3.4	3.8		0.5	0.5
O			1.38	1.08	4.6	9.7		0.1	0.1
P	0.15	0.23	1.1	1.05	3.6	4.5	0.51	0.31	0.31
Q	0.15	0.23	1.38	1.08	4.6	9.7	1.31	0.095	0.095
R	0.17	0.23	1.38	1.08	4.6	9.7	1.28	0.10	0.10

9.4 Berm Breakwaters

Conventional rubble mound breakwaters are statically stable structures. They are expected to stay in place without change in shape and with little damage. Generally, N_s is less than 4. If N_s exceeds 4 substantial damage will occur to the armor layer and the breakwater may fail. If the whole structure were built of the same size stone, however, there would not be failure, but simply a reshaping of the breakwater profile. This is the principle behind the *berm breakwater*. Steep rock slopes are allowed to deform into a stable "S" shape and form a "berm". For berm breakwaters, $3<N_s<6$. Of course there is a continuum of "granular structures". For natural rocky beaches, $6<N_s<20$, for gravel beaches $15<N_s<500$, and for sand beaches, $N_s>500$, according to Van der Meer (1987).

An alternative to the conventional (statically stable) rubble mound breakwater is therefore a rubble mound breakwater that is dynamically stable. The individual armor stones will be moved around by the waves, to form a stable mass similar to a beach, where individual sand or gravel grains may move about substantially, but the total beach is stable against wave attack. Such a dynamically stable berm breakwater is built up of stones that are not specifically sorted or built up in various layers.

Construction is much simpler and there are other advantages:
- the breakwater is very porous and hence much of the incident wave energy is absorbed,
- small armor stone can be used,
- all the quarried rock is used.

Van der Meer has developed a computer program **BREAKWAT** to design these structures, but for most purposes, the guidelines developed by Hall (1993) are sufficient. A breakwater cross-section, as in Fig. 9.13, will eventually deform into a natural profile as shown with the dotted line[4].

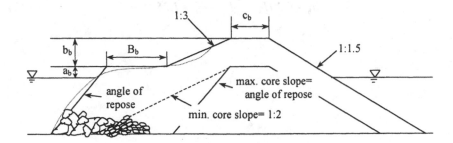

Figure 9.13 Berm Breakwater

The most important design parameter is the berm width B_b

$$B_b = D_{a,50}\left[K_b + 7.5\left(\frac{D_{a,85}}{D_{a,15}}\right) - 1.1\left(\frac{D_{a,85}}{D_{a,15}}\right)^2 + 6.1P_r\right] \qquad (9.50)$$

where

$$K_b = -10.4 + 0.5\left(\frac{H_s}{\Delta_a D_{a,50}}\right)^{2.5} \qquad (9.51)$$

and P_r is the fraction of rounded stones.

---

4 The dynamic nature of this breakwater may also result in alongshore transport of the armor units

It is usual to divide the armor coming out of a quarry into core and armor. Figure 9.13 shows the use of a core (the inner mass of stones) and armor units (the outside). That is better construction than placing the rock indiscriminately. The diameter of the core material is less than a certain size and armor is larger. The division between the two is a function of the volumes of rock that the quarry produces in the different sizes. The volume of the core and the armor is about 50% of the total volume of stone. Construction methods define the elevation of the top of the core and the berm (a_b). Usually these breakwaters are constructed with land-based equipment and hence the core and berm are built above water to allow safe passage of the equipment during construction. The crest width and elevation prevent most overtopping. Typical values are

$$a_b = 0.2\,H_s \; ; \quad b_b = c_b = 1.25\,H_s \qquad\qquad (9.52)$$

10. Introduction to Coastal Management

10.1 Introduction

The management of coastal resources is integrally related to every country's *economy*. In developing countries, the coastal zone has great economic value, as a resource for food production and basic industry and it has probably also become an income generator through *tourism*. In developed countries, the coast has often been the machine that drives a country's economy, but there also coastal tourism has become an important generator of wealth. In developed countries, struggle for survival is no longer a primary concern and therefore, in addition to economic considerations, there is also an interest in *quality of life*. This is expressed primarily by private citizens and environmental movements and often pits them against government and business. The value of the coast is summarized in Table 10.1.

Table 10.1 Value of the Coast

Developing Countries	Developed Countries
Tourism	Tourism
Production of Foodstuffs	Economic Machine
Basic Industry	Quality of Life (Environment)

Historically, coastal management has been synonymous with coastal engineering. Managing the coast (essentially to maximize its economic value) involved design and construction related to personal safety, military defense and transportation

Two early coastal management/engineering examples are:
- The Netherlands: Life of a very dense population depends on maintenance of the coast as a barrier against intrusion by the sea, and therefore coastal management (essentially coastal engineering) has existed there for a long time.
- The United States: For historic reasons, military defence of its shores, was considered of paramount importance and hence the United States Army Corps of Engineers has been active for a long time managing its coastal resources.

More recent priorities are related to quality of life. These are sustainable development, environment, water quality and recreation. These new concerns are clearly part of modern coastal management.

10.2 The Coast under Pressure

Most coastal areas around the world are under pressure from economic causes, summarized in Table 10.2. Initially people who lived near the coast formed an integral part of a coastal fabric. They were fishers, sailors, dockhands, etc. and they lived in a tenuous *balance* with the coastal resources.

Recent large migrations of "outsiders" to the coast, however, have resulted in *stress and overload*. People turn to the coast for recreation and quality of life. Destination resorts, for example, along the Mediterranean Coast, South Florida and the Caribbean Islands have seen large increases in tourism and recreation. New resorts are opened up every year. But other coastal areas, such as the West Coast of North America have seen large influxes of new settlers, because people find the lifestyle associated with the coast to be important.

One constraint on the coast is that it is essentially linear; it is a narrow strip of land along the coast. As new areas become popular and open up for development, the additional coastal zone is measured in kilometers while the newly developed land resources are measured in square kilometers. This invariably causes high pressure on land prices and recreational facilities along the shore. The coastal zone is essentially a scarce commodity. Finally, if the coastal zone were robust, there would not be so many problems. However, the coastal zone is fragile and there is a world-wide tendency for coasts to erode, as shown, for example in Ch. 11 and Bird (1984, 1985). This puts high priority on protecting and maintaining what little is there, particularly because real estate values along the coast are so high.

Table 10.2 Pressures on the Coast

Population Density
- Historically, population densities were high along the coasts.
- 50% of the population of the United States lives near the coast.
- 80% of the population of Australia lives near the coast.
- >80% of the population of Canada lives near its oceans or the Great Lakes.
- Most of the world's major cities are near the coast.
- These numbers are increasing

Recent Migration
- Younger and more affluent people value the lifestyle afforded by coastal areas.
- Redevelopment and high real estate values result in high-density development.
- Many people can now afford to live near the coast in spite of high real estate values.

Tourism
- People take vacations in far away coastal areas.
- There has been a large increase in air traffic and package vacations to destination resorts.
- A wave of (younger) retirees seeks to live near the coast in warmer climates.

Linear
- The coast is always a narrow, linear strip of land.
- If a new "coastal area" is developed, the focus is always on the coastal strip.

Erosion
- Most of the world's coasts are eroding.

10.3 Conforming Use

Traditionally, the coastal zone has had many uses, which compete for limited space and may or may not conflict with each other. Some of the more important uses are listed in Table 10.3 and coastal management may be defined as the management of these uses of the coastal zone. Since there is much demand for the limited space along the coast, the first step in coastal management would appear to be to define conforming use. To be classified as a conforming use, it must be *necessary* for a project to be situated along the coast. Examples are swimming beaches, fishing ports and marinas. If a project does not need to be specifically along the coast, such as for example casinos, theaters and car parking areas, it is not a conforming use and does not automatically belong on the coast. As a next step, building permits would only be granted to projects that have a conforming use. This is not simple. For instance, a harbor or marina is normally required by law to supply sufficient infrastructure to support its operation. Therefore a car park suddenly becomes a conforming use, if it is needed in support of a conforming use. Similarly, money generators, such as casinos are usually permitted to locate along the coast by the local authorities, regardless of conforming use.

Table 10.3: Typical Uses of the Coastal Zone

– Tourism
– Residential
– Recreational
– Industrial and Commercial
– Agricultural
– Transportation
– Waste Disposal
– Aquaculture
– Fishing
– Nature Reserves
– Military and Strategic

In the past, many factories, railway lines, highways, commercial harbors and military bases were situated directly along the coast. Coastal defense was a matter of national priority. Manufacturing was the most important generator of wealth and commerce was highly dependent on transportation by water. Factories also needed to discharge effluents and for that reason needed to be close to the "receiving water". Historically, before mechanized transport, most goods were moved by ship, then transported by horse-drawn carts or carried on foot. The connecting roads needed to be as short as possible and therefore ran close to the shore, where the ports and industrial areas were. Later, highways, railways, telegraph, etc., followed the same transportation corridors. Although there was little, if any, serious planning involved, all the above were high priority, conforming uses of the coastal zone and shaped the infrastructure of the coastal areas of many countries over the past 300 years.

How times have changed! Except for bulk goods, transportation by water and railroad has essentially been replaced by air and road transport. Military bases and factories have closed, rail lines and docks have been abandoned. With a few exceptions, many of the original high priority, conforming, economically justified uses are no longer high priority or economically justified. Agriculture has also become a non-conforming use of the coast. What about the roads? Transport by automobile and truck is a most important mode of transportation. The initial cartage paths along the shores have grown, first into highways earlier in the previous century and later into expressways. But do these transportation corridors need to be near the water? Obviously not and therefore with the exception of local access, highways are also now a non-conforming use of the coastal zone. Table 10.4 shows (for developed countries) how priorities of the uses listed in Table 10.3 are changing.

Table 10.4: Changes in Priorities as Conforming Use

Higher Priority	Lower Priority	Changed Priority
Tourism	Industrial and Commercial	Fishing
Residential	Agriculture	Waste Disposal
Recreational	Transportation	
Nature Reserves	Military and Strategic	
Aquaculture		

If legislation permits private residential development, it is also a conforming use. Clearly, there is a major conflict between exclusive use of shore sections by individuals and the common enjoyment of the shore by all. As a result, private ownership is subjected to limitations. Aquaculture has increased in priority simply because it is new. It needs to be carefully integrated with other coastal activities, since it conflicts in a major way with many other uses. Fishing, close to the coast, has changed from commercial fishing to sport fishing in many developed countries and waste disposal is changing from dumping anything at any time to very restricted ocean disposal and to disposal of treated rather than raw sewage and industrial effluents.

Slowly, abandoned older facilities are taken over by the new interests and the newly non-conforming facilities such as factories, ports and rail lines are slowly converted to new, conforming uses. Most major port areas are now being re-developed from busy, dirty, dangerous work areas with limited access into people places with housing, parks and walkways. Loading terminals are being converted into apartment buildings as in the Old Ports of London and Rotterdam and in the Port of New York.

Abandoned rail lines become hiking trails and small commercial ports become marinas, in many cases surrounded by housing and condominiums. This poses unique opportunities, but also unique responsibilities. How are we going to make this wholesale change? Will the next generation commend our foresight, or will it condemn us for being shortsighted and only interested in short-term profits?

Tourism was identified in Table 10.1 a very important economic force for development of the coastal zone. The term as used here includes not only foreign visits, which generate wealth, or long-distance visits, but also the many visits originating in the local vicinity. There is a negative impact from tourism in locations where there is little concern for the environment. Developers are permitted to be very careless with the coastal resources and tourism (or the economic gain from tourism) drives some insane misuses and over-uses of the coastal zone. On the other hand, in environmentally conscious regions, tourism can be a positive influence. Proper coastal management is seen there as enhancing the intrinsic value of the area for tourism.

Table 10.5: Changes in Tourist Requirements

Standard Tourist	Renaissance Tourist
– Airports	– Dunes
– Highways	– Wetlands
– Hotels	– Clean Water
– Fast Food	– Birds
– Alcohol	– Fish
– Parking Lots	– Bicycle Paths

The definition of intrinsic tourist value of the coast is also changing, as summarized in Table 10.5. There are areas that are mainly destination resorts (where people fly in, sit on the beach, soak up the sun and the alcohol and then leave a week later to make room for the next group). But tourists are also becoming more interested in nature and physical activities such as hiking, biking, birding, boating and fishing. The coastal environment preferred by these *renaissance tourists* is much more natural (dunes, beaches, wetlands, clean water, abundance of animals) than the traditional tourist environment (airports, highways, hotels, fast food restaurants and parking lots, all close to the water's edge). Nevertheless, under both scenarios, tourism development is clearly driven by economics. Income is normally considered first and environment second.

From the above discussion it should become clear that ideals such as enhanced environment, ecosystem planning, etc. are laudable, but they must be presented within an economic framework. Co-operation with the economic mindset, not confrontation, will be the only way to place environmental considerations on the agenda of governments and business.

10.4 Conflict and Compatibility

One basic management tool is the compatibility matrix. Examples may be found in Carter (1988). A compatibility matrix for the conforming-use categories in Table 10.4 is given in Table 10.6. Compatibility is measured there on a scale of -2 (bad) to +2 (good). Each of these categories has within it, its own set of conflicts. For recreation, the areas of conflict are shown in Table 10.7.

Table 10.6: Compatibility Matrix

		a	b	c	d	e	f	g	h	i
a	Residential	x								
b	Recreational	-1	x							
c	Nature Reserves	-1	2	x						
d	Aquaculture	-1	-2	-1	x					
e	Fishing	1	1	-2	0	x				
f	Waste Disposal	-2	-2	-2	-2	-2	x			
g	Industrial and Commercial	-2	-2	-2	0	0	2	x		
h	Agriculture	-2	-2	-2	1	0	1	-1	x	
i	Transportation	-1	-1	-2	0	0	0	2	1	x
j	Military and Strategic	-2	-2	-2	0	-1	0	1	-1	1

If coastal management is the management of the uses of the coast, it must be primarily the management of conflicts. That requires introduction of legislation and enforcement of the proper use of this precious, narrow strip of land along our shorelines, all the time keeping in mind the economic framework that normally outweighs ideals. Coastal management pre-supposes technical skills to be able to make informed decisions. These skills must be based on geological, biological, legal, engineering and other training. They also involve political savvy and skills in communicating with everyone from government officials to children (or adults) building sand castles. Although coastal management is inter-disciplinary, it is the engineers who are asked to make the crucial technical decisions. For this, engineers

234	*	*Introduction to Coastal Engineering and Management*

need to be properly informed and need to establish necessary and appropriate networks with the other disciplines.

Table 10.7: Partial Compatibility Matrix for Recreational Activities

		a	b	c	d	e	f	g	h	i	j	k
a	Swimming	x										
b	Sailing	-1	x									
c	Motor boating	-2	-1	x								
d	Jet-Skiing	-2	-1	2	x							
e	Windsurfing	-1	-1	-2	-1	x						
f	Fishing	-1	0	2	-2	0	x					
g	Diving	2	-1	-2	-2	-1	-2	x				
h	Snorkelling	2	-1	-2	-2	-1	-2	2	x			
i	Kayaking	2	1	-1	-1	2	0	0	0	x		
j	Bird Watching	0	0	-2	-2	0	0	0	0	2	x	
k	Camping	2	2	2	2	2	2	2	2	2	2	x
l	Rock Concerts	0	-2	0	0	-2	-2	0	0	-2	-2	-2

10.5 Management Strategies

Townend (1994) presents some management principles and management issues. These are summarized in Tables 10.8 and 10.9.

Table 10.8: Management Principles
(after Townend, 1994)

> – The coast is *dynamic* and policies must reflect this.
> – Management boundaries should reflect *natural* processes.
> – Conflict cannot always be resolved, requiring planning and legislation.
> – Conflicts change with time, requiring a flexible management framework.

Table 10.9: Management Issues
(after Townend, 1994)

Frameworks (Conceptual and Computational)
–   Geographic Information System

Tools
–   Zoning
–   Regulations and Enforcement
–   Public Awareness and Consultation

Responsiveness
–   Legal Considerations
–   Economic Considerations
–   Social Considerations
–   Other Scientific and Technical Disciplines
–   Many Jurisdictions involved

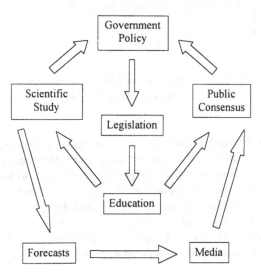

Figure 10.1 Decision Making Process
(after Townend, 1994)

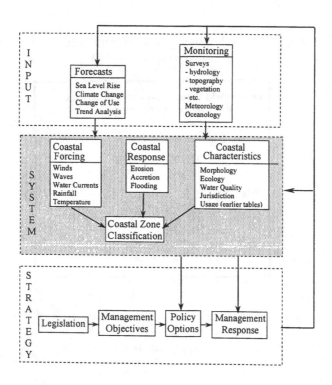

Figure 10.2 Responsive Management Framework
(after Townend, 1994)

Townend and others recommend the use of Geographic Information Systems (GIS) as the conceptual/computational framework. It forms the management database. On this geographic database are stored all pertinent data such as locations of buildings and infrastructure, coastal protection structures, sewerage outfalls, property ownerships, legal jurisdictions and physical conditions such as flood and erosion hazards, sediment sources and sinks, etc. The tools of management are: Zoning, Regulation Enforcement, Public Awareness and Consultation. These tools should be carefully selected and sharpened, showing sensitivity to the projects and the physical environments involved. They need to be incorporated into an appropriate decision making process and a responsive management framework such as shown in Figs. 10.1 and 10.2. Legal, economic and social considerations and the involvement of many disciplines require a responsiveness to and cooperation with others who may not think the way we do.

10.6 Coastal Management in Spite of the Odds

Jurisdiction over the coast varies from country to country, within countries and even within regions. Our detailed discussions will refer to Canada in general, and to the Great Lakes and the province of Ontario in particular. Since Ontario and the Great Lakes area have an advanced coastal management strategy, this will be a good example of what may be expected.

In many countries, the jurisdiction over the coast is badly fragmented between and within several levels of government. When considering any project along the Canadian shore of Lake Ontario, for example, it is necessary to take into account at least the agencies in Table 10.10 (Waterfront Regeneration Trust, 1995).

Table 10.10: Agencies Involved in a Project on Lake Ontario

> − International Joint Commission
> − Great Lakes Water Quality Board
> − International St. Lawrence River Board of Control
> − Canadian Federal Government Departments, such as:
>   − Fisheries and Oceans
>   − Energy
>   − Transport
>   − Heritage
> − Provincial Ministries, such as:
>   − Municipal Affairs
>   − Natural Resources
>   − Energy and Environment
>   − Culture, Tourism and Recreation
>   − Conservation Authorities
> − Municipal Government(s)

Donnelly (1991) presents a partial list of items of legislation by various governments, relevant to the Canadian shores of the Great Lakes. This is reproduced in Table 10.11. Although the 1978 meeting of the Canadian Council of Resource and Energy Ministers (CCREM, 1978) decried such fragmentation, very little has been done about it. CC-SEA (1991) describes some of the above environmental regulations. Carter (1988) describes the jurisdictional situations in several countries (United States, United Kingdom, Netherlands, France, Australia, Canada and Israel). This makes interesting reading. An over-riding concern everywhere is the fragmentation of jurisdictions by political boundaries (municipalities, states, provinces and countries) and real or imagined physical boundaries. For example, in many cases one agency is in charge of areas above high water, while another agency has jurisdiction below high water.

Table 10.11: Legislation Pertaining to a Project on Lake Ontario

− Public Lands Act	− Planning Act
− Conservation Authorities Act	− Ontario Water Resources Act
− Aggregate Resources Act	− Canada Petroleum Resources Act
− Lakes and Rivers Improvement Act	− Endangered Species Act
− Environmental Assessment Act	− Road Access Act
− Fisheries Act	− Surveys Act
− Navigable Waters Protection Act	− Shoreline Protection Act
− Environmental Protection Act	− Land Titles Act
− Municipal Act	− Boundary Waters Treaty Act

Integrated Coastal Zone Management (ICZM) is the strategy that is used to deal with the many of disciplines involved with the coast as well as the various of laws, regulations and jurisdictions. Here, the concerns (such as physical, environmental and biological) are considered together. Regulation is administered by agencies to which decision making power has been delegated over many different statutes governing the coast. Thus ICZM intends to integrate disciplines as well as jurisdictions. One step of ICZM is the lead agency, an agency that is approached first and where the integration of jurisdictions resides. A further step of integration provides one window for all information and regulation. Ontario has come a long way in such integration of coastal jurisdictions.

10.7 Management of Coastal Lands

The management of coastal lands is difficult because of the pressures expressed in Table 10.2. Since there is a world-wide tendency for coasts to erode, protecting and maintaining them has a high priority, particularly because land values along the coast are so high. The problem of protection and maintenance of the coast is often under-estimated, however. In simple terms, it is never economically feasible to protect shorelines, unless there is very dense development or extensive tourism. Areas such as Miami Beach in the United States, Gold Coast in Australia, Scheveningen in the Netherlands, Copacabana in Brazil and the Chicago and Toronto waterfronts on the Great Lakes are prime candidates for coastal protection. On the other hand, agricultural areas, cottage country and areas of single-family residential properties will never be. Thus decisions about what to protect, beyond the obvious, are painful and rising sea levels (Ch. 6) make the decisions even more difficult.

An interesting picture of the economics for Miami Beach is provided by Houston (1995, 1996). Table 10.12 shows first of all that Miami Beach attracts many more tourist visits than any other major tourist attraction in the United States. This is the same for the beaches of most countries. Hundreds of people visit museums and national or historic sites, but thousands go to the beach. Secondly, Houston shows that for this internationally popular beach, the money spent on protection - artificial beach nourishment in this case, is US $ 3 Million/yr, while the wealth generated by the Miami Beach economy (from foreign visits) is US $ 2 Billion/yr. Not only is it economically justified to protect this beach, it is extraordinarily good economics.

Table 10.12: Annual Tourist Visits to Popular U. S. Sites

Miami Beach	21 Million
Grand Canyon	4 Million
Yosemite Park	3.3 Million
Yellowstone Park	2.6 Million

Louisse and Kuik (1990) present some insight into the future coastal management strategy of the Netherlands. It is of some interest to note that the Netherlands, a densely populated country in which the very life of its citizens depends on coastal protection, has come to realize that it may have difficulty maintaining its shoreline as it would like. Fig. 10.3 indicates the absolute necessity of protecting the Dutch shoreline. It shows the length of shoreline threatened with flooding (translate as people drowned and billions of dollars in damage) and with destruction of drinking water reservoirs, if no further protection is provided. The total length of Dutch shoreline is 350 km and the three rates of sea level rise considered are 2, 6 and 8.5 mm/yr. The same study presents four strategy alternatives: withdraw (except from areas where further erosion would result in loss of life), selective erosion control, full erosion control, and expansion in the seaward direction by artificial beach nourishment where the coastal defense is considered to be weak. Full erosion control and perhaps expansion in threatened areas would appear to be almost a necessity for this densely populated country that could flood by up to 2/3 its area. Yet, the political choice made in the Netherlands in 1990 was only to exercise full erosion control. They will not to expand and perhaps may withdraw from some areas. This policy is reviewed every 5 years. At present, full control involves placing 6 to 10 million m^3 of nourishment sand along the coast annually.

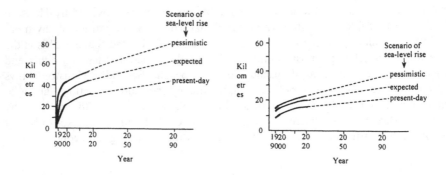

Figure 10.3 Length of Dutch Coast in Danger of :
a)   Flooding b) Loss of Fresh Water Resources
(after Louisse and Kuik, 1990)

With respect to the efficacy of shore protection, McKeen (1995) states:

"Historically, human responses to natural hazards have primarily involved the construction of various forms of protection or remedial works. In the shoreline areas, for example, these works are often installed in an *ad hoc* fashion or largely ignore natural processes and environmental impacts. In a significant number of cases, rather than protecting against flood or erosion damages, the failure or improper selection, design or installation of protection works have often created new hazards resulting in marked increases in property damages, losses of land, social disruption and environmental damage."

Similar statements may be found elsewhere in the literature. They are made against backgrounds of years of observation. They are indictments to engineers who designed the structures and to the property owners who commissioned them. All erosion control is driven by economics (protection of property, water supplies, etc.). It will change the environment (by definition). Clear guidelines need to be developed to evaluate the net overall benefits of erosion control. Erosion control structures should not only produce economic gain, but also a net positive impact on the environment. Both these aspects are difficult to define adequately and unequivocally. The management and protection of the Great Lakes - St Lawrence shoreline will be introduced in Section 10.9 as an example of development of such guidelines and of management of the land-water interface. Coastal protection, in general, is discussed in Ch. 15.

10.8 Management of Coastal Waters

10.8.1 Groundwater

Fresh water is a precious resource in a maritime coastal region. It has two sources: the fresh water in rivers and lakes, and precipitation. Both of these sources feed fresh water into the groundwater reservoir. The sea feeds salt water into the reservoir. Since the flow rates of groundwater are very small, there is little mixing of the salt and fresh groundwater and the lighter fresh water overlies the heavier salt water as shown in Fig. 10.4. The groundwater level is higher than the surrounding sea level and the density difference results in a floating fresh water lens that extends down to about 40 times the difference in elevation between the groundwater table and the sea. Clearly, any lowering of the groundwater table has a multiplier effect (40 times) on the volume of fresh water in the lens and hence any fresh water aquifer needs very careful management. The fresh water reservoirs of small island communities are very susceptible to damage. Some of the common disturbances that occur partly as a result of phenomenal increases in coastal populations are given in Table 10.13.

One serious consequence of pumping a coastal fresh water aquifer for water supply is that the depletion of the aquifer causes subsidence of the land mass, which in turn, results in increased flooding. Similar subsidence is caused by extraction of oil and natural gas. An example of such subsidence is the (now) regular flooding of Venice by the Aqua Alta discussed in Ch. 6.

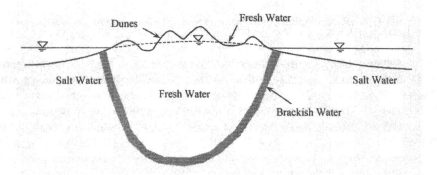

Figure 10.4 Fresh Water Coastal Aquifer

Table 10.13: Common Disturbances of the Fresh Water Aquifer

– pumping the aquifer for fresh water supply
– lowering of the land mass by cutting away the dunes – lowers the water table
– construction of buildings, roads and parking lots – prevents the recharge of aquifer
– dredging rivers and creeks for navigational improvements - result in salt water intrusion (salt water comes further upstream and hence further into the coastal fresh water aquifer).

10.8.2 Waste Water

Coastal waters have traditionally been used for waste water disposal. Sewage and chemical effluents have polluted these waters, particularly since World War II. A model of what was to come with pollution of the oceans may be found in the history of pollution of the Great Lakes. These huge water bodies were thought to be unlimited receptacles for wastewater and, being smaller than the oceans, they began to show serious overload and eutrophication in the 1960s. Of course, sometime later the same process fouled many seas and ocean basins.

The story did not end for the Great Lakes in the 1960s. The 1970s saw a concerted effort to clean up the lakes. The many different agencies (from two countries - Canada and the United States) quarreled and negotiated. Some cleanup was effected, but there appears to be little collective will or funding to do it properly.

Today, the Great Lakes continue to be heavily contaminated with toxic chemicals. Even though many of the offending chemical dump sites have been cleaned up, the toxic chemical content of the fish, for example, is still virtually the same. Cleanup of the oceans will be much more complex since it involves many rich and poor nations and the prevailing philosophy is still very much "out of sight out of mind", when it comes to pollution of some maritime coastal waters.

Although sewage on the Great Lakes is now treated before discharge into the lakes, many maritime locations still discharge raw sewage, often into the nearshore zone. An area is indeed fortunate, if the sewage outfall consists of a many kilometers long pipe, with appropriate diffusers, as at Sydney, Australia. Many outfalls are just open channels to discharge storm water and short pipes to discharge sewage relatively close to shore. The toxic chemicals and nutrients are trapped by currents and wave action, causing high levels of pollution near the shore. Why would people who are otherwise reasonable, discharge raw sewage into the sea? That is again a matter of economics. Tertiary treatment costs 10 times as much as dumping raw sewage; incineration costs 15 times as much.

Figure 10.5 shows the situation around the British Isles in 1981. One international incident involves the dumping of raw sewage by the city of Victoria, Canada. The neighboring smaller towns on the United States side of the border complain, because they pay for secondary and in some cases tertiary sewage treatment to protect the environmentally sensitive Puget Sound, while the largest city in the area continues to pollute.

10.8.3 Other Forms of Pollution

Another major source of pollution of coastal waters is the dumping of solid waste. Although ocean dumping legislation severely limits the dumping of solid waste, the oceans have been the recipient of all sorts of waste, such as hospital waste, contaminated dredge spoil and nuclear waste.

Runoff from farming is also harmful to the coastal waters particularly for bays and lakes. The fertilizers that promote the growth of agricultural products, also encourage algae, weeds, etc. Pesticides introduce high levels of toxic substances, such as heavy metals. Even without the chemicals, the runoff from farming can be undesirable. In the Netherlands, for example, the density of the cattle population is so high, that their manure causes high levels of pollution to both the groundwater and the surface waters. The runoff of fine sediment materials from soil erosion resulting from converting forests to agricultural land has also caused problems for many marine organisms. In fact, the dying of the coral reefs in many tropical countries can be attributed, at least in part to the sediment that has entered the water column since the land was cleared for agriculture, as early as in the 18th century in some cases.

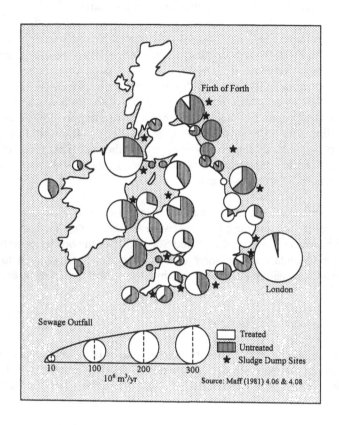

Figure 10.5 Pollution around British Isles
(after Carter, 1988)

Oil spills, resulting from transportation and exploration of oil close to shore have been the cause of well-known disasters such as the spill from the Exxon Valdez in 1989. Areas along major shipping routes are particularly vulnerable, as witnessed by the spills of the Torrey Canyon in 1967, the Amoco Cadiz in 1978 and the Erika in 1999, all in the same geographic area – the northwestern coast of France. Although such major disasters cause public anger and some government action, much larger volumes of oil are released, virtually unheralded, into the oceans each year by leakage from ships, production platforms and refineries.

10.9 Example: Management of the Great Lakes - St Lawrence Shoreline

Ontario has an advanced shore zone management policy that will be used as an example. Details may be found in McKeen (1995) and Sullivan and Davidson-Arnott (1995). The management of the shoreline is outlined in the "Natural Hazard Policies: Great Lakes - St. Lawrence River System" (McKeen, 1995). It hinges on a so-called Regulatory Shoreline, which is explained in Table 10.14.

Table 10.14: Regulatory Shoreline in Ontario

Regulatory Shoreline is the furthest landward limit of:

– *Regulatory Flood Standard*, (Fig. 10.6). The sum of:
  – 100 year flood level
  – flood allowance, which consists of the sum of:
    – wave uprush
    – allowance for other water-related flood hazards
– *Regulatory Erosion Standard for Eroding Banks or Bluffs*. (Fig. 10.7). The sum of:
  – stable slope allowance – three times the height for the bank or the bluff
  – 100 times the average annual recession rate, or 30 m if the recession rate is not known
  The minimum setback from the crest of the bank or bluff must be 30 m (Fig. 10.8).
– *Regulatory Dynamic Beach Standard for Eroding Beaches or Dunes* (Fig. 10.9). The sum of:
  – regulatory flood standard
  – 100 times the average annual recession rate
  – allowance for the dynamic fluctuations of the beach

Some notes with Table 10.14 are:
–   the 100 year flood level is the peak instantaneous water level, combining still water level and wind setup, which is equaled or exceeded with a 1 % probability (Sullivan and Davidson-Arnott, 1995).
–   the standard allowance for other water-related flood hazards is nominally:
  –   15 m for the Great Lakes
  –   5 m for the connecting channels

–   When wave uprush overtops the banks, the flood hazard allowance must include the distance over which the water ponds behind the shore (Fig. 10.10).

–   The allowance for the dynamic fluctuations of the beach is nominally 30 m.

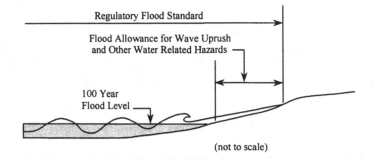

Figure 10.6 Regulatory Flood Standard

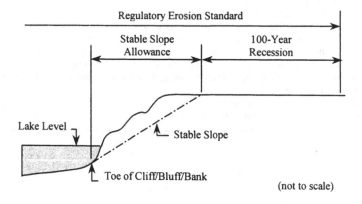

Figure 10.7 Regulatory Erosion Standard for Bluffs

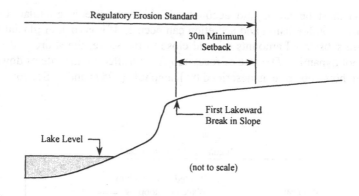

Figure 10.8 Minimum Setback from Bluff Crest

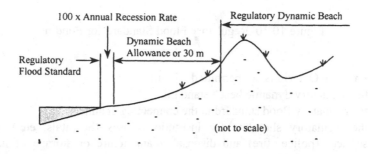

Figure 10.9 Regulatory Dynamic Beach Standard

The regulatory shoreline developed above is conservative and all the nominal values can be re-defined for a site using "accepted engineering principles". This essentially forces every project to be properly designed, since the regulatory shoreline as defined in Table 10.14 is not a very attractive option.

To distinguish between eroding bluff that commonly occur along the Great Lakes and dynamic beach shorelines, the nearshore substrate, (which controls the erosion process) must be taken into account. If this substrate is granular, the shore is "dynamic" (both erosion and accretion can occur). If it is rock or glacial till, even if there are substantial amounts of sand close to the shore, the shore will only recede and is not dynamic. The rate of recession is controlled by the rate of downcutting of the nearshore substrate as described by Kamphuis (1987) and in Section 11.6.

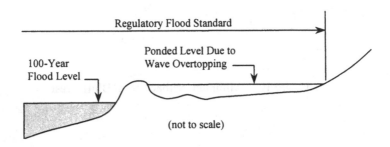

Figure 10.10 Regulatory Flood Standard for Ponding

Development in Ontario is not permitted within:
–   the regulatory dynamic beach standard;
–   the regulatory flood standard in the connecting channels;
–   the regulatory shoreline for institutional uses (hospitals, etc.), essential services (police, fire) and disposal, manufacture or storage of hazardous materials and sewage.

Development is defined (McKeen, 1995) as:

> "The construction, erection or placing of a building or structure of any kind; or the making of an addition to a building or structure that has the effect of increasing the size or usability thereof; and includes such related activities as site grading and the placing of fill."

Development must meet the standards, described in Table 10.15 and shoreline management practices are summarized in Table 10.16. Prevention is generally cost effective and results in the least damage to the environment. Since protection structures only alleviate erosion, the protection credited to the structure must be related to its useful life, as shown in Fig. 10.11. The 30 m nominal or the residual flood/erosion allowance is always required for access.

The impacts on the physical environment, caused by various types of shore protection practices have been summarized by Kolberg (1995) from which Table 10.17 was adapted.

Table 10.15: Ontario Standards for Development

– flooding and erosion hazards must be safely addressed
– new hazards must not be created
– no adverse environmental effects must result
– access must exist for vehicles and people, particularly during times of flooding and erosion

Table 10.16: Shoreline Management Practice

prevention
– land use planning
– regulation of development

protection
– *non-structural*:
   – relocation
   – bluff drainage
   – dune enhancement
These methods have a minimum impact on the environment.

– *structural* (described in Ch 15):
   – filling and diking
   – revetments and seawalls
   – beach nourishment
   – groins
   – artificial headlands
   – detached breakwaters

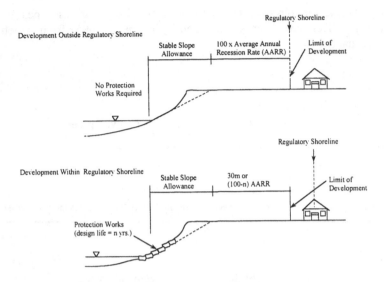

Figure 10.11 Protection Credited

Table 10.17: Impact of Structural Shore Management Practices

Management Procedure								Influence	Impact
Nourishment	Fill	Dike	Revetment	Seawall	Groins	Headlands	Offshore Breakwaters		
			x	x	x	x	x	Decreases sediment supply	Increases long-term erosion
x								Increases sediment supply	Decreases long-term erosion
x					x	x	x	Impedes littoral transport	Accretion/erosion
					x	x	x	Diverts sediment to deep water	Increases erosion downdrift
						x	x	Impedes cross-shore transport	Decreases beach movement
			x	x	x	x	x	Increases reflection	Localized scour
			x	x	x	x	x	Concentrates wave action	Localized scour
						x	x	Decreases wave exposure	Decreases circulation
x			x	x	x			Occupies backshore	Alters backshore topography
x					x	x	x	Occupies nearshore	Alters nearshore topography
	x	X						Occupies onshore	Alters onshore topography
x	x	X	x	x	x	x	x	Alters onshore terrain	Alters onshore topography
	x	X	x	x				Disrupts drainage	Alters water flow

10.10   Example: Management of Coastal Ecosystems

The environmental regulations vary greatly between the various jurisdictions. The regulations applying to the Great Lakes - St. Lawrence system will be given here as an example (Harker and Mortson, 1995).
Environmentally sound design requires an understanding of:
- – environmental sensitivities,
- – biological impacts,
- – environmental policies.

Environmental sensitivities are concerned with habitat, which is defined as:

"The combination of living and non-living things that provide a particular species with the resources it needs to complete its life cycle."

Diversity of habitat is essential to accommodate many species. The wildlife strategy for Ontario (1991) includes birds, reptiles, amphibians, fish, invertebrates, plants, algae, bacteria and other wild organisms. Wildlife is not only large animals that are hunted or trapped. The shore ecosystem includes upland and terrestrial wildlife and habitat, wetlands, and aquatic wildlife and habitat (which includes spawning grounds, nursery, rearing, food supply and migration areas).

Biological impacts are summarized in Table 10.18. Their impact can be assessed according to Table 10.19. Table 10.20 describes the many impacts resulting from various influences on the physical processes.

Although this summary of environmental policy is brief, we should note that Ontario policy ascribes *equal* weighting to physical and environmental considerations. Structures will simply not be built if there is a net negative environmental impact.

Table 10.18: Description of Biological Impact

Importance or Significance:
– habitat
– species
– diversity

Spatial Extent:
– immediate site
– off the site

Duration:
– only during construction
– during operation and/or post-design life of the structures

Recovery:
– how long before the impact to disappears
– is the impact irreversible?

Mitigation:
– Can the impact be mitigated?
– Is compensation for the displaced habitat possible?

Cumulative Impact:
– Small additional stress may have devastating effect on a
  shore that is already highly stressed.

Table 10.19: Assessment of Impacts

Minor Impacts:
– can be mitigated
– are of short duration
– have a high rate of recovery
– and/or are local

Major Impacts:
– impact important (significant) habitat or species
– have long-term or permanent impacts
– have a low rate of recovery
– and/or affect critical habitat or species

Table 10.20: Biological Impact and Mitigation

Influence	Biological Impact	Mitigation
Decrease in sediment supply	May change substrate, e.g. granular to cohesive Impacts fish spawning areas Impacts plant growth Impacts turbidity levels	Avoid disruption
Increase in sediment supply	May cover cobble, gravel (preferred spawning substrate) with sand	Minimize intrusion of sand in the nearshore Re-establish aquatic plants elsewhere
Local erosion	May deposit fines on existing substrates	Avoid
Alteration of backshore	Clearing vegetation may disturb habitat	Preserve vegetation where possible Replant after construction Avoid nesting periods Do not remove trees
	Increased runoff may increase suspended sediment in the water course	Plant disturbed areas, Maintain a strip of vegetation along shore Construct silt curtains and traps Stockpile materials away from shore
	Wildlife access to water/land may be restricted	Provide access areas Avoid times when access is important (e.g., breeding)
Alteration of nearshore	May increase of suspended sediment during construction	Use silt curtains
	May change in topography may alter light reaching the bottom	Establish vegetation in areas with more light
	May cover spawning substrate or change habitat	Mitigate at adjacent site
	Removal of boulders, logs, etc., which form feeding areas and protection from predators for small fish	Add more natural surfaces to shore protection structures
	Removal of aquatic vegetation	Replace vegetation in the vicinity
	Burial of benthic colonies	Do nothing, they will re-colonize quickly
Alteration of onshore topography	Removal of vegetation	Avoid and re-vegetate
	Soil compaction and increased erosion	Stabilize
Alteration to surface or groundwater drainage	Increased sedimentation	Use drainage
	Disruption of spawning activity	Avoid critical times
	Drainage changed	Avoid critical times Avoid affecting wetlands

10.11   Concluding Remarks

Coastal management is at the same time very simple and very complex. It is based on simple, common sense principles, such as "live and let live". The complexity comes when such principles need to be carried out in a complex, high pressure environment that has many competing uses and which is governed by over-riding economic considerations. The coastal zone is subject to pressures from populations that are too large to sustain. This results in very high land values. Return on investment in the coastal zone (for example, from tourism) must match these land values. From a strict business sense, in comparison to these economic driving forces, what do a few fish or dune grasses matter? But they do matter!

In such a scarce, busy and valuable environment, it is small wonder that there are many competing uses for the coastal zone. Some of these such as safety from flooding, defense against enemies, factories, waste disposal and roads have historically been high priority uses. But priorities are rapidly changing. Factories, farms and transportation corridors have little business being in the high-pressure coastal zone and will make room for parks, beaches, nature reserves and private residences. In this rapidly changing environment, it is very important that things are done right. This is both a daunting task as well as a unique opportunity.

The management of this change, the refereeing between competing uses, the setting of priorities all fall into the task known as Coastal Management. This task involves concerted effort from the whole community (business, politicians, land owners and the public - both young and old). It involves education, particularly of the young people and in the schools. It concerns various disciplines (such as geology, biology, engineering), it requires a communal willingness to overcome a legislational - jurisdictional jungle and finally it needs to take place within a political, social and especially economic framework that was not developed to provide for good coastal management.

The coastal engineer is at the centre of this task, partly because coastal management has developed from coastal engineering historically, but more so because we have the appropriate background to synthesize many diverse ideas into coherent working systems. However, engineers have lost the confidence of the general public because they have made mistakes in the past (and still make them today). In order to carry out their task, engineers must know their trade well and must learn to deal effectively with all the elements of the community, with the other disciplines involved with the coast and with the relevant legislation. They must also be familiar with basics of coastal biology, geography (planning), geology, politics and economics.

Coastal management is difficult because of the tension between economy and environment. This tension has the motion of a *pendulum*. Initially interest in the coastal zone was strictly economic, as if someone pulled the pendulum in the economic direction. Of course this was undesirable and the pendulum was released to take into account environmental concerns. In many countries this pendulum has swung over to the environment. For example, in 1999 the Dutch government considered issuing permits to pump natural gas from under a complex of offshore islands and tidal flats called the Waddenzee. In preparation, many engineering studies were performed to determine the net impact on the fragile tidal environment, which supports a prolific wildlife. In particular, it was important to determine rates of lowering of the tidal flats and design mitigation measures. The environmental lobby's activity concentrated on the uncertainties in the engineering estimates. In effect, engineers were caricatured as not knowing enough to proceed and as natural risk takers. As a result, the Dutch government did not issue the permits. It is clear that with the pendulum in the position, as it was in the Netherlands and in some other developed countries in 1999, environmental movements can essentially stop any coastal project with the same argument. Engineers can only do so much with the available data and tools, and major uncertainties, that form a thread throughout this book, will continue to be with us.

Such an environmental checkmate, however, is no victory for the environmental side, because public reaction will cause the pendulum to swing in the other direction. For example, the enlightened coastal management policy for Ontario, discussed above was developed in 1995. Subsequently the public in reaction to earlier, more liberal governments elected a conservative government. As a result, the economy rose to the top of Ontario's political agenda and the environment (and effective coastal management) took a back seat. On another front, the oil-pollution-from-ships-disasters pendulum will one day swing in favor of the large numbers of animals that are killed, every time there is an oil spill. That was clear from the reaction to the breakup of the Erika off the coast of France in 1999.

Our task as coastal managers and engineers is to try to stop the swing of the pendulum by creating designs that reasonably and technically balance economy and environment. Considering the myriad of coastal problems and the diversity of rich and poor countries, this is a difficult task. We are (once again) building bridges - not concrete and steel bridges, but technological bridges linking business interests and political expediency with true sustainable development, dreamt about by nature lovers and social activists. We must learn to explain our knowledge clearly to governments, property owners and elementary school children alike. Finally we must learn from our mistakes and educate ourselves, and fellow engineers who can only think in terms of sheet pile, concrete and rock.

11. Coastal Sediment Transport

11.1 Introduction

Although there are many important aspects to coastal zone management, such as the environment, transportation, economics, biology, etc. the most important consideration and ultimate design criterion in a design for the coastal zone is often the movement of sediment. The remainder of this lecture will deal with the proper management of coastal sediment transport. The general characteristics of sediment transport will be discussed here; details and equations will be given in Ch. 12.

Sediment, moved by waves and wind, may be academically divided into cross-shore and alongshore sediment transport. These are discussed in Sections 11.3 and 11.4. Sediment movement can result in *erosion* or *accretion* (removal and addition of volumes of sand). Erosion normally results in shoreline *recession* (movement of the shoreline inland); accretion causes the shoreline to move out to sea.

There is a *fundamental* difference between granular shores (consisting of sand and gravel) and cohesive shores (soft rock, till and clay). Granular shores will be discussed in Sections 11.2 to 11.5; cohesive shores in Section 11.6.

11.2 Dynamic Beach Profile

The shape of a beach (its water depth as a function of distance offshore) is called the beach profile. It responds to the environmental conditions (waves and water levels)

257

imposed upon it and if such environmental conditions are constant, the beach profile should remain relatively constant (equilibrium profile). In some model tests at Queen's University, however, the beach profile still changed after 1500 hours (2 months) of continuous testing with a single wave and water level in a hydraulic model,. Hence, an equilibrium profile can only be approached.

There is normally an overriding (often annual) storm-calm cycle of beach profile change (Fig. 11.1). The high and steep storm waves that usually occur in autumn, winter and spring move material offshore from the upper beach. This erodes the nearshore area and forms one or more bars near where the waves break. These bars are generated by the breaking waves, and in turn they cause the waves to break. Smaller and not so steep waves, occurring during the calm periods (usually in summer) move beach material back onshore. The bars are reduced in height at that time and the visual beach above water is widened again, re-forming a berm on the upper beach. The changes both in the location of the sediment and the position of the shoreline due to these storm-calm cycles can be quite substantial, but the net changes over several years may be quite small. The beach is said to be in dynamic equilibrium if the mean beach profile does not move in the cross-shore direction. This cyclical dynamic profile mobility gives rise to the requirement for a Regulatory Dynamic Beach Standard in Ch. 10[1].

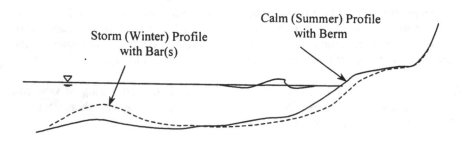

Figure 11.1 Annual Change in Beach Profile

1. These storm-calm cycles require that the fetch be substantial. For small fetches, the waves will be more or less the same fetch-limited, small (and steep) waves at all times.

If the combined cross-shore and alongshore movement of beach material in a system results in a net loss of beach material, the shore in the system is said to be eroding. The mean beach profile will retreat landward causing a net recession of the shoreline. Over the long term, most of the world's beaches are eroding, and actively accreting beaches only exist because of unique conditions. Figure 11.2 and 11.3 show such an unusual accreting beach and dune system on the Atlantic Coast of Denmark. Extensive dunes can be seen on Fig 11.2 (looking landward) and a wide beach is seen on Fig 11.3 (looking seaward from the same position). This is a special location where sand arrives from both the North and the South.

Bird (1993) identifies 20 causes of long-term beach erosion. The causes may be combined into four important groups:
- Decrease in sediment supply,
- Comminution,
- Submergence,
- Human interference.

The first three are natural causes of erosion that have taken place over geological time. Only in the last decades have shores become subject to human interference, but it is usually not the only or even the major cause of general beach erosion.

Figure 11.2 Accreting Dunes

Figure 11.3 Accreting Beach

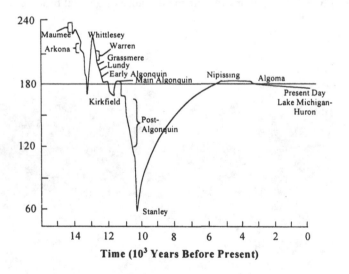

Figure 11.4 Recent Water Levels in the Great Lakes
(after Chapman and Putman, 1984)

Decrease in sediment supply may be explained as follows. Beaches are formed when very large volumes of granular material are moved about. That happens when rivers carry large amounts of sand to their deltas and when ice-age glaciers retreat, leaving large volumes of sand to be moved by large flows of melt water. Large fluctuations in water level will redistribute this sediment to form sandy shorelines. Particularly water level rise will form new beaches. Figure 11.4 shows the latest (Wisconsinan) ice age. It caused a large decrease in water levels in the Great Lakes area between 15,000 and 10,000 years ago (Chapman and Putnam, 1984). Over the following 5,000 years the water levels increased by 125 m. These large fluctuations in water levels with the accompanying flows resulted in very large fluctuations in sediment movement and formed the present day beaches. Figure 11.4 also shows that over the past 5,000 years, the water levels have been quite constant. Section 6.7.2 discusses similarly large water level fluctuations for sea levels that have also leveled out over the past 3000 years. As a result, the beach-forming conditions of the past no longer occur and existing beaches are essentially relict beaches from processes that took place at these earlier times. The beaches thus formed have lost their supply of material thousands of years ago and are now slowly starving of sand and eroding under the present environmental conditions.

Beach sediment is not the final size of material resulting from the hydro-geological processes that turn solid rock into silt and clay. Shingle, cobble and sand are only intermediate steps and the violent coastal climate will continue to decrease the size of the beach materials through a grinding process called comminution. As perfectly stable beach materials are reduced into finer materials, they will no longer be stable as beach building materials and waves will carry the fine sediment into deep water or along the shore to an area with less violent wave action. Wind may also carry it inland. The disappearance of these smaller grain sizes from the active coastal system results in erosion.

Net coastal submergence (resulting from a relative sinking of the land or a rise in water level) was discussed in Ch. 6. It causes an increase in water level at the coast, which results in greater wave action close to the shore. This is another cause of erosion, as shown in Ch. 12.

11.3 Cross-shore Transport

In the interest of simplicity, it is assumed in this section that sediment transport in the alongshore direction is negligible.

11.3.1 Dune-Beach Utopia

Section 11.2 showed that beach profiles respond to storm-calm cycles by shifting sand in the cross-shore direction, forming a dynamic equilibrium. But any beach profile will need additional material during times of high stress, such as during infrequent combinations of high wave action and storm surge resulting from exceptionally deep depressions, hurricanes, etc. Nature has provided for such emergencies by stockpiling large quantities of sand in dunes. The dunes are a long-term protection against coastal erosion, because they provide adequate elevation of the land contours to prevent flooding and form emergency reservoirs of sand.

During emergencies such as major storms and storm surges, sand on the upper beach is moved offshore. This allows the waves to come further into shore and they will attack the foot of the dunes causing them to become unstable and deposit large amounts sand close to the shore, compensating for the sand moved offshore by storm waves. When the emergency has passed, most or all of the sediment moved offshore will normally return onshore, forming expansive dry beaches. Winds will then blow the dry sand inland to replenish the dunes. Ideally, a dune-beach system can thus take care of emergency situations for millennia to come (or at least until there is no more sand in the dunes). In practice, the situation is complicated by alongshore transport, offshore bar formations, canyons, etc., which may prevent some of the sediment from moving back onshore after a storm. But it is not difficult to see that modern engineering design and coastal management should:
- not disturb existing dune-beach systems,
- encourage growth of dune-beach systems, and
- emulate dune beach systems wherever possible.

11.2.3 Dune-Beach Disturbance

Greed and ignorance have disturbed many dune-beach systems. Hotels have been built on the most seaward (and most vulnerable) dunes, or on sites that were first cleared and graded, thus destroying the dunes. The hotel guests can now see the sea from their room and can walk straight onto the beach without having to climb over the dunes first. Roads (normally called "boulevards") have been built on or seaward of the dunes so people can drive along the sea. Developers sell building lots and condominiums in dune areas so that people can live very near the water. Dunes are also regularly paved into parking lots. Such practices have invariably resulted in disaster and examples are given in the figures below. Figure 11.5 shows Surfers' Paradise, Australia, 1973, Fig. 11.6 shows Miami Beach (what beach?), 1969.

Figure 11.7 shows a beach road in California. There was a 4 m drop-off where the asphalt ends. Figure 11.8 shows a (now useless) boulevard near Perth (Australia). Figure 11.9 shows the slowly eroding natural beach and low dunes at Dalvay Beach on Prince Edward Island, Canada. Figure 11.10 shows the "protection" used at Dalvay Beach for a paved parking lot and small building. Needless to say, this "protection" disappeared in a few years. Fortunately, up-to-date coastal management philosophy and regulations, as presented in Ch. 10, do much to prevent this type of destruction, to understand the fragility of the coastal system and to repair existing problem areas.

Figure 11.5 Surfers' Paradise, 1973

Figure 11.6 Miami Beach, 1969

Figure 11.7 Beach Road in California

Figure 11.8 Boulevard in Australia

Figure 11.9 Dalvay Beach

Figure 11.10 Dalvay Beach "Protection"

11.3.3 Dune-Beach Encouragement

It is possible to encourage a dune-beach system. First, we ensure that indigenous vegetation covers the dunes. Lists of indigenous plant types as well as planting and maintenance instructions may be found, for example, in CERC (1984).   The vegetation cover decreases the wind velocity in the boundary layer above the sand, thus preventing the loose sand from being blown inland. Examples may be found in Figs. 11.11 and 11.12. Dune vegetation is fragile and traffic (walking, riding and motor vehicles) can easily destroy such a cover. For that reason, all-terrain vehicles, automobiles and trail bikes do not belong in the dunes. Lighter traffic such as pedestrians and horses can be channeled on specially reinforced paths. Any removal of the fragile vegetation cover leaves exposed sand. When this sand is blown away, a depression is formed and adjacent plants will be undermined, increasing the size of the damaged area and allowing further removal of sand to form a blowout (Fig. 11.13).   Special paths can be designed (Fig. 11.14) that are reinforced by larger material that is not removed by the wind, to prevent blowouts. Dune growth can also be encouraged by placing fences parallel to the shore in strategic locations, (CERC, 1984). The fences will slow down the wind velocity, so that blowing sand is deposited behind the fences.

Figure 11.11 Dune Vegetation

Figure 11.12 Dune Vegetation

Figure 11.13 Blowout

Figure 11.14 Reinforced Paths

11.3.4 Soft Protection

If the dune-beach system is really a utopia, then our protection designs would do well to emulate it.   Instead of building a seawall that reflects wave action and provides no emergency sand reservoir, we can protect the coasts by an artificially placed dune-beach system. Such soft protection has many advantages, but the most important is that it can be used as a recreational space. Blowing sand can be a problem because of a sudden abundance of dry sand, immediately after placement. Careful planning and immediate covering of the dune by vegetation are important to keep the sand in place and prevent sandblasting facilities landward of the nourishment.

Both Miami Beach and Surfers' Paradise have used this method to solve their immediate erosion problems. The late 1970s nourishment at Surfer's Paradise is shown in Fig. 11.15. This figure may be compared with Fig. 11.5 taken at the same location. Nourishment has been repeated several times at Surfer's Paradise, which demonstrates an inherent property of nourishment schemes. They normally need to

be repeated and should be considered beach maintenance, rather than beach re-
construction. In many instances, soft protection is placed in the form of a beach
only. This is obviously less desirable than a complete dune-beach system and will
need more regular maintenance.

Figure 11.15 Nourishment of Surfers' Paradise

To put the concept of artificial beach nourishment into perspective; most major
tourist destination beach resorts are regularly replenished. When the hotels replace
the dunes, the beautiful beaches quickly erode to the point that no beach is left. The
hotels and other structures are then endangered and are often protected by seawalls
and groins (Ch. 15). Fewer tourists will now come to sunbathe (on the seawalls).
The hotels need both protection and recreational beach and this can be provided by
artificial beach nourishment. The North Sea shore of the Netherlands has been
repaired in this way and future maintenance of the Dutch coast will be by artificial
nourishment. Nourishments can also be reinforced by structures and Ch. 15
discusses combinations of artificial nourishment with groins, artificial headlands and
offshore breakwaters.

11.4 Alongshore Transport

Section 11.3 assumed for ease of discussion that all the sediment transport takes place in the cross-shore direction. This is of course not true. When waves approach a shoreline at an angle, alongshore transport (also often called littoral transport) takes place. It is often the most important design consideration.

11.4.1 The Process

Waves approaching the shore at an angle will move sediment along the shore in the direction of wave propagation. There are two mechanisms (Fig. 11.16): beach drifting in the swash zone and transport in the breaking zone. Beach drifting is not difficult to understand conceptually. The wave action pushes sand up the beach in the wave direction. When the wave retreats, the water and sediment particles are accelerated by gravity and travel down the steepest incline, perpendicular to the beach. To express this process in volumes of sand transported, however, is difficult.

Transport in the breaking zone is also conceptually easy to understand. The turbulence in the breaking zone stirs the material into suspension and it is carried by an alongshore current, generated by the momentum of the breaking waves. The same turbulence and current also transport sand as bedload along the bottom. To quantify the suspension process and to determine universal expressions for the velocity of the alongshore currents and the amounts of material moved alongshore is once again difficult.

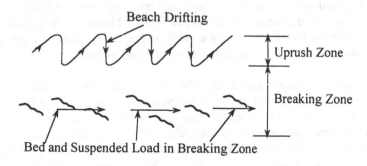

Figure 11.16 Alongshore Sediment Transport

Referring to Fig. 11.17, sediment transport due to the various incident waves from the left can then be added up to yield a sediment transport rate to the right (Q^+). Similarly we can define sediment transport rate to the left (Q^-). The sum of these two is called the gross sediment transport rate and the difference is the net sediment transport rate. This net rate has a direction and the terms updrift and downdrift are relative to the direction of the net sediment transport.

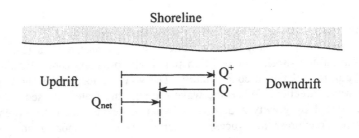

Figure 11.17   Sediment Transport Rate and Direction

11.4.2 *Measurement of Littoral Transport*

Alongshore sediment transport rate may be obtained by using tracers, by measuring differences in deposited volumes of sand, or by integration of suspended sediment measurements. Measurement of littoral transport rates using tracers is based on following quantities of native sand or similar material that have been labeled by fluorescent or radioactive materials. Sediment transport rate is difficult to derive from the movement of tracers, because tracers become buried or lost, so that in most studies the recovered tracers only form a small portion of the tracers actually placed. Tracer studies are also short-term tests, covering hours to weeks.

Deriving sediment transport rate from differences in measured volumes of sand is also not easy. To obtain good results, large, new structures should be built or large trapping areas dredged out especially for this purpose. Extensive bathymetric surveys need to be made at frequent intervals. Ideally, the survey intervals should be short enough to distinguish different seasonal changes and should also identify the effects of individual major storms. Since most alongshore transport measurements are a by-product of construction projects, the results are usually quite inaccurate. The structures can be bypassed by sand that moves seaward or over the top of the structure during major storms, and surveys are often not accurate or frequent enough to define volumes adequately.

It is also possible to measure suspended sediment concentrations quite accurately, using acoustic or optical equipment. Generalization of such point measurements into suspended sediment transport rate is possible in theory, but because of extreme variations of concentration in time and distance, there are many practical problems. It is still not possible to obtain good long-term sediment transport rates via this method.

The conclusion about measurement of sediment transport rate by these three methods must be that measurement is difficult and expensive and that all measurements contain large uncertainties.

11.4.3 Computation of Littoral Transport

Perhaps it is possible to calculate sediment transport rates using theoretical expressions, calibrated to long- and short-term experimental results published in the literature or obtained specifically for this purpose. Because parameters vary with time and distance, the calculation is normally not simple and may involve numerical modeling discussed in Ch. 13 and 14. Any calculated values of alongshore sediment transport rate necessarily contain larger uncertainties than the measured values on which they are based.

A *detailed* sediment transport calculation incorporates many carefully measured wave, current, beach and sediment parameters into a numerical model to determine actual detailed (in time and space) sediment transport rates. Several efforts have been made worldwide to collect the necessary field data to formulate such models properly. The ultimate success with a detailed computation depends on data. Field data are required, along with data from hydraulic models to provide controlled and repeatable data sets. One application of such a detailed sediment transport calculation may be found in Briand and Kamphuis (1993 and 1993a). For the present, our limited stock of good, sufficiently detailed calibration data and our inadequate understanding of the detailed sediment transport processes normally leads us to simpler, bulk volume computations discussed below.

The *bulk* sediment transport method relates total alongshore sediment transport rates to a few simple wave and beach parameters. The calibration data for this method are simpler to obtain but the answers are less sophisticated. Examples of bulk sediment transport expressions are the CERC formula (CERC, 1984) and the Kamphuis (1991) expression, discussed in Ch. 12.

11.5 Complications

Throughout the previous discussions it was assumed that
- infinite amounts of beach material are available for sediment transport,
- alongshore sediment transport is essentially the net transport, which takes place in one direction,
- the effects of individual storms can easily be averaged into long term littoral drift quantities.

11.5.1 Limited Amounts of Beach Material

Many coastal areas exhibit wide beaches backed by substantial dunes and thus have virtually unlimited amounts of sandy beach material available, and all the usual expressions were developed for such areas. But many coastal areas do not have unlimited quantities of sand, because erosion processes meet either man-made or natural formations that do not contain sufficient sand to supply the cross-shore or alongshore sediment transport potentials. Examples of this are:
- depleted beaches near seawalls and structures (Figs. 11.5 and 11.6),
- discontinuous sand cover at cohesive bluffs (Section 11.6),
- artificially nourished beaches, with a limited amount of sand.

In such cases, it is necessary to distinguish between *potential* sediment transport rate (calculated from formulas) and *actual* rate. The actual rate is less than the potential rate and can only be determined by a sediment budget calculation, which takes account of all the sediment inflows and outflows, and all the sediment sources and sinks of a system. The actual rate is normally viewed as a long-term average rate and it is most often considered to be a simple fraction of the long-term average potential rate.

Both potential and actual sediment transport rates play an important role on beaches with limited sand supply. Potential rate is approached during short periods of time when a large supply of material does exist. For example, a beach may have accumulated at the actual rate over a whole summer or over a number of calm years. Such a beach may disappear in hours during a relatively small storm, as this storm can transport material at the potential sediment transport rate (alongshore, as well as offshore) for the time that sufficient material is available. Only when the beach has eroded back to existing structures or when most of the beach material has been removed from a rocky or cohesive sublayer will the actual sediment transport rate become substantially smaller than the potential rate. Similarly a beach formed slowly at the actual rate will respond to waves from an unusual direction very rapidly at the potential sediment transport rate.

Thus the potential sediment transport rate, which cannot exist over the long term, will be reached during single storms or storm segments and short-term erosion and accretion rates, even in areas with relatively little beach material, are closely related to potential rate. For this reason sediment transport rate should be expressed over short time spans of hours or days rather than years. In areas of short sand supply, potential sediment transport rate is large, relative to available sand. This leaves the impression that erosion and damage is always very rapid. One storm can remove material (at the potential rate) that took years to build up (at the actual rate). Short term actual sediment transport rates can approach potential rates while the long-term average actual rates are considerably smaller and a function of supply and loss of sand.

11.5.2 Sediment Transport in Two Directions

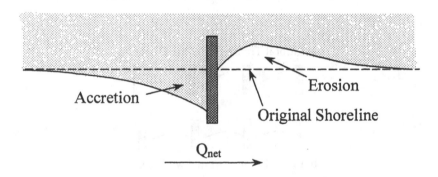

Figure 11.18 Accretion-Erosion around a Groin

One day a property owner said to me: "Build some groins; they attract sand - everybody knows that!" This statement seems funny and naive. It is well known that groins interrupt littoral transport and that sand will build up (accrete) on the updrift side and be taken away (erode) from the downdrift side (Fig. 11.18). One can find plenty of illustrations in publications on coastal engineering. But! When a groin was built on that site, it attracted sand as the person said. The site had little sand to begin with and the long-term average actual littoral transport in one direction was almost equal to the rate in the other direction. The gross transport (the sum of the two) was large and the net transport (the difference) was small. The contours around the groin, which had been placed on an initially straight, eroding shoreline are shown in Fig. 11.19. They quite clearly show accretion with radically differing beach profiles on both sides of the groin. At the waterline, however, the classical pattern of Fig. 11.18 prevails, fooling the casual observer or the person who depends solely on air photos.

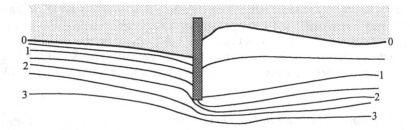

Figure 11.19 Transport in Two Directions

a. Littoral Transport in One Direction

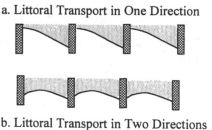

b. Littoral Transport in Two Directions

Figure 11.20 Transport in Two Directions in Groin Field

We cannot simplify designs so that they only take (apparent) net longshore sediment transport into account. One example of a design difference is shown in Fig. 11.20

where it may be seen that same length groins will collect more sand (and damage adjacent areas more) if the alongshore transport is in two directions. They will also not be flanked[2] as easily in case of predominantly bi-directional transport.

11.5.3 Short Term Littoral Transport

Even when sand supply is not limited, a few days of storm will normally move as much or more material than the relatively small waves move during the remainder of the year[3]. That is, in fact, why the accretion around the groin in Fig. 11.19 is different on the left side and the right side. The left side, where the slopes are gentle, accretes most of the year as a result of small waves. The right side, with steeper slopes, accretes during short periods of high wave energy. Thus, we cannot only think in terms of annual littoral transport but must be careful to consider short-term storms, etc. For example, the groins in Fig. 11.20b appear quite safe. However, one storm could remove all the sand, flank the groins and destroy them if they are located in an area where single storms account for a large proportion of the total sediment transport or are capable of moving most of the available material.

11.6 Cohesive Shores

Cohesive shores are *essentially different* from sandy shorelines. They can be broadly classified as soft and hard shorelines. *Soft* cohesive shores consist of soft, unconsolidated cohesive materials, recently deposited on deltas, tidal flats and coastal wetlands. The deposited material is the very fine, ultimate product of weathering and erosion that is found at the seaward end of long river systems. It is in the form of silt and clay that can travel in suspension for very long distances from a river mouth, before coming to rest in sheltered water. The wave climate, where the deposition takes place is gentle, but it does not need to be gentle all the time. Once the particles have settled, they are difficult to remove, even by large tides and waves, because of their cohesion. The transport of the silt and clay also interacts with the movement of sand through the same area, making the sediment transport in such areas very complex.

Coastal management of soft cohesive shores is also not simply concerned with sediment transport and prevention of erosion, but with management of a very

---

2. The waves erode around the landward end of the groin.
3. Often the storm waves are accompanied by higher water levels due to storm surge, permitting larger waves to come closer into shore.

complex and fragile system. These areas are usually valuable and unique habitat. The annual growth of marine plants and the dense dendritic channel patterns form prolific feeding, breeding and nursery areas for wildlife of all sorts and their proper maintenance is essential, even to animals that may spend most of their life thousands of kilometres away. Detailed discussion of the management of soft cohesive shores is beyond the scope of this book. Further details may be found in Black et al. (1998), Dyer (2000), and Van Rijn (1998).

Hard cohesive shores consist of consolidated cohesive materials that were deposited thousands of years ago and have been cemented or compressed together into rock-like, hard masses. Although they were originally deposited under benign wave conditions, they are now attacked and eroded by wave action that usually occurs at a totally different water level.

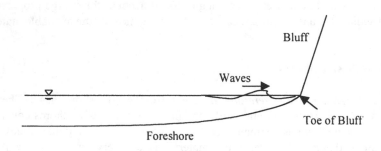

Figure 11.21 Cohesive Shore Profile

An eroding, hard cohesive shore consists of a steep bluff (or cliff) and a foreshore as in Fig. 11.21. Because the system erodes, the bluff is oversteepened (steeper than it would be if subjected to normal weathering). The foreshore of an eroding bluff normally has the same composition as the bluff itself, since the foreshore was itself the bluff, years ago. The bluff is attacked at its toe by waves, which undercut the bluff and result in oversteepening, slumping, sliding and mass wasting of the bluff material onto the foreshore. There, any large pieces of cohesive material are broken up by the waves into the original grain sizes of the cohesive mass. Sandstone becomes sand, mudstone becomes mud and till becomes a mixture of gravel, sand, silt and clay. The fine materials are removed permanently (in suspension) into deep water and this material never returns to shore. Except in the case of sandstone, the erosion debris will not form protection from the waves. The profile never achieves a (dynamic) equilibrium with the incident wave conditions, as it does for a granular beach (Fig 11.1). A hard, cohesive shore, unless it consists of mainly sand and gravel will always erode.

The mass wasting of the bluff results in its recession (movement inland) and it is popularly assumed that bluff recession rate is controlled by the wave action on the toe of the bluff and by geo-hydrological processes, such as seepage and drainage. In order to identify the real reason for bluff retreat, Fig. 11.22 shows a foreshore that remains constant (suppose it is made of concrete). While the waves cause the bluff to recede, the waves will eventually be unable to reach the toe of the bluff. Bluff recession will stop and geo-hydrological processes will cause the bluff to stabilize at a flatter slope. For bluffs to continue to recede, it is necessary that foreshore to be lowered by the erosion process. In fact, the whole profile, above and below water will remain the same, as the shore recedes (Fig. 11.23). Evidence of this may be seen in areas where the water levels have been lowered in recent geologic time. Old shorelines are visible, high above the existing water level and these shorelines exhibit the same bluff and foreshore profiles as the eroding bluffs at the present water level.

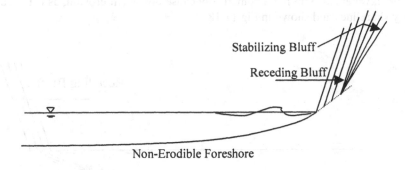

Stabilizing Bluff

Receding Bluff

Non-Erodible Foreshore

Figure 11.22 Bluff with Non-Erodible Foreshore

Comparison of Figs. 11.22 and 11.23 indicates that the rate of bluff recession is controlled by the rate of downcutting of the foreshore. Kamphuis (1983, 1987 and 1990) shows that the rate of downcutting of the foreshore is a function of wave action but that the wave-generated bottom shear stresses are usually insufficient to remove the hard rock-like material from the cohesive foreshore. Yet the bluff recedes. The erosion is made possible by a discontinuous layer of granular material over the foreshore. This granular material can come from erosion of the cohesive foreshore itself, for example, when till erodes. It can also be moved into the area by alongshore transport from nearby formations. The coarser materials (gravel) will collect at the shore and the finer materials (sand) will remain over the foreshore. Since the layer of sand is thin (usually < 0.3 m) and discontinuous, it does not provide protection against scour. Quite the opposite - this sand is moved around by the waves and scours the foreshore, as an abrasive. This scouring action readily removes the hard cohesive material to bring about the erosion, as shown in Fig. 11.23. The consequence of this scouring process is that the foreshore profile will be similar to a sand profile. If the cohesive profile is locally lower than the sand profile, the sand will fill in and protect such a depression from further erosion. Similarly, a bump in the cohesive profile will protrude above the sand and will be subjected to intensive scour, lowering it to the level of the surrounding sand profile (Kamphuis, 1987).

Understanding this complex foreshore erosion process is the key to understanding the erosion and possible protection of bluff or cliff shorelines. For example, it explains why bluff erosion is not simply a result of toe erosion at times of high water. Erosion of the foreshore at times of low water is equally responsible. Groins and breakwaters also do not automatically cause downdrift erosion, as is the case on sandy shorelines and shown in Fig 11.18.

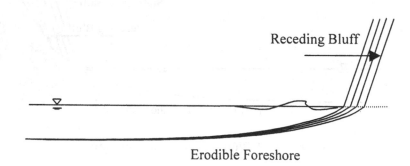

Figure 11.23 Receding Bluff Profile

12. Basic Shore Processes

12.1 Introduction

In Ch. 11, the general characteristics of sediment transport were described to give the reader an impression of how beach processes work and how we can recognize and deal with certain problems. In this chapter some of the details will be filled in to complement the general impressions of Ch. 11.

12.2 Nearshore Current Patterns

Nearshore current patterns are a combination of longshore currents, rip currents and undertow. For a large incident wave angle, alongshore momentum generated by the wave breaking process sets up strong longshore currents (Fig. 12.1). Smaller incident wave angles generate weaker longshore currents. The forward flow of the water particles in the breaking waves also "pumps" water across the breaking zone, increasing the water level there. The onshore momentum of the waves holds some of this water close to shore, causing an elevated water level near shore called wave setup. However, most of the water flows from the shore to deeper water in the form of undertow (a general offshore flow, everywhere) and rip currents (concentrated strong local currents) as shown in Fig. 12.2.

The rip currents occur at locations of least resistance such as breaks in the offshore bars. This is a feedback system - rip currents also increase the size of shoreline irregularities and breaks in the bars. They can also be triggered by local irregularities in the shoreline direction, such as at coastal structures or beach cusps[1]. If there are no specific irregularities, the locations of rip currents may be determined by a pattern of edge waves - alongshore standing waves consisting of wave energy that is trapped near the shore. The nearshore current patterns shown in Fig. 12.2 are therefore a complex combination of several different water levels and types of flow. The current velocities cannot easily be estimated. Undertow and rip currents are, however, substantial and dangerous, and many swimmers are carried to sea and some are drowned as a result of strong undertow and rip currents.

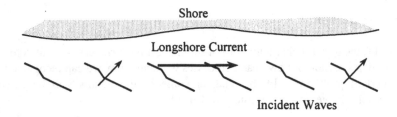

Figure 12.1 Circulation for Large Incident Wave Angle

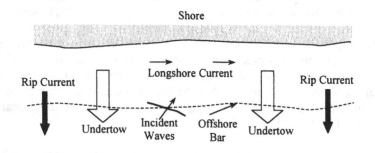

Figure 12.2 Circulation for Small Incident Wave Angle

---

1. Beaches are generally straight or slowly curving. Sometimes, this general pattern is modified by a secondary, rhythmic pattern of shallow bays, separated by rather sharp points. These points are called cusps.

From small amplitude wave theory, according to Table 2.2 the horizontal component of wave orbital velocity is

$$u = \frac{\pi H}{T} \frac{\cosh\{(k(z+d)\}}{\sinh(kd)} \sin(kx - \omega t) \qquad (12.1)$$

In the breaking zone, this may be simplified. Since

$$\sinh\left(\frac{2\pi d}{L}\right) \rightarrow \left(\frac{2\pi d}{L}\right) \ , \quad \cosh\left(\frac{2\pi d}{L}\right) \rightarrow 1 \quad and \quad C = \frac{L}{T} \rightarrow \sqrt{gd} \qquad (12.2)$$

the maximum value of the orbital motion near the breaker is

$$\hat{u} = \frac{\pi\,H\,L}{T\,2\pi d} = \frac{H}{2d}C = \frac{H}{2d}\sqrt{gd} \qquad (12.3)$$

Equation 7.32 shows that at breaking, the ratio H/d is constant. This ratio is called the breaker index.

$$\gamma_b = \frac{H_b}{d_b} \qquad (12.4)$$

and Eq. 12.3 may be written as

$$\hat{u}_b = \frac{\gamma}{2}\sqrt{gd_b} = \frac{1}{2}\sqrt{\gamma\,g\,H_b} \qquad (12.5)$$

There are several expressions to calculate the potential velocity of longshore current (current for an infinitely long, straight beach). One commonly used expression is shown in CERC (1984) and is based on Longuet-Higgins (1970)

$$V_L = 20.7m\sqrt{gH_b}\,\sin 2\alpha_b \qquad (12.6)$$

where m is beach slope.

12.3 Littoral Materials

Littoral materials vary in size from boulders to clay. They may be classified according to size, based on the median grain diameter D_{50}. Figure 12.3 shows two such common classifications. The grain sizes in any sample are usually log-normally distributed (the log of the particle size forms a straight line on a probability plot). A parameter Φ is defined as

$$\Phi = -\log_2 D \tag{12.7}$$

These Φ sizes are used as basis for the Wentworth Classification shown in Fig. 12.3.

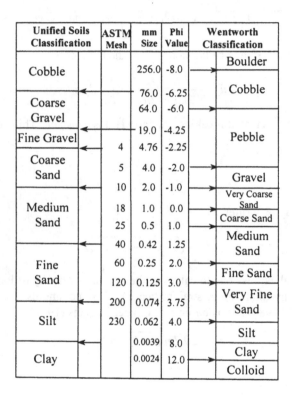

Unified Soils Classification	ASTM Mesh	mm Size	Phi Value	Wentworth Classification
Cobble		256.0	-8.0	Boulder
		76.0	-6.25	Cobble
Coarse Gravel		64.0	-6.0	
Fine Gravel		19.0	-4.25	Pebble
	4	4.76	-2.25	
Coarse Sand	5	4.0	-2.0	
	10	2.0	-1.0	Gravel
Medium Sand	18	1.0	0.0	Very Coarse Sand
	25	0.5	1.0	Coarse Sand
	40	0.42	1.25	Medium Sand
Fine Sand	60	0.25	2.0	Fine Sand
	120	0.125	3.0	
	200	0.074	3.75	Very Fine Sand
Silt	230	0.062	4.0	Silt
		0.0039	8.0	Clay
Clay		0.0024	12.0	Colloid

Figure 12.3 Grain Size Classification

Grain size distribution may be defined by median grain size.

$$D_{50} \quad or \quad \Phi_{50} \tag{12.8}$$

where the subscript refers to the percentage of grain sizes that are smaller than this size. Mean grain size is defined as

$$M_\Phi = \frac{\Phi_{16} + \Phi_{50} + \Phi_{84}}{3} \tag{12.9}$$

Standard deviation of the grain size distribution is

$$\sigma_\Phi = \frac{\Phi_{84} - \Phi_{16}}{2} \tag{12.10}$$

and its skewness is

$$\alpha_\Phi = \frac{M_\Phi - \Phi_{50}}{\sigma_\Phi} \tag{12.11}$$

Since sediment transport involves the dynamics of particles under water, it is also common to use the settling velocity (fall velocity) of particles in still water to describe the sediment. CERC (1984) publishes a fall velocity curve. For natural grains

$$w_f = \left[\left(\frac{\rho_s}{\rho} - 1 \right) g \right]^{0.7} \frac{D_{50}^{1.1}}{6 v^{0.4}} \tag{12.12}$$

$$for\ 0.13 \cdot 10^{-3} \leq D_{50} \leq 1.6 \cdot 10^{-3}\ m$$

and for larger material

$$w_f = 1.05 \left[\left(\frac{\rho_s}{\rho} - 1 \right) g D_{50} \right]^{0.5} \tag{12.13}$$

$$for\ 1.6 \cdot 10^{-3} \leq D_{50} \leq 8 \cdot 10^{-3}\ m$$

where v is the kinematic viscosity of water.

12.4 The Beach

12.4.1 Beach Slope

A beach is often characterized by its slope, which is related to grain size. Larger grain sizes generate steeper beaches as shown, for example in CERC (1984). Kamphuis et al (1986) show that beach slopes through the breaker zone are a function of the ratio (H/D) which represents the ratio of disturbing wave forces to restoring particle forces

$$m = 1.8 \left(\frac{H_b}{D_{50}} \right)^{-1.2}$$ (12.14)

Equation 12.14 may be explained as follows: Steep beach slopes result in a large energy dissipation rate (the breakers tend toward plunging and collapsing breakers as shown in Ch. 7). This results in more concentrated disturbing forces. Thus smaller grain sizes are readily removed from steeper beaches and the larger sizes remain.

12.4.2 Beach Profile

Beaches do not consist of just a simple, planar slope. It is usual to describe beach shape as a beach profile, as introduced in Section 11.2. It was shown in Ch 11 that in practical terms, equilibrium profiles do not exist, but the concept of equilibrium or stable beach profile is still useful since beach profiles tend to approach an equilibrium. According to Bruun (1954) and Dean (1977) the underwater portion of long-term average profiles may approximated by

$$d = A_p x^{2\,3}$$ (12.15)

where d is the depth of water and x is the distance offshore of the still water line. The profile coefficient A_p is mainly a function of grain size and Fig. 12.4 summarizes the relationship proposed by Moore (1982) and Dean (1983). From this curve, it is possible to define relationships over ranges of grain sizes. For example

$$A_p = (1.04 + 0.086 \ln D)^2 \quad for \ 0.1 \times 10^{-3} \le D \le 1.0 \times 10^{-3} \, m$$ (12.16)

Simpler exponential relationships may also be defined for smaller ranges. Suppose we wish to define the relationship for $0.1 \times 10^{-3} < D < 0.2 \times 10^3$ m. A straight line between $D = 0.1 \times 10^{-3}$ and 0.2×10^{-3} m has a slope of 0.63 and an intercept of 1.3, which means that

$$A_p = 20 \, D^{0.63} \quad for \ 0.1 \times 10^{-3} \, m \le D \le 0.2 \times 10^{-3} \, m$$ (12.17)

Dean (1983) also proposes a simple relationship between A_p and fall velocity

$$A_p = 0.50 \, w_f^{\,0.44}$$ (12.18)

where w_f is in m/s. Equation 12.14 shows that Eqs. 12.16 to 12.18 are overly simplistic in that they do not include wave height. Indeed, the equations were derived by averaging many beach profiles and thus they represent average values of H.

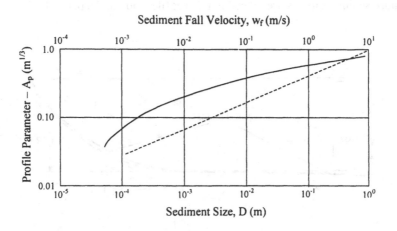

Figure 12.4 Beach Parameter A_p as a Function of Grain Size
(after Dean, 1993)

The profile depth, according to Eq. 12.15 increases indefinitely with x. That is unrealistic. A beach profile has a practical seaward limiting depth, where the wave conditions can no longer change the profile. Sediment will still move back and forth, but there is no perceptible change in depth. Hallermeier (1981) discusses this critical depth or closure depth and (CUR, 1990) approximates it as

$$d_c = 1.6\ H_{s,12} \qquad\qquad (12.19)$$

where $H_{s,12}$ is significant wave height which occurs 12 hrs/yr on average.

In the simplest terms, a long-term average beach profile must satisfy both Eqs. 12.15 and 12.19. Figure 12.5 shows the two equations and an assumed profile that contains a short transition section. Equation 12.19 forms a horizontal base line and the assumed profile moves over top of this base line. Erosion means that the profile moves landward along the base line, accretion means the opposite. This combination of a profile and a horizontal base line is of course a simplification. For example, long-term erosion means that the profile is flatter than average. Since most beaches are slowly eroding, Eq. 12.15 probably best represents a slowly eroding profile. An accreting profile would be steeper. An actual beach profile will also slope very slowly to deeper water. The slope is a function of offshore supporting rock formations, currents, tides and gravity. Offshore sediment losses are related to this profile slope in deeper water.

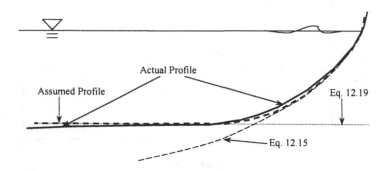

Figure 12.5 Beach Profiles and Closure Depth

12.5 Cross Shore Sediment Transport

Cross-shore sediment transport described in Section 11.3 takes place when an existing beach profile changes. If the beach profile is close to its equilibrium with the existing environmental conditions, little cross-shore sediment motion will take place. Should the environmental conditions change, however, substantial cross-shore sediment transport must be expected in order to come to the new equilibrium that accompanies the new conditions. The rate of cross-shore transport is normally assumed to be proportional to the difference between the existing beach profile and the equilibrium profile that matches the new environmental conditions (Bakker, 1968). This means cross-shore sediment movement is large, immediately after a change in environmental conditions and subsequently slows down. As a result, shoreline change also begins rapidly and then slows down in time.

An example of the cross-shore sediment transport process may be found in Ch. 11 for the annual beach profile change from a summer berm profile to a winter bar profile (Fig. 11.1). The summer berm is more or less in equilibrium with the small summer waves and when the higher and steeper winter waves come, the beach seeks a new equilibrium. Similarly, at the beginning of summer, the winter equilibrium beach must respond to the more gentle summer wave climate and the beach adjusts itself again. In both cases, cross-shore motion is the main sediment transport vehicle. In the same way, temporarily higher water levels such as storm surge permit larger waves to come closer into shore, and the beach will respond suddenly and dramatically through extensive offshore transport.

Research to determine which wave conditions produce offshore and onshore sediment movement indicates that a fall velocity ratio

$$\frac{H_o}{w_f T} \cong 1 \tag{12.20}$$

approximates a critical condition. If the ratio exceeds 1, sediment moves offshore (tends to produce a bar profile); if it is less than 1, sediment moves onshore to produce a berm. Further information may be found in Kraus et al (1991).

If, over the long-term, the material that is moved offshore does not all come back onshore, the beach will erode. Erosion of the beach (decrease in volume of beach material) will cause recession of the beach profile (movement landward). One classic example of beach recession results from sea level rise. Higher water levels allow larger waves to come closer into shore, resulting in erosion of the top portion of the profile to adjust to the more severe wave conditions. It is possible to estimate the net beach recession accompanying sea level rise by assuming that the wave climate remains the same and the beach profile retains its shape. This beach profile eventually must rise with the water level and the volumes of sand required to raise the profile in the foreshore must come from a landward movement of the profile (Fig. 12.6). This results in Bruun's Rule.

$$R = \frac{x_c h}{(d_d + d_c)} \tag{12.21}$$

where R is the recession, h is the water level rise, d_c is the closure depth and d_d is the dune height. This expression is very approximate because
- The distance out to closure depth, x_c (or the width of the active profile) is computed from Eqs. 12.15 and 12.19, which makes x_c very sensitive to d_c,
- The final slope AB will be very flat and hence the triangle ABC will contain

quite a volume of sand that is ignored in this calculation,
- There is no offshore sediment movement as a result of currents, tides and gravity
- An eroded volume of sand is expected to produce the same volume when deposited.

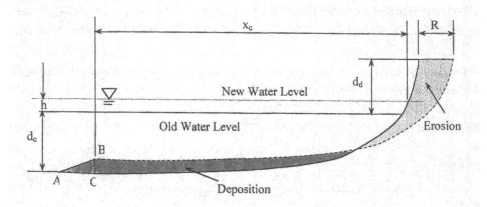

Figure 12.6 Beach Profile Recession from Water Level Rise

12.6 Alongshore Sediment Transport Rate

Alongshore sediment transport takes place by beach drifting and transport in the breaker zone (Ch 11). Usually a beach consists of graded (finer as well as coarser) material. Generally coarser sediment is found close to shore, on the steeper part of the profile. It is moved along the shore by beach drifting. The finer material further offshore is moved along any offshore bars by the alongshore current patterns. This normal transport pattern can easily be disturbed by shore protection structures as shown in Ch. 15. Unfortunately, the effects of grain size distribution and offshore diversion of coarser sediment by structures ignored in most designs.

The alongshore sediment transport rate may be computed by using a detailed or a bulk sediment transport expression as Section 11.4.3. A detailed sediment transport calculation requires computer programs and much data to calibrate such programs. Because such data are normally not available, bulk sediment transport calculations and measurements are normally used for practical engineering solutions. These bulk expressions simply relate the total alongshore sediment transport rate to some easily measured wave and beach parameters. Two such bulk expressions are the CERC expression (CERC, 1984) and an expression developed in Kamphuis (1991). These expressions are derived below.

12.6.1 Alongshore Component of Wave Power

Sediment transport along the shore is related to the wave-generated momentum or energy gradient. The energy flux or wave power between wave rays, according to Fig. 7.5 and Eq. 7.14 is

$$\overline{P} = nCEb \tag{12.22}$$

and the average wave power per unit length of beach is

$$\overline{P'} = \frac{nCEb}{b/\cos\alpha} = nCE\cos\alpha \tag{12.23}$$

The alongshore component of this wave power (a misnomer since P is a scalar) is

$$P_a = (nCE\cos\alpha)\sin\alpha = \frac{1}{2}nCE\sin 2\alpha \tag{12.24}$$

In the breaking zone

$$n \to 1, \quad C_b \to \sqrt{gd_b} \quad and \quad E_b = \frac{1}{8}\rho g H_b^2 \tag{12.25}$$

and

$$P_{ab} = \frac{1}{16}\rho g^{3/2} H_b^2 d_b^{l\,2}\sin 2a_b \tag{12.26}$$

Using the breaker index of Eq. 12.4 results in

$$P_{ab} = \frac{1}{16}\frac{\rho g^{3/2}}{\gamma_b^{1/2}} H_b^{5/2}\sin 2\alpha_b \tag{12.27}$$

For irregular waves, H_s is used to define the component of wave power (in S.I units) as

$$P_{asb} = \frac{1}{16} \frac{\rho g^{3/2}}{\gamma_{sb}^{1/2}} H_{sb}^{5/2} \sin 2\alpha_b \qquad (12.28)$$

or

$$P_{asb} = 2.0 \cdot 10^3 \frac{H_{sb}^{5.2}}{\gamma_{sb}^{1/2}} \sin 2\alpha_b \qquad (12.29)$$

12.6.2 CERC Expression

The best known equation for bulk sediment transport rate is found in the CERC (1984).

$$I_s = 0.39 \, P_{asb} \qquad (12.30)$$

where I_s is the underwater weight of sediment transported.

Assuming a dense sand with $\rho_s = 1800$ kg/m^3 and porosity, n = 0.32, Eq. 12.30 may be converted to m^3/yr as

$$Q_c = 2.2 \cdot 10^6 \frac{H_{sb}^{5.2}}{\gamma_{sb}^{1/2}} \sin 2\alpha_b \qquad (m^3/yr) \qquad (12.31)$$

For a flat beach (m → 0), Eq. 7.32 yields $\gamma_{sb} = 0.56$ and hence

$$Q_c = 2.9 \cdot 10^6 \, H_{sb}^{5/2} \sin 2\alpha_b \qquad (m^3/yr) \qquad (12.32)$$

or

$$Q_c = 330 \, H_{sb}^{5/2} \sin 2\alpha_b \qquad (m^3/hr) \qquad (12.33)$$

where H_{sb} is in meters and Q is a function of H and α only.

12.6.3 Kamphuis (1991) Expression

Kamphuis (1991) derives an expression that includes the effects of wave period (or wave steepness), beach slope and grain size (Fig. 12.7)

$$\frac{Q_s}{\rho \, H_{sb}^3 / T_{op}} = 1.3 \cdot 10^{-3} \left(\frac{H_{sb}}{L_{op}} \right)^{-1.25} m_b^{0.75} \left(\frac{H_{sb}}{D_{50}} \right)^{0.25} \sin^{0.6} 2\alpha_b$$

$$(12.34)$$

This reduces to

$$Q_s = 2.27 \ H_{sb}^2 \ T_p^{1.5} \ m_b^{0.75} \ D^{-0.25} \sin^{0.6} 2\alpha_b \qquad (12.35)$$

where Q_s is in kg/s underwater. This may be converted to

$$Q_k = 6.4 \cdot 10^4 \ H_{sb}^4 \ T_{op}^{1.5} \ m_b^{0.75} \ D^{-0.25} \sin^{0.6} 2\alpha_b \qquad (m^3/yr) \quad (12.36)$$

or

$$Q_k = 7.3 \ H_{sb}^4 \ T_{op}^{1.5} \ m_b^{0.75} \ D^{-0.25} \sin^{0.6} 2\alpha_b \qquad (m^3/hr) \qquad (12.37)$$

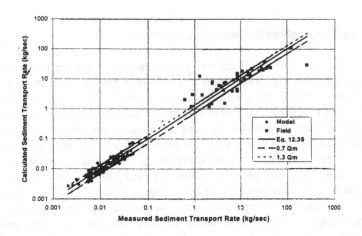

Figure 12.7 Development of Kamphuis (1991) Expression

Equation 12.34 to 12.37 were derived from small-scale hydraulic model tests and were found to be valid (without further calibration) for available field results. They over-predict transport for gravel beaches because they do not include a critical shear stress (they assume that particles move even for small wave conditions, which is true for sand but not for gravel). Equations 12.32 and 12.36 are compared in Table 12.1. The CERC expression has been criticized for over predicting Q, particularly at high-energy wave conditions.   Table 12.1 shows that the Kamphuis (1991) expression gives smaller values.

Table 12.1 Comparison of Bulk Expressions

| | H_{sb} | m | α_b | D | T | Q (m^3/yr) | |
	(m)		(deg)	(mm)	(sec)	CERC	K(1991)
Model	0.1	0.1	4	0.1	1	1.3×10^3	0.4×10^3
Field	1	.02	4	0.2	8	0.4×10^6	0.2×10^6
Field	2	.02	4	0.2	8	2.3×10^6	0.8×10^6
Field	3	.02	4	0.2	8	6.3×10^6	1.8×10^6

12.7 Actual Alongshore Sediment Transport Rate

The above sediment transport expressions assume that there are infinite amounts of sand along the shoreline. They imply infinitely long beaches with sandy profiles that extend far offshore. At most locations, as discussed in Ch. 11, the assumption of an infinitely long beach with unlimited amounts of sand is not valid. It is necessary to distinguish between *potential* alongshore sediment transport rate (resulting from the above calculations) and *actual* rate (the amount of sand actually moving along the shore). The actual alongshore sediment transport rate is calculated by examining the various inflows, outflows, sources and sinks of sand. Such a calculation is known as a sediment budget. Sand sources are the supplies of sand provided by rivers, erosion products from dune or bluff erosion and the lowering of the foreshore that accompanies shoreline recession. Common sediment sinks are offshore losses into deeper water, onshore losses when wind blows the sand inland so it can no longer be reached by the waves, and man-made losses resulting from construction, dredging and sand mining.

As a simple example of actual and potential sediment transport rate and the interaction between waves, sediment transport rate and sand mass available for transportation, consider the alongshore sediment transport process for the section of sandy shoreline (littoral control volume) in Fig. 12.8. The sediment is supplied at the actual sediment transport rate. The outflow of sediment is a function of the amount of sand available for wave-sand interaction in this section and could theoretically reach the potential rate. If the outflow of sand exceeds the inflow, erosion will occur in the section resulting in shoreline recession. If the sand volumes are limited, there will also be a decrease in the amount of sand overlying any hard offshore layers. The decrease in the amount of sand, on the beach and the foreshore, will mean that less sand is available to interact with the waves. That will decrease the outgoing actual sediment transport rate. Over the long term, the beach size and sand mass will adjust to approach an equilibrium with the incident wave climate and the incoming (actual) sediment transport rate so that the sediment outflow is exactly in balance with the incoming (actual) rate. Clearly the short-term sediment transport rates differ from the long-term rates and that is why we need to take both into account.

12.8 The Littoral Cell

A littoral cell is defined as a reach of shoreline in which all sediment transport processes are related. In theory, it has zero alongshore sediment flow past its updrift and downdrift boundaries. It may contain several sand sources and sinks. Consider a littoral cell as in Fig. 12.9. From the upstream boundary, sediment transport continues to be zero until some sediment is supplied to the cell by a creek or river, or as a result of shore erosion. A short distance into the littoral cell one might expect a small actual transport rate as shown in Fig. 12.9. The sediment transport rate will continue to increase along the cell as more creeks deposit their sand, more bluffs erode, etc.

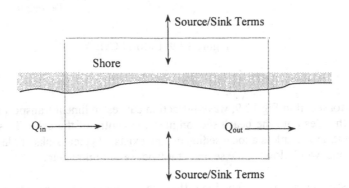

Figure 12.8 Littoral Control Volume

Local offshore conditions may cause the local potential sediment transport rate to be less than the overall actual supply. For example, an obstruction or a shallow, erosion-resistant foreshore may not allow larger waves to come close to the shore. In that case, some of the sand brought in by the actual sediment transport rate may be deposited as a small local beach, as shown in Fig. 12.9. Such beaches will become larger as one proceeds further into the cell, since the actual alongshore sediment transport rate will continue to increase. When local conditions become unfavourable for the formation of a beach (such as a deeper foreshore), the local sediment outflow rate will be increased locally and there will be no visible beach. A beach will not re-appear until conditions become such that a beach can form again. If conditions remain favourable for beach formation, the beach will continue to increase in size with distance along the littoral cell.

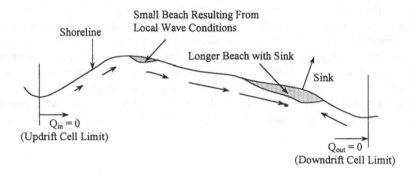

Figure 12.9 Littoral Cell

In the littoral cell in Fig 12.9, wave refraction causes sediment transport to converge from both sides onto the beach section near the centre of the cell. This beach will continue to grow, unless a local sediment sink exists. Typical sinks are large offshore depressions, wind blowing the sediment inland, sand mining, etc.

Coastal structures (Ch 15) function differently, depending on where they are located within such a littoral cell. Understanding the dynamics of a littoral cell means that the engineer or manager knows about how much sediment moves, where it moves, what the influences of the foreshore and offshore conditions are, where the sediment sources and sinks are, etc. In short, it permits proper management of the sediment, a primary consideration for shore protection design and shore management. For example, shore protection on a beach will normally impact the immediate surroundings. But, if shore erosion (such as bluff erosion) is the major source of sediment, shore protection to prevent erosion will decrease the sediment transport rate downdrift throughout the littoral cell and impact the whole downdrift area of the cell.

12.9 Uncertainty

Uncertainty was defined in Section 3.7 and the uncertainties in the basic wave parameters, H, T and α were presented in Sections 3.7, 5.4 and 7.6. These values are summarized in Table 12.2. It is clear that derived values such as velocities and sediment transport rates, particularly if they involve breaking wave conditions and wave angle will contain large uncertainties.

Table 12.2 Uncertainties of Basic Parameters

Parameter	Section	σ	σ'
H,T	3.7		0.05 –0.10
H,T (Hindcast)	5.4		0.25 –0.30
H_b	7.6		0.45
γ_b	7.6		0.10
α_o	3.7	2°	0.2*
α_o (Hindcast)	5.4	8°	0.8*
α_b	7.6		1.0

* assuming an incident angle of 10°.

For the CERC expression, σ' for P_{ab} is calculated to be 0.8, because of σ' values for H_b and α. The uncertainty in Eq. 12.30 approaches 0.5 and hence the total σ' in Q_{CERC}, in theory, is 1.3. Clearly the concept of a normal distribution defined by uncertainty is inadequate. The Kamphuis (1991) relationship itself has a σ' of about 0.3 as shown in Fig. 12.7. The right side of Eq. 12.37 contains $\sigma'=0.75$. Thus, σ' for Q_k may be estimated as 1.0.

This is depressing. How can we work with such uncertain data and expressions? First, many results lie very much closer than one standard deviation from the mean. Second, these are the only tools we have. This discussion about uncertainties is not introduced to discourage, but to remind us that uncertainties exist, that they are serious and that we must analyse them carefully.

13. Coastal Design

13.1 Introduction

Coastal engineering and management are fields for which there is no code of practice, no proven design techniques, no manuals and no general systems of computer-based design programs. We also cannot define input conditions with sufficient accuracy and are uncertain of the "strength of our materials". Some standard procedures exist, but any application of standard procedures is limited because solutions are generally site specific. Thus, most projects become unique challenges that need application of ingenuity and common sense. Frankly, that is why most of us are coastal engineers and managers.

To make a design, we have the following tools at our disposal:
- Field measurements
- Knowledge (theory and experience)
- Models

Discussion about the coastal design process was begun in Ch. 1 and continued in Ch. 8 and 9. In this chapter the design process and the use of modeling, associated with design will be explained further. This chapter in a way draws together the other chapters of this book. The ideal design process is shown in Fig 13.1. We obtain the necessary data about the prototype, using the latest field measurement techniques. These are surveys from land, from vessels, with divers and via remote sensing, often involving specially designed equipment. Field measurements are difficult and expensive to obtain, because the equipment needs to be delicate enough to measure small quantities and small changes in conditions, while at the same time being rugged enough to withstand the violent wave breaking environment. It is possible to design projects, using only data and existing *knowledge* (theory and previous experience). Such studies are referred to as *desk studies*. Feasibility studies are done that way. The experience must be truly coastal experience in order for the design to be safe and effective. Experience in other related fields, such as in structural or geotechnical engineering, or in river basin management is not adequate, unless supplemented by experience with waves, water levels and other subjects discussed in this book. Reading textbooks or handbooks may provide theory, but does not provide appropriate experience. Some experience must also be directly related to the site. That is why local historical experience and impressions of the local residents must be integrated with the experience of the engineers and managers.

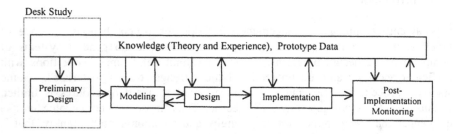

Figure 13.1 Ideal Design Process

A design based only on data and knowledge will be conservative[1]. The inherent uncertainties in the data and design assumptions require substantial margins of safety to ensure that the design meets the required specifications (Ch. 8 and 9). These safety margins increase the cost of the project, which can only be reduced if the uncertainties are reduced.

One method to reduce the margins of safety (and cost) would be to design by *trial and error*. We use small safety factors and hope the structure stays up. If it collapses or turns out wrong, do it again, until the design is satisfactory. This approach is obviously unacceptable, because of the very large (social and economic) costs of failures. It is possible, however, to use exactly such trial and error methods in *models* (numerical or small-scale physical simulations[2] of the prototype). First, a preliminary design is made. Its sensitivities to various changes in the input

---

1. It is quite possible that a desk study design is entirely incorrect.
2. These are small-scale physical models, but not small physical models. Typical surface area dimensions are 100 to 1000 m^2.

parameters and design layouts are then assessed in a model, the design is modified and eventually a final design is chosen and implemented, as in Fig. 13.1. Essentially, this is trial and error. The final design will now be less costly than a design from desk studies alone. It is certainly less costly than a prototype failure if the original desk study was wrong. Modeling is usually economically justified, because the cost savings as a result of the modeling exercise are greater than the cost of the modeling. Important and larger coastal projects are therefore designed using models.

An important and final step in the design process is to monitor the completed project. This permits us to make final corrections, alerts us to problems and most importantly, it provides the only route by which we can learn from our past projects and improve our art.

Two basic types of models can be distinguished in the literature - *physical* models and *numerical* models. These are discussed in Sect. 13.2 to 13.4. The remainder of this chapter addresses the shortcomings of our design tools and how to improve our designs. Each tool must be used, so that it brings its own strengths to the design and its weaknesses are minimized. We discuss field measurements in Sec. 13.5, uncertainties in Sec. 13.6 and how to reduce the uncertainties in Sec. 13.7. Section 13.8 discusses how our knowledge is used in interpreting model results. Some possible ways to optimize the output and increase confidence in our designs for the future are investigated in Sec. 13.9, and Sec. 13.10 discusses composite modeling.

The challenge of coastal engineering design is to use the available tools to provide the best product. The challenge of coastal management is to recognize what coastal design really produces, be aware of its limitations and place coastal design within the larger social, environmental, political context. This chapter and Ch. 14 are rather detailed, compared to the other chapters of this book. This does not simply reflect the author's area of research interest. It is crucial that everyone understands the strengths and limitations of our tools and how they should be used. This understanding should not be left to a select few "modeling experts". Everyone in the design chain must at least be familiar with design procedure and modeling.

13.2 Model Classification

13.2.1 Time-Space Classification

Models can be classified according to the time and space (area) covered. Figure 13.2 identifies Short Term and Small Area Models, Medium Term and Medium

Area Models and Long Term and Large Area Models. These lengthy descriptions are introduced to identify the models exactly, but we will shorten the designations to S, M and L models. The exact definitions of S, M and L will obviously be a function of the problem to be solved. Some typical definitions are: S-models cover prototype durations of hours (or less) and areas of 1 to 100 m^2. Coastal applications are models of bedforms (ripples and dunes), breakwater cross-sections, local scour near structures, water intakes, sewerage outfalls and diffusers.

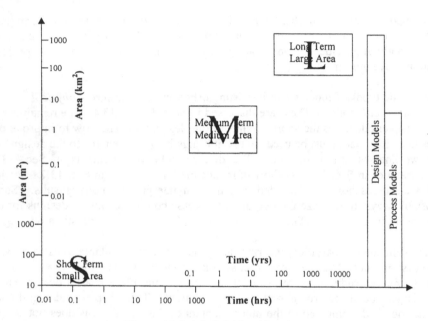

Figure 13.2 Model Classification

M-models typically cover prototype areas of several km^2 and durations of years. Coastal applications are models of shore sections (littoral cells), harbors, inlets, estuaries or portions of estuaries, and shore protection with offshore structures. This category also includes fluid flow models (waves and currents) that cover medium areas, although they may only represent a short duration. Examples are refraction and diffraction of a single wave condition. We include them under M-models because their outcome is normally applied to medium term problems such as wave agitation in a harbor and coastal morphology.

L-models typically cover areas greater than 100 km^2 over centuries or even millennia. Coastal L-models extend in the cross-shore direction from behind the dunes to the continental shelf. To simulate long durations and the slow, long-term erosion of coasts, they must account for the (very slow) sediment transport processes that exchange sediment between the active coastal zone, the continental shelf and the backshore. Processes on a geologic time scale, resulting from water level changes, such as in Fig. 11.4 must be modeled. Examples are models representing the shoreline evolution of sections of the Great Lakes since the last ice age (11,000 years ago), or the development of river deltas over centuries.

The icons in Fig. 13.2 are quite far apart. They delineate the most common domain of activity of S, M and L models. It shows that there are substantial gaps in the usual modeling activity (and knowledge) between these classes and that translation of the results from one type to another may be difficult.

13.2.2 Classification by Purpose

Kamphuis (1991d, 1996, 2000, 2000b) distinguishes between two purposes of physical models: design models and process models. These same purpose categories can also be applied to numerical models.

The *design* model simulates actual complex prototype situations in order to provide specific information that can be used directly in design or in retrospective study of failures. The model is as close as possible to a small scale replica of an actual prototype situation. Models to determine the effects of proposed construction such as dams and navigation channels on estuarine flows, salinity and sedimentation; models of accretion and erosion near harbor entrances; outfall design models simulating effluent plumes, perhaps in stratified tidal flow and models of breakwater stability are all examples of design models. These models simulate specific prototypes with a defined geometry and boundary conditions.

Process models study a physical process; they do not model a specific prototype. Examples are: how does stratification affect sedimentation in an estuary, how are bedform ripples related to vortices, how do vortices move sediment up into the water column, or how do wind waves cause mixing and influence dispersion. The processes to be modeled and the boundary conditions are specified abstractly and thus the model boundary conditions and the scales can be defined to minimize the laboratory and scale effects, discussed below. Since models of details of processes are normally S-models, which represent small prototype areas, they can be built at much smaller scales[3] than design models.

Process models can also be larger, M-type models. These simulate combinations of processes, such as how sand islands erode under a single wave and water level condition (Sec. 13.10), or how inlets change under one defined combination of waves and currents.

Design models can be either S or M-models[4]. M-type design models refer to the classical hydraulic models that study, for example, the effects of construction on estuaries, or shore morphology near harbor entrances. S-type design models are, for example, outfall design studies or breakwater stability studies.

Numerical process models are usually S-models, such as classical studies of refraction-diffraction over a shoal, wave breaking over a bar, sediment transport driven by Boussinesq waves, or beach profile response to waves. Numerical design models can be either S or M-type models, like their physical counterparts.

13.3 Physical Models

13.3.1 General

Physical modeling has been discussed in many publications. Some examples are Chakrabarti (1994), Hughes (1993), Ivicsics (1980), Kamphuis (1975, 1985, 1991d, 1996, 2000, 2000a, 2000b), Langhaar (1951), Martins (1989), Shen (1990), and Yalin (1971). Summaries of the historical development of physical modeling of coastal areas may be found in Kamphuis (1996 and 1999a). Physical modeling has three attributes that commend it:

a) Qualitatively, physical models are close simulations of the prototype.
b) Viewing a physical model in operation adds to the physical understanding of the problem.
c) Physical models can be used, even if not all the details of the relevant processes are clearly understood.

Point a) occurs because simulations are normally carried out with media that are similar to those in the prototype. For example, for coastal models, the fluid is

---

3. The scale is small when large models represent small prototype areas (Eq. 13.2).
4. Kamphuis (1991d, 1996, 1999a) used different terminology. S-models were called short-term design models; M-models were called long-term design models.

normally water, as in the prototype. The flow of this water is subjected to the same gravitational force. Therefore, water will flow downhill, as in the prototype. In the case of sediment transport, the model sediment will be granular, like the prototype material and hence, currents and waves, which are also similar to the prototype will move the sediment as in the prototype. Thus, qualitatively, the physical model results will be very similar to the prototype.

Since the behaviour of the model is quite similar to the prototype, but at a different scale, it is possible to see aspects in the model that may not be obvious from prototype observations – Point b). For example, large circulations, refraction and diffraction all are more obvious in the model than in the prototype. In addition, response of a system under extreme stress and over long durations of constant conditions can be tested in a model. Finally, trial and error changes that would be costly or impossible to make in the prototype are simple and inexpensive in the model. For example, model breakwaters can be reshaped in minutes to indicate how certain changes in breakwater layout affect wave penetration in a harbor.

Point c) states that the model simulates the prototype closely and therefore a physical model provides qualitative results, also for problems for which the processes are not well understood or not well described by theory. Complex non-linear physical processes, for example, can be reproduced in a well-designed physical model. Multi-phase phenomena involving combinations of air, water, other fluids of different densities, and/or particles ranging from rock to mud to pollutants can all be studied with success in physical models, even though we do not know the equations that govern the interactions.

As a result of points a, b and c, physical models constitute a major step from a coastal problem to its solution. In some cases, they are the only possible step. To translate the qualitative model results into quantitative results is, however, a difficult task. Although the model water flows downhill, its actual velocity will probably not be correct. The model sediment transport may be in the right direction, but the actual volumes of sediment transported will not be correct and the sediment will begin to move at the wrong times. Thus, model results cannot be incorporated directly into designs. They must be *interpreted*. That crucial step is discussed Section 13.8.

13.3.2 Scaling and Scale Effect

To design a physical model, we must determine the scales for the various model parameters. Model scales may be derived either from the governing equations or by dimensional analysis. Both methods have their strengths and limitations and hence both should be used. Neither method can completely describe a physical model since a model simulates the prototype better than either equations or dimensional analysis. That is why physical models are used.

There are many types of coastal models such as models of wave agitation in harbors, of wave interaction with structures or of spreading of contaminants by currents and waves. Extensive examples may be found in the literature, such as in Sharp (1981). Kamphuis (1996, 1999a) describes the scaling of the coastal mobile bed, sediment transport and morphology model. That is one of the most difficult of all physical hydraulic models, but it is of vital interest because its results impact almost all coastal problems. It is the most difficult to comprehend and interpret, but will be essential in the future development of coastal modeling.

It is never possible to achieve complete similarity between model and prototype, because some quantities cannot be scaled down. For example, gravity is the same for model and prototype and the only practical model fluid is water. This means gravity, fluid viscosity and density are the same in model and prototype, and certain trade-offs need to be made. The most common example of such a trade-off is that the model cannot simulate gravity-driven and fluid viscosity-related phenomena simultaneously. Since waves and currents are gravity-driven, we usually elect to simulate gravity properly. This means that viscosity effects will not be properly reproduced. The effects of such non-similarity are called *scale effect*.

Using either equations or dimensional analysis, and simulating gravitational forces correctly, the scaling of fluid flow in a coastal physical model may be summarized as:

$$n_x = n_y = n_z = n_d = n_k = n_\eta = n_H = n_L = n_\tau = n$$
$$n_u = n_v = n_w = n_U = n_V = n_{u'} = n_{v'} = n_{w'} = n_{u_u} = n_{v_w} = n_C = \sqrt{n}$$
$$n_t = n_T = \sqrt{n}$$
$$n_g = n_\rho = n_\mu = 1$$

(13.1)

Here n_x is the model scale for x, defined as the prototype value of x over its model value:

$$n_x = \frac{x_p}{x_m}$$

(13.2)

and k is bottom roughness, τ is shear stress. The unsubscripted parameter, n, represents the general model scale. Velocities in the x, y and z directions are u, v and w; U and V are the depth-averaged velocities in the x and y directions, u', v' and w' are turbulent velocity fluctuations, u_w and v_w are the wave generated velocities. The first line in Eq. 13.1 states that all lengths are scaled down by the same factor n. The second and third lines state that all velocities and durations are scaled down by $\sqrt{n}$. The last line indicates that gravity, fluid density and viscosity are (out of necessity) the same in model and prototype[5].

We can use these basic scales to derive related scales by substituting into known equations. For example, area is the product of two horizontal distances:

$$A = x\,y \tag{13.3}$$

To calculate the area scale:

$$n_A = \frac{A_p}{A_m} = \frac{x_p y_p}{x_m y_m} = \frac{x_p}{x_m} \cdot \frac{y_p}{y_m} = n_x n_y = n^2 \tag{13.4}$$

It is seen that simple substitution of the scales for the actual values in Eq. 13.3 could have derived the scale relationships. Similarly

$$n_{Q_f} = n_V n_A = n^{5/2}; \quad n_M = n_\rho n_x n_y n_z = n_\rho n^3 \cong n^3$$

$$n_F = n_M \frac{n_V}{n_t} = n_\rho n^3 \cong n^3; \quad n_p = \frac{n_F}{n_A} = n_\rho n \cong n^3 \tag{13.5}$$

where Q_f is fluid discharge, M is mass of water, F is force and p is pressure.

Equation 13.1 may also be used to derive the scales of often-used similarity numbers, denoting common force ratios. Table 13.1 shows that when gravity is modeled correctly (Froude Number scale = 1), there are serious scale effects that distort the viscous, compressive and surface tension forces. For example the Reynolds number scale is $n^{3/2}$, instead of 1. For a model with n=50, it means that viscous forces are exaggerated by a factor $50^{3/2}$=350.

---

[5] A small (often negligible) exception occurs when sea water is modeled by fresh water.

Table 13.1 Derivation of Scales for Common Similarity Numbers

Similarity Number	Force Ratio	Algebraic Expression	Scale of the Number
Reynolds	Inertia/ Viscosity	$\dfrac{Vx}{\mu/\rho} = \dfrac{Vx}{v}$	$n^{3/2}$
Froude	Inertia/Gravity	$\dfrac{V}{\sqrt{gx}}$	1
Mach	Inertia/Compressive	$\dfrac{V}{\sqrt{K/\rho}}$	$n^{1/2}$
Weber	Inertia/Surface Tension	$\dfrac{V}{\sqrt{s_t/\rho x}}$	$n^{3/2}$

V is a typical velocity, x is a typical length; K and s_t are the compressibility and the surface tension of the fluid, which are assumed to be the same in model and prototype.

Example 13.1 Model Scaling

We need to design a model of a rubble mound breakwater to scale 30. This will be an S-type model. Equation 13.1 indicates that for the model dimensions, all prototype lengths must be divided by 30. Therefore a prototype breakwater that is 10 m high will be 10/30 = 0.33 m high in the model. A 2 m prototype wave height becomes 67 mm in the model. All velocity and time scales are $\sqrt{30} \doteq 5.5$. Thus a prototype velocity of propagation of 1.0 m/s becomes 1.0/5.5=0.18 m/s in the model, a 5.5 second wave period becomes 1 second and 1 hour prototype duration becomes 60/5.5 = 10.9 minutes in the model. Armor stone mass is calculated with Eq. 9.33 and it can be used to derive the scale for armor stone mass.

$$n_{M_a} = \frac{n_{\rho_a}\, n_H^3}{n_{K_D}\, n_{\Delta_a}^3\, n_{\cot\theta}} \tag{13.6}$$

Assume that the model represents a breakwater on a maritime shore (ρ_p = 1035 kg/m³), and assume that $(\rho_a)_p$ = 2650 kg/m³ and $(\rho_a)_m$ = 2400 kg/m³. If $(K_D)_m$ and θ_m are the same as the prototype, then

$$n_{\rho_a} = \frac{2600}{2400} = 1.10; \quad n_{\Delta_a} = \frac{(2650-1035)/1035}{(2400-1000)/1000} = \frac{1.56}{1.4} = 1.11 \tag{13.7}$$

and

$$n_{M_a} = \frac{1.10 \, n^3}{(1)(1.11)^3 \, (1)} = 0.80n^3 = 21,700 \qquad (13.8)$$

If our model tests determine that a stable armor stone size is 0.23 kg, then the prototype would require 0.23 (21,700)/1000 = 5.0 tonne armor stone.

The choice of n=30 in this example is not arbitrary. It is based on prototype dimensions that need to be modeled in available laboratory space and with equipment that can only reproduce parameters within certain ranges. Table 13.2 presents typical prototype values and limits to the model parameters. The possible scales are obtained by dividing the prototype value by the possible model values.

Table 13.2 Parameters Governing Choice of Model Scale

	Prototype	Model		Possible Scale	
	Typical	Max	Min	Max	Min
H_{mo} (m)	3.8	0.13	0.02	190	29.2
T_p (sec)	8.4	2.0	0.7	144	17.6
d (m)	9	1			9

To minimize scale effect, we choose the smallest possible scale that can reproduce all the relevant parameters. That is n=29.2, which is rounded to 30.

M-type design models must simulate large surface areas in a limited laboratory space and hence require large values of n_x and n_y. If we must model a 10 km long shoreline in 50 m of laboratory space, then $n_x=n_y=200$ and Eq. 13.1 requires that $n_z=200$. In that case, a 1.0 m wave prototype height would be only 5 mm in the model and a depth of water of 1 m would be reduced to 5 mm. Such small model quantities will lead to large viscous and surface tension scale effects (Table 13.1) and for that reason, M-type physical models are often geometrically distorted, so that the vertical scales are less than the horizontal scales. Kamphuis (1996, 1999a) has shown that such geometric distortion leads to large scale effects in the model results. The large scales (large scale effects) and the model distortion (more scale effect) essentially mean that M-type physical design model results are very difficult to interpret and that alternative methods should be attempted.

To determine the scales required for the mobile bed portion of a coastal model, the sediment transport is assumed to be a function of shear stress on the bottom. Since there is a lower limit on the model sediment size, in order for it not to go into continuous suspension, the grain size cannot be properly scaled down, unless the prototype grain size is large. That results in scale effects related to sediment transport rate, bottom friction, percolation into the beach, etc. The bedform patterns in the model and prototype are also not properly reproduced, and therefore it must be expected that the sediment phase of a physical model exhibits further large scale effects. Mobile bed models also contain what could be termed an (unavoidable) natural distortion. The beach in a model is formed by its waves and currents and will take on its own profile that is a function of these conditions and the composition of the beach. The modeler is not free to set the beach profile to produce an undistorted model. That again introduces additional scale effect. Clearly, sediment transport models may produce results that look like the prototype, but they need careful interpretation before they can be used in design.

13.3.3 Laboratory Effect

The boundary conditions of a model connect it with the outside world. The model boundaries should simulate the prototype conditions as closely as possible, but we can never achieve a perfect simulation. For example, a wave climate contains many years of waves of ever varying wave heights, coming from many directions. Such a wave climate is normally simulated in a physical model by a few representative wave combinations of H_s, T_p and α, and only the most sophisticated models introduce directional wave spectra. Currents are normally introduced as bulk volumes of water and the velocity distributions in the model will be only approximate. The difference in response between the model and the prototype, resulting from such simplified boundary conditions is called *laboratory effect*.

13.3.4 Implications for Physical Modeling

Physical models are clearly plagued by substantial scale and laboratory effects. However, points a), b) and c) of Sec. 13.3.1 continue to pique our interest in physical modeling. Although the qualitative results provided by a physical model do not correspond directly to prototype reality, at least they are based on a degree of physical similarity so that many of the complex processes, their interactions and the complicated boundary conditions are all reasonably modeled. Yet, a model study cannot be quantitatively successful unless the modeler understands the shortcomings of the model and works around them. Only a modeler who understands which parameters are important and which scale effects result from not adhering to what scaling laws, a modeler who goes through detailed reasoning of scales and scale effects will be successful in obtaining valid quantitative answers.

Clearly, the future for physical modeling lies in reducing the scale and laboratory effects. To reduce scale effects, future physical models will inevitably tend away from the M-type design model with its geometric distortions and large scales and toward S-type (or sometimes M-Type) process model abstractions of the problem. This leads to the concept of *Level 2000* models (Kamphuis 2000a, 2000b), discussed in Section 13.9.

Process model results do not provide direct solutions to practical problems. They must be seen as abstract building blocks that can be used as steps toward a solution of a practical problem. Usually many such building blocks are required to solve a practical problem. This introduces the concept of *Composite* Modeling (Kamphuis, 1995, 1996, 2000, 2000a, 2000b). It combines process modeling, analysis of the model results and computations as shown in Section 13.10.

13.4 Numerical Modeling

13.4.1 General

If physical models produce mainly qualitative solutions, perhaps we could use numerical models. With the advent of computers and with the incredible explosion of size and speed of computers, sophistication of software and development of information technology, numerical modeling would seem to be a natural choice. In 1960, computers consisted of large rooms filled with vacuum tubes and switches. Today, every desk is graced with a much more powerful processor that comes as a small box. Input is via CD's or the Internet and output consists incredible color graphics. Obviously this is the tool of the present and the future. It is on a rapid upward curve of sophistication, and technology that is regarded as impossible today, will be routine in a short time.

Is this relatively new tool cost effective? With respect to both physical and numerical models, there is an initial investment. Laboratories for physical modeling need large real estate properties and must construct specially equipped buildings and infrastructure. Numerical models and associated software must also be purchased or developed at substantial cost, but can be operated in standard office space. Physical models have high operating costs associated with them. The large laboratories need to be maintained. There are also the costs of model construction, equipment, such as pumps and wave makers, measurement instrumentation, and provision of water and power. To operate a physical model also needs a substantial staff with technical backgrounds varying from backhoe operators to instrument makers. A numerical model needs computing equipment and a small, homogeneous group of people familiar with the computer programs. Both types of models need coastal experts to evaluate and interpret the model results (Sec. 13.8). Clearly numerical modeling is more cost effective and lower operating cost is a major advantage for numerical modeling. As a result, small firms with licensed software can carry out numerical model studies. Physical modeling normally requires large laboratory facilities that are either government agencies or heavily supported by government grants.

However, numerical models also have disadvantages. For example,

a) They can give spurious solutions that show little similarity to the prototype,
b) They do not add to the modeler's understanding; they essentially reflect the modeler's input,
c) A problem must be clearly understood before a model can be properly formulated.

The first two points can be simply illustrated. If the sign on the gravity term in the equation of motion in a numerical model is incorrect, water in the numerical model will flow uphill. The modeler knows, however, that water flows downhill and looks for the reason why the water flows uphill. Once the mistake in sign is found and corrected, water will flow in the right direction. The resulting model, however, still only reflects the modeler's innate knowledge.

Contrary to physical models, in numerical modeling, a problem must be clearly understood before a model can be properly formulated so that it produces valid solutions. Equations governing the processes, numerical methods, transfer functions and calibration coefficients must all be known, at least approximately, from the outset. To continue the above simplistic example, the modeler must know the value of gravitational acceleration, before the model can produce reasonable results. Since gravity has been determined accurately, that should not be a problem, but many other constants and coefficients used in numerical models are not so well known. For example, we only know dispersion coefficients within one or two orders of magnitude, most of the time. Thus, because of uncertainties in the equations and the coefficients, and because of approximations made in the numerical simulation of the equations, numerical modeling exercises can only produce qualitative results, at best. Interpretation of such qualitative results into quantitative estimates is the major task of numerical modeling (Sec. 13.8). It requires a thorough understanding of the coastal processes, the applicable equations, the interactions between variables and the shortcomings in evaluation of the coefficients. In addition, just as the physical modeler needs to know about scaling, scale effects and laboratory effects, the numerical modeler needs to understand the implications of simplifications brought into the model, and needs to know about numerical modeling methods with its pitfalls, such as instabilities, numerical diffusion and dispersion, smoothing, etc.

For some problems we know enough about the equations and coefficients to model them numerically with some confidence. Such tractable problems as fluid flow with relatively simple boundary conditions, can be modelled using either physical models or numerical models, combining long waves, short waves and currents. Once numerical models can be successfully applied to solve a type of problem, the use of physical models for such a problem declines, and eventually, mainly physical process models will be used to simulate unknown details.

One aspect about numerical models is very important. We have seen that the results from M-type physical coastal models contain large uncertainties and using L-type physical models would be out of the question. Thus M and L-type modeling is a unique niche for numerical models. The development of M-type numerical models is well underway; work on L-type models is beginning.

13.4.2 Simplifications of Three Dimensional Models

Coastal numerical models couple a transport model (sediment, pollutant, etc.) with a hydrodynamics model that represents wave action, water levels and currents. The model calculates hydrodynamics over a certain region (domain) using the equations of motion and continuity. The calculation domain is connected to the outside world through boundary conditions (also equations). In three-dimensional (3-D) models, the hydrodynamics equations are written in three dimensions. Finite difference models, for example, schematize the domain over a 3-D grid, as shown in Fig. 13.3. A complete 3-D representation involves much computer time and memory but some fully 3-D models have been successfully developed.

Three-dimensional models can be simplified into two-dimensional (2-D) models. Using vertically integrated values for the fluid flow, results in a 2-D model with a horizontal computational grid (2-DH model), shown in Fig 13.4. Such a fluid flow model can be applied to solve M-type transport problems. Shortcomings result from errors brought about by vertically integrating fluid velocities. These shortcomings are directly comparable to scale effects in physical models. Extensive reviews of such models may be found in Basco (1983), Horikawa (1988) and de Vriend et al. (1993 and 1993a). The external forces normally include gravity, pressure, bottom friction, lateral exchange of momentum and wave induced forces. They could also include Coriolis effect, wind stress, tidal fluctuation, etc. The output of the 2-DH hydrodynamics model is water levels and depth-averaged current velocities.   A coupled transport model would essentially describe conservation of mass (of sand, pollutant, etc.) and use the currents and shear stresses calculated by the hydrodynamics model to entrain and transport the material. For sediment transport, the change in sediment volume in a calculation cell ($\Delta x \; \Delta y$) results in a local change of the sea floor level in that cell.

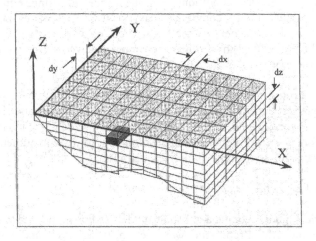

Figure 13.3 Three Dimensional (3-D) Modeling

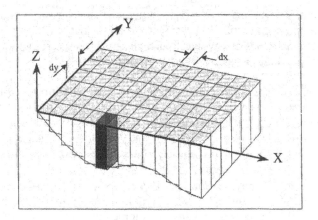

Figure 13.4 Two Dimensional (2-DH) Modeling

Another simplification of the 3-D model can be achieved by ignoring all alongshore variations in water levels, fluid velocities and their derivatives. This results in a cross-shore model calculated over a 2-D vertical grid (2-DV model), shown in Fig. 13.5. Examples may be found in Stive and Battjes (1984), Stive (1986), Steetzel (1987, 1990) Watanabe et al (1980), Watanabe and Dibajnia (1988), Roelvink (1991), Sato and Mitsunobu (1991) and Broker et al (1991). These models are essentially S-type models with some promise toward future M-type modeling.

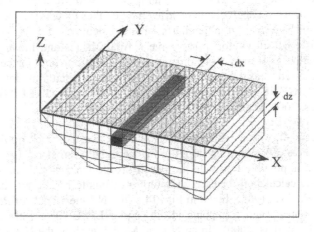

Figure 13.5 Two Dimensional (2D-V) Modeling

To overcome the shortcomings of the 2-DH models, it is possible to replace its depth-averaged values by assumed distributions of velocity, shear stress and concentration or with a version of the 2-DV model, thus combining the two concepts. Such a model is called a quasi-three-dimensional model (Q3-D). Simple, early examples of Q-3D sediment transport and morphology models are Briand and Kamphuis (1993a, 1993b) and Katopodi and Ribberink (1992). Work on Q3-D models is ongoing and this type of model looks promising as a practical M-type tool in that it can perform sophisticated computations in reasonable time on desktop computers and workstations.

13.4.3 One Dimensional Models and their Extensions

At the other end of the spectrum of coastal models, is the one-dimensional (1-D) model, presented in detail in Ch 14. The 1-D coastal model does not pretend to be able

to convert detailed fluid dynamics into detailed sediment transport rates. Its key ingredients are:
- a bulk sediment transport expression, such as Eqs. 12.33 or 12.37. It translates wave and beach parameters directly to overall sediment transport rate,
- a global conservation of sand mass equation.

The 1-D model assumes that when erosion and accretion take place, a typical beach profile simply moves cross-shore over an imaginary horizontal plane located at the closure depth of the profile (Figs. 12.5 and 14.2). Closure depth is defined as the depth below which no appreciable profile change takes place. It is derived from the bathymetry or from hydrographic charts, or it is calculated (Eq. 12.9). Such a 1-D model implies that all contour lines have similar shapes and move landward and seaward together as if there were only one single contour line, hence this model is also often referred to as a 1-Line model. Examples of such a model are Perlin and Dean (1983), **GENESIS** (Hanson and Kraus, 1989) and **ONELINE** (Kamphuis, 1993; Dabees, 2000; Dabees and Kamphuis, 1998, 1999).

One major concern with a 1-Line model is the fact that the beach profile remains the same as it moves offshore or onshore during accretion or erosion. To introduce some 2-D aspect, it is possible to divide the beach profile into vertical subsections and to perform a 1-D calculation for each. Examples of this technique are the early 2-Line approach of Bakker (1968) shown in Fig. 14.5, the N-Line models of Perlin and Dean (1983) and Johnson and Kamphuis (1988) and **NLINE** (Sec. 14.8, Dabees, 2000; Dabees and Kamphuis, 2000). In an N-Line Model, each of the N subsections of the beach profile is related to the others by a cross-shore transport calculation. The cross-shore sediment transport rate is related to the difference between the existing profile and an equilibrium profile; thus the beach profile tends toward equilibrium throughout the calculation. The resulting computation is on a 2-D grid which flexes since the locations of the grid points move in the cross-shore direction. The cross-shore distribution of alongshore sediment transport rate is also needed for these calculations.

Many 1-Line and N-Line models make small angle assumptions, in which either the incident wave angles with respect to the shoreline or the changes in shoreline orientation or both must be small. This becomes a major problem when calculating for large incident wave angles (such as on the Great Lakes) and for large amounts of accretion or erosion (such as for long duration, L-type models). Calculation schemes valid for larger angles are presented in Ch 14.

13.4.4 Performance of Coastal Models

Péchon et al (1997) compare a number of commonly used, wave-driven current models. These models only simulate the fluid mechanics (waves and currents). They find that the models provided a good prediction of the wave field, but forces, bed shear stress and turbulence produced by such waves need more work. In other words, as input to beach change and morphology models, the existing fluid mechanics models are deficient.

Roelvink and Broker (1993) compare the performance of five commonly used 2-Dimensional Vertical (2-DV) or profile change models. The models were expected to simulate the results of two simple physical model tests of beach profile change, conducted in a large laboratory wave flume. Comparison of the results (a pure 2-D situation) indicated that these models can at best only predict short-term coastal profile change. They function best in the central part of the surf zone, for spilling breakers, for "quasi-uniform" conditions, and to compute erosion, although the authors think that the models could simulate beach recovery. These models do not perform well for steep beach profiles. They also do not include long wave activity and sediment transport in the swash zone. The comparison showed that there were wide variations in answers and in some cases, none of the models came close to the measured values.

De Vriend et al (1993a) discuss a similar comparison for (2-DH) M-type coastal area models. The results were compared to physical process model tests of a simple semi-circular bay and of a river outflow. De Vriend et al concludes "significant achievements have been made, but more research over a wide area is needed in order to validate these models and to make them robustly applicable to practical situations".

De Vriend et al (1993) discusses L-type numerical models. For such long durations and large dimensions, the relationships between the input and the response parameters cannot be simply produced by an integration of the detailed physical responses, such as reproduced by M-type models. That would cost too much computing time and the uncertainties of the numerical process simulations would produce uncertain long-term results. Also, long-term response over large areas is mostly a response to low-intensity, second-order stimuli, such as long-term water level fluctuations, climate change, large scale resonant systems of small-order currents, sediment exchange between the active beach and the offshore, etc. Therefore, in addition to the detailed physics, the behavior of the system in response to this low-intensity forcing must be modeled correctly. Such behavioral modeling is a new direction.

Davies et al (1997) compare four numerical models of sediment transport over a plane bed to physical model tests carried out in an oscillating tunnel. It was concluded that net sediment transport was modelled well, but that suspension of sediment and quantities of suspended sediment were not modelled well.

Nicholson et al (1997) compare five commonly used M-type coastal area numerical models to a physical model experiment and a prototype situation involving an offshore breakwater. Their conclusion is that major morphological features like tombolos and salients can be reproduced satisfactorily, but that the results were greatly affected by the sediment transport expressions used. They also found that even for M-type models, higher orders of approximation are needed to reproduce subtle interactions.

The common theme in these comparisons is that the best numerical models still need much development and detailed testing against information obtained from physical models and prototype measurements. The good news is that since numerical modeling is the obvious direction of the future, and costs are such that many (small) players can participate in development of this relatively new technology, there is much work being done to improve the numerical models.

13.5 Field Measurement and Data Models

Since modeling cannot simulate prototype processes exactly, why not use field experimentation? Obviously field results are better than physical model results that contain scale and laboratory effects or numerical diffusion. Indeed, field experimentation needs to be pursued and modeling directly from prototype data, using neuro network, data assimilation and data reduction techniques needs to be developed. But prototype measurement has some serious drawbacks, such as:
- Field experiments cannot be designed (only instrument deployment is designed, but the actual test conditions must be left to nature),
- Close observation of details of the processes, particularly underwater processes, is difficult,
- Field experiments are expensive and therefore field tests are few and short,
- Boundary conditions cannot be manipulated in field experiments,
- Field results are favoured toward fair weather and low-energy wave and current environments. Experimentation in poor weather and high energy environments is too difficult,
- Long-term steady conditions cannot be imposed in the field,
- Field tests cannot be repeated.

For those reasons, prototype input in the foreseeable future, except for a few large, international, co-operative process studies, will consist of gathering limited input data for design and calibration of physical and numerical models. The above limitations also mean that field measurements contain large uncertainties. There is a unique role for physical process modeling here. It is capable of filling in gaps in the field data. It can provide designed test conditions, close observation of processes, response to extreme conditions, etc. that cannot be done in the field. Particularly very large physical models discussed in Section 13.9 will be useful to extend limited field data series. For example, fluid velocities, shear stresses and sediment entrainment processes can all be studied with ease, for real, measured wave trains. The results will be very good in large models with scales of 5 or less.

To obtain sufficient field data, it is also necessary to reduce the cost of field measurement substantially. We need to develop new and better equipment to collect field data. That equipment should be simple, rugged and transportable. We need to take full advantage of remote sensing and video imaging technology. Mobilization costs of field experiments must be greatly reduced, so that field measurements can be made much more often and field experiments can become more focussed.

13.6 Uncertainty

The above review of the available tools, shows that both physical and numerical models produce only qualitative results (results with large uncertainties). In the simplest terms, any model can be viewed as a black box with inputs, outputs and dials to make adjustments. It is like a radio that uses high frequency radio waves as input, produces music as output and has dials to adjust the volume and tone of the output. If the input signal is weak, the output music will be poor. If it is a poor quality radio or it in a bad state of repair, it will produce poor music, regardless of the quality of the input signal. Similarly, the output from a model will contain uncertainties, resulting from the input values and from the quality of the model.

We have discussed the uncertainties of wave-related prototype data in earlier chapters; these are summarized in Table 13.3. For such uncertain primary data, we found that derived quantities, such as sediment transport rate, contain uncertainties of at least 0.5, if measured wave data are used, and closer to 1.0 if calculated from breaking wave data. Any model output, based on such uncertain input and derived quantities will contain large uncertainties. These uncertainties cannot be removed simply by using larger physical models with more sophisticated boundary conditions, or by using more esoteric numerical formulations. For example, a Quasi 3-Dimensional (Q3-D) model will theoretically calculate more realistically than a 2-Dimensional Horizontal (2-DH) model, which in turn will outperform a 1-

Dimensional (1-D) or 1-Line model. But are the answers of the Q3-D model better than 1-D model results?

Table 13.3
Uncertainties in Wave Parameters

	Measured	Hindcast	Wave Climate	Breaking
Wave Height (H)	.075	0.25	0.3	0.45
Wave Period (T)	0.1	0.3	0.3	0.3
Wave Angle (α)*	0.2	0.8	0.9	1.0

(*) Assuming a wave angle of 10°.

If both are formulated correctly and up-to-date models are used, the answer is "not necessarily". First, a weak input signal will produce poor music regardless of the quality of the radio. But secondly, computation time for a sophisticated model is greater and hence, sophisticated models generally use more simplistic and less extensive input data sets than simpler models. A sophisticated Q3-D calculation will use only one or a few wave climate bins (combinations of H_s, T_p, α,) to represent a complete wave climate, whereas a simple 1-D model can use, for example, a complete time series of 3-hourly combinations of (H_s, T_p, α) covering many years. This is entirely similar to laboratory effects in physical models; unsophisticated boundary conditions introduce additional uncertainties. Thus it is quite possible that a simpler model produces more useful results than its more sophisticated counterparts.

13.7 Reducing Uncertainty

The uncertainties in model results must be reduced. As a first step, this can be done immediately through proper model *validation*, which consists of
–   Benchmarking,
–   Calibration,
–   Verification.

Any numerical model, regardless of its sophistication, should be properly *benchmarked*. The model must be run for simple boundary conditions and with simplified equations to be compared with analytical solutions. For example, a 1-D shoreline model must be able to simulate diffusion-type solutions of Pelnard Considère (1956) and Larson et al. (1987), before it can be successfully applied to more complex situations (Sec. 14.5).

In *calibration*, the model parameters are adjusted so that the model reproduces measured prototype values. The concept of model calibration is based on the fact that directly measured prototype parameters contain less uncertainty than the output of the model, which is based on the combined uncertainties of the input data and the model. For example, sediment transport rate calculated by a numerical model can easily contain uncertainty of 1.0. At the same time, Kamphuis (1999) estimates that sediment transport rate computed directly from prototype bathymetric measurements contains an uncertainty of 0.4. Thus it makes good sense to calibrate the model sediment transport rate using prototype bathymetry measurements. Calibration can consist of several intermediate stages. A coastal morphology model can first be calibrated for current directions and velocities, velocity distributions, and sediment transport rates and their distributions, before being calibrated for final changes in morphology.

Model calibration inherently assumes that a calibrated model can correctly extrapolate existing field conditions. For example, a coastal model that was calibrated against shoreline change data can predict future shoreline change. An estuary model calibrated with changes in shoals and channels can predict changes in shoals and channels. But, consider a beach that was only interrupted by some shore-perpendicular structures, such as groins or jetties. A model calibrated with the historical conditions along such a beach cannot be used to design offshore breakwaters to protect this shore. Even a carefully calibrated model will not include wave diffraction, the major new influence introduced by the shore-parallel structures. The model cannot be used, because a design with shore parallel breakwaters is not a simple extension of existing prototype conditions. Such a model can only be useful for the design of offshore breakwaters, if it is benchmarked against simple diffraction solutions and compared to known, similar prototype situations.

To gain further confidence in the generality of a model, the calibrated model must be *verified* against additional prototype data that were not used in the calibration. Thus, one key to success is extensive prototype monitoring to obtain as much information as possible, about the input parameters (waves, tides, currents), but especially also about the output parameters (new wave and current patterns that resulted from the design, shoreline change, sediment volume changes).

In the past, calibration and verification of physical models consumed most of the time required for a model study. It was a major cost item, but it was, in fact, only a fraction of the total cost, considering the other large costs of physical modeling. The cost of a numerical model study is more directly related to the running time of the study. Thus lengthy validation will not just increase the cost of a model study by some fraction, but by a factor (perhaps 2 to 4). This makes adequate validation of numerical models problematic and in the recent past we have moved away from proper validation. Verification is often degraded to quick and simple *comparisons* of model results with the sparse field data. Even worse, results are sometimes only validated by textbook concepts such as "the model shows that accretion and erosion occur, where expected". On the other hand, Section 13.4.4 indicates that subjective terms such as "function well", "significant improvement" and "satisfactory" may be the best we can do, at this time, to describe numerical model results.

The costs of proper validation are large, but the benefits are immediate. Proper validation is the only way to achieve acceptable solutions. In fact, neglecting the basics of model benchmarking, calibration and verification and working with grossly inadequate field data is very deceptive. Neither physical nor numerical models give any indication that its answers are uncertain and that they need careful interpretation before the results can be used. On the contrary, the qualitative impact of physical models and the graphic output of numerical models produce unfounded confidence in the results.

13.8 Model Interpretation

Models do *not* provide direct *quantitative* answers. A modeler needs to interpret the qualitative model results to produce quantitative answers and that is a very difficult step. Because of this difficulty, it is, first of all, a common practice to use models just for their qualitative results. They are often used simply to gain some preliminary insight, prior to the real study. They are simply used to understand problems that cannot be solved another way. Such qualitative model testing is invaluable, but everyone must clearly understand the purpose and validity of the results. These qualitative results cannot be simply translated into quantities.

Interpretation to produce quantitative results requires close interaction between the modeler, the model and prototype data. Since a physical model is inherently a reasonably close simulation of the prototype, the physical modeler has a good base from which to begin quantifying the qualitative model results. In the case of numerical modeling, there is no such guarantee of reasonable simulation of the prototype. Thus quantification of qualitative numerical modeling results is more difficult. It requires critical assessment of the equations and constants used and of numerical properties, such as numerical diffusion, dissipation and the effects of smoothing, in addition to extensive validation. It is, therefore, essential that the interpreter of a numerical model be expert in numerical analysis as well as coastal engineering, or that a team of coastal and numerical experts works closely together.

It is clear, from the above, that we cannot think of a physical or numerical model without the modeler. The modeler is an *integral* part of the model.

Often in numerical modeling, it is not the modeler (the person who developed the model) who interprets the results, but a user (the person who operates the models). This user may know little about the actual computer code of the model or about the underlying relationships. Numerical models contain smoothing to suppress irregularities generated by physical non-linearities or by the numerical scheme. Does the user really understand these smoothing functions and how they affect the results? The "better" numerical models are even made "user-friendly" through attractive interfaces, specifically designed so that the user will not need to be concerned about the internal vagaries of the model. Hence, it is unlikely that a user is even aware of the smoothing in the model or of other simplifications introduced to solve the basic equations. Since the numerical routines may affect the final results in a major way, this is unfortunate.

Numerical model studies are sometimes done by users who are neither familiar with numerical simulation, nor very experienced in coastal engineering. Such users will, however, readily begin to feel that they are coastal experts, simply because they are expert at operating a coastal numerical model and may have done several similar studies. A further false sense of security is introduced in both the user and the client by the very attractive computer-generated output graphics. Because of the attractive and very clever representations of reality by full-colour graphics and animations the answers appear to be indisputable and are readily accepted as correct. If such attractive graphical results are based on questionable models, run by users who are not experts in coastal engineering, and are based on data with large uncertainties, what do the output graphics really mean?

The task of a coastal manager in model interpretation is to understand the interactions between the various people and disciplines, to understand the purposes of the model studies, and to be aware of the modeling limitations. They should, for example, realize that a Q3-D model does not necessarily provide better answers than a good 1-D model, that expensive field measurements are essential for model input and validation, and that attractive color graphics or realistic impressions from physical models do not necessarily represent good answers.

13.9 The Future

At the end of this chapter, we will take a brief look at the future of coastal design and modeling. We will need some changes in direction and in the use of our existing methods. Future design must be based on improvement and integration of our tools. The ultimate goal is to develop numerical models and numerical modelers who can produce acceptable results. Much physical modeling and field measurement will be necessary to produce better numerical simulations and to define the constants that go into the various equations. Until we have developed (much) better numerical models, it is most likely that a combination of numerical and physical modeling, along with informed use of prototype data will be required to provide immediate answers.

Physical modeling will need to be improved. We will need to design models, strictly according to appropriate scale laws and to limit scale effects and laboratory effects. That will require a better understanding of scaling as well as sophisticated, costly facilities. To limit scale effects, *Level 2000* process models will be large, and the prototype sections to be modelled need to be small to permit physical modeling at the smallest possible scale factors. Some future physical models should be very large to permit scales of 1 to 5. Such scales are already possible in oscillating tunnels and wave flumes, but we need to extend this possibility to wave basins. To limit laboratory effects, such Level 2000 facilities will need very sophisticated boundary conditions. Computer-controlled inflow-outflow, wave generation, sediment feeding, etc., based on sensors at the model boundaries and within the facility will be needed. Since the sophisticated boundary conditions are computer generated and controlled, the Level 2000 model is essentially an integration of numerical and physical modeling with field measurements. The near field (close to the points of interest) is modelled physically and with minimum scale effect; the far field is modelled numerically. This method is not new. So-called Hybrid Models have been used since the late 1960s. The method must simply be overhauled and we must improve every aspect of this technique. Even though Level 2000 facilities will be very large, they will still not be used as M-type design models to solve complete problems. That would not be cost effective and modeling large prototype sections at scales of 1 to 5 would still not be possible, even in very large facilities.

The future of physical modeling will consist of simulating small sections of the prototype in the largest possible facilities. Future models will therefore be mainly S-type process or design models. Such models are generic and do not refer to any specific prototype. To draw practical design conclusions from them would be impossible and that introduces the next category of models – the composite model.

13.10 Composite Modeling

Because process model results do not provide direct solutions to practical problems, they must be seen as building blocks that can be used as steps toward the solution of a problem. The mortar that will hold these building blocks together is computation. The complete modeling task produces a composite model consisting of three distinct phases - a process modeling phase, the analysis of the process model results and a computation phase to use these results in production of a useful output (Fig. 13.6).

Composite modelling is possible using either physical or numerical models as the process models, but process models, by definition, cannot be calibrated. They are based on simple, chosen boundary conditions that cannot be imposed on the prototype at will. When we use physical models as process models, we count on the innate similarity between the model and the prototype, or use multiple test series to determine scale effects. Numerical models can only be considered as process models, if they have been fully validated − models with a proven track record that can be applied with confidence to simple boundary conditions.

The computational portion of the composite model can be a numerical model, in which case the process model results simply provide appropriate coefficients and transfer functions. It is more likely, however, particularly for engineering studies that need to provide useful answers within a limited time and budget and for situations with complicated boundary conditions, that the computational module will be relatively simple. Perhaps it is a statistical summation of a number of process modeling results or of relatively simple empirical relationships derived from such model results. Non-linear interactions between the various building blocks need to be investigated through study of the various processes as functions of time.

An example of a composite model study using physical process modeling with waves, currents and sediment transport will now be given. Circular artificial islands were used for oil exploration in the Canadian Beaufort Sea (Kamphuis et al, 1986, 1987). Since there was no rock or even gravel available as construction material, these islands were built entirely out of dredged sand. They were typically located in 20 m of water and their surface area was a 100 m diameter circle, 5 m above still water level. With average stable side slopes of 1:10, their volume was 6-10 Million m^3. During the winter (October to June) the ice surrounding the island provides shore protection against erosion of the sand by the waves and currents. During the open water season (July, August and September) the huge mass of sand itself formed the only insurance against erosion of the centre of the island where the buildings and the drilling rig were located. To determine the complex, three-dimensional erosion and deposition of sediment under waves and currents, 52 M-type process model tests were run. The test layout is shown in Fig. 13.7 and the test parameters were varied as in Table 13.4. Testing was performed on full models for n≥100 and on half models for scales n≤100. Figure 13.7 shows the two n=100 test layouts used. The four different model scales and three model grain sizes were used to provide different scale series to determine scale effects.

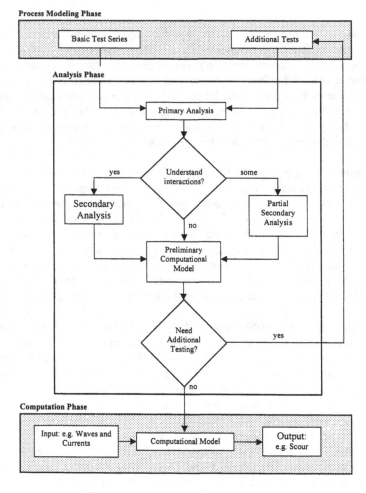

Figure 13.6 Composite Modeling

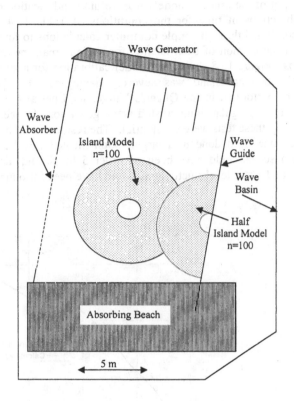

Figure 13.7 Layout of Physical Model Tests of Sand Island Erosion

Table 13. 4
Process Model Tests of Artificial Islands
Summary of Test Parameters

Model Scales – n	200	100	75	50
Model Particle Sizes - D_{50} (mm)	0.56	0.18	0.11	
Prototype Wave Heights (m)	6.5	4.75	3.0	
Prototype Wave Periods (sec)	8.0	10.0		
Wave Types	Regular	Irregular		

The primary output of the process models was volume and location of erosion and deposition, as functions of time for the conditions of Table 13.4. The erosion volumes were combined through simple computer calculations to form a composite model to predict the erosion of such islands for various real, medium-term wave climate scenarios. Figure 13.8 shows the model calibration for a known, prototype storm, in which such an island was severely damaged. The process model experiments were performed in the Queen's University Coastal Basin, which is a small facility with an irregular, long-crested wave generator. Hence, the scale and laboratory effects in these tests were substantial. The results would have been much improved if they had been done in a larger, sophisticated Level 2000 modeling facility, in which scales could have been of order 5 to 10 and directional wave spectra, currents and water level fluctuations could have been introduced.

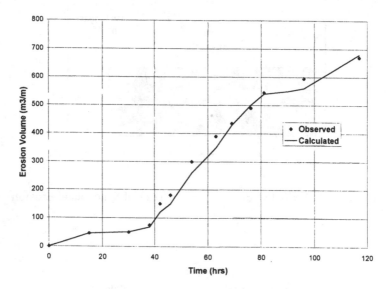

Fig. 13.8 Calibration of Island Model Results

Composite modeling, using physical process models, has many distinct advantages over either physical modeling or numerical modeling. Because scale and laboratory effects are limited in large process models, the main drawbacks of physical modeling have been reduced. Because physical modeling results are included in numerical calculations, the output goes far beyond simply echoing input equations and coefficients, as would be the case in a pure numerical model. Because both modeling concepts are combined, the method is immediately useful for problems that cannot be solved by either, by drawing on the strengths of both methods.

Several aspects of composite modeling also make it economically attractive. The physical process models are relatively simple (with respect to scaling relationships), inexpensive and easy to understand. They are repeatable and because the tests are very similar to each other, the experience gained with the first studies is immediately used in the later, similar studies resulting in a high efficiency. The process model results used in composite modeling are generic which means they are not very site specific and could be used to solve many similar problems for totally different layouts and locations. One could visualise, in time, complete libraries of such process modeling results that can be combined computationally to solve many different problems, greatly reducing the number of new model tests actually required to solve any particular problem.

An interesting aspect is that model validation takes place within the computation phase. This permits extensive calibration, verification and "what-if" scenarios at low cost. Finally, the physical modeling and the computation phases of such a composite model study need not be carried out by the same organisation. For example, a client who is well informed about the study area can do the scenario computations, once a laboratory has provided the generic building blocks.

An application of composite modeling, using a numerical process model occurs when a problem is basically 1-D, but has some sections that are too complex for 1-D analysis. Wave diffraction and currents for the complex sections can be computed with a 2D-H numerical model for many typical (simple) input conditions. These 2-DH process model studies would be simple and repetitive. The computational phase then interprets and generalizes these 2D-H building blocks and integrates these into an overall 1-D model that uses several years of hourly wave data. Again, all the validation and what-if scenarios are done in the less costly computation phase of the composite model.

13.11 Summary

The discussion may be summed up as follows:
- Coastal management and design must understand the strengths and limitations of all elements of the design process, recognize the uncertainties and evaluate the design results objectively.
- Physical models are plagued by scale and laboratory effects, and high operating

costs, but are the only tool that can produce certain results.

- The existing trend to use S-type design and process models will continue. In fact, all physical models will become more abstract and less of a direct simulation of a particular prototype problem.
- Reducing scale and laboratory effects will require the development of some new, larger laboratory facilities with more sophisticated boundary controls ("Level 2000" facilities). In parallel with this, a thorough understanding of scaling and modeling techniques, and scale and laboratory effects is needed to make use of these superior tools.
- Many processes cannot yet be modelled numerically and relevant constants vary over large ranges. Numerical modeling needs to be improved substantially to reduce the uncertainties in its results.
- All models need to be properly validated (benchmarked, calibrated and verified) in spite of the costs involved.
- Prototype experimentation and measurement is expensive. Its contributions to engineering solutions will consist mainly of providing quality input and validation data for numerical and physical models.
- Prototype monitoring should provide valuable post-construction verification data to further our understanding about our tools and our results.
- Success in the future will only be ensured only if prototype measurements, physical modeling and numerical modeling are integrated. We need to use such combinations of physical and numerical modeling, and field results that minimize cost and uncertainty of the results.
- Physical modeling can play a pivotal role between field observations, which are necessarily spot measurements of uncontrolled and uncontrollable physical events, and the full understanding of the problem needed for a numerical model.
- Direct integration of numerical models and field data is possible through neuro-network, data assimilation and data reduction techniques.
- Integration of physical modeling with field data and numerical modeling will in part consist of the use of Level 2000 facilities. Some of these need to be large enough to approximate field conditions (that can be controlled, closely observed and repeated), while being controlled by sophisticated computer programs, that will resemble and be based on existing numerical modeling techniques.
- Using composite models can control the cost of modeling.
- In physical composite models, many relatively simple, repetitive, physical process model results are integrated by a relatively simple computer program. Low cost is achieved since all the physical models are simple and relatively similar, and all the lengthy calibration and what-if scenarios are carried out in the less expensive computational phase.
- Numerical models must have a proven track record before they can be used as process models.

14. One-Dimensional Modeling of Coastal Morphology

14.1 Introduction

This chapter discusses the one-dimensional (1-D or 1-Line) coastal morphology model. It is the simplest of all coastal numerical models. At the same time it is the most important, since any study involving coastal change normally involves a 1-D computation somewhere. A relatively detailed discussion on 1-D modeling is presented, since the ideas and principles can be used to solve other problems. The examples will be based on the 1-D general model **ONELINE** (Kamphuis, 1993; Dabees and Kamphuis, 1998, 1999; Dabees, 2000). Other one-dimensional models are Perlin and Dean (1983) and **GENESIS** (Hanson and Kraus, 1989). Section 14.8 introduces an extension of the 1-Line model, called the N-Line model (Perlin and Dean, 1983; Johnson and Kamphuis, 1988; Dabees, 2000; Kamphuis and Dabees, 2000).

The coordinate axes used are shown in Fig. 14.1. A 1-D model solves two simple 1-D simultaneous equations. The first equation, the 1-D Morphology Equation expresses conservation of (sand) mass and calculates shoreline change as a function of distance along the shore. The second equation is the equation of (sand) motion. It is a bulk sediment transport rate formula that expresses alongshore sediment transport rate as a simple function of relevant wave climate and beach parameters.

14.2 The 1-D Morphology Equation

The development of the 1-D morphology equation assumes that a beach profile of constant shape slides along a horizontal base located at closure depth d_c, as introduced in Ch 12. This is shown in Fig 14.2. Closure depth is the depth at which beach profiles are not changed by normally occurring wave conditions. This closure depth may be measured from beach profiles or hydrographic charts. It is the more or less constant depth offshore of the active profile. It can also be estimated from the wave climate (Eq. 12.19).

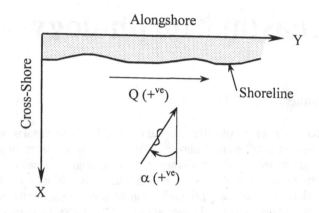

Figure 14.1 Coordinate Axes

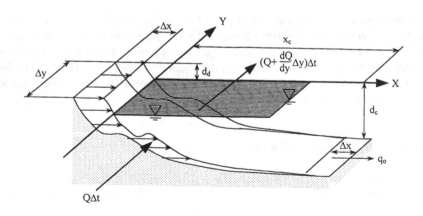

Figure 14.2 Conservation of (Sand) Mass

Erosion causes the profile to move landward and accretion moves it seaward. Since the profile remains the same, all the contours move the same distance and one single contour line can represent the complete beach movement. Hence this method is also known as a 1-Line model[1]. Expressing conservation of (sand) mass in the alongshore direction as shown in Fig. 14.2 results in

$$\frac{dx}{dt} = -\frac{1}{d_p}\left\{\frac{dQ}{dy} - q_o\right\} = -\frac{1}{(d_d + d_c)}\left\{\frac{dQ}{dy} - q_o\right\} \qquad (14.1)$$

where x is the distance to the shoreline from the y (alongshore) axis, d_p is the total profile depth consisting of a dune (or berm) depth (d_d) and the closure depth (d_c), Q is the bulk alongshore sediment transport rate and q_o is the net cross-shore gain of sand per unit distance in the alongshore direction.

14.3 Sediment Transport Rate

14.3.1 Potential Sediment Transport Rate

Alongshore sediment transport rate is computed using the bulk expressions, developed in Ch. 12. These equations integrate all pertinent fluid flow and sediment entrainment properties into simple sediment transport expressions involving a few wave and beach parameters. Because of the simplicity of these expressions, many calculations can be made without involving large computation times. Hence, many years of wave data can be introduced into a 1-D model,. The simplification, using bulk sediment transport expressions is justified, since the beach parameters and wave data normally contain considerable uncertainties.

For bulk sediment transport rate we could use the CERC expression (Eqs. 12.31 to 12.33) or the Kamphuis (1991) expression (Eqs. 12.35 to 12.37). Since computations involve time series of waves defined over short intervals, such as hours, we will use Eqs. 12.33 and 12.37 in which Q is calculated in m^3/hr.

If a longshore gradient in wave height exists, such as in the shadow of structures, we take this into account by changing the wave angle term in Eqs. 12.33 and 12.37 to

---

1. Actually the development of Eq. 14.1 assumes that the shore is a vertical line that moves in the cross-shore direction over a depth d_c, but the same equation can also represent a profile of constant shape, as discussed here.

$$\left[\sin^{c_1} 2\alpha_b - \frac{c_2}{m_b} \cos\alpha_b \frac{dH_{sb}}{dx} \right] \tag{14.2}$$

where c_1 is 1 for the CERC expression and 0.6 for the Kamphuis expression. The coefficient c_2 is a matter of discussion (Gourlay, 1978; Ozasa and Brampton, 1980; Kraus and Harikai, 1983). Hanson and Kraus (1989) suggest values of c_2 between 1 and 2. We retain this coefficient as a calibration coefficient in the model.

14.3.2 Actual Sediment Transport Rate

The sediment transport expressions calculate potential sediment transport rates. In most practical cases, storm conditions vary rapidly and the sand is of limited extent and volume. As a result, the above equations will overestimate the actual sediment transport rate. Although the differences between the two rates are a complex function of many parameters, as shown in Ch. 12, we follow the usual (simplistic) practice

$$Q_a = C_Q \, Q_p \tag{14.3}$$

where Q_a is the actual rate, Q_p is the potential rate and the coefficient C_Q is retained as a second calibration coefficient, where $0 \le C_Q \le 1$. The assumption that actual sediment transport rate is a simple fraction of potential rate is essentially valid only for long term computations and cannot really be used for individual storms as discussed in Ch. 12.

14.4 Wave Transformation Computation

14.4.1 Wave Shoaling, Refraction and Breaking

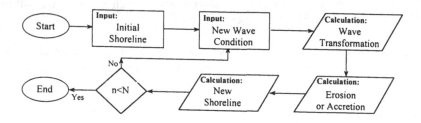

Figure 14.3 Computation Scheme

To solve beach morphology problems, a series of computations is repeated for each incident wave condition, as shown in Fig 14.3. First a shoreline is defined at the initial time t_I. Wave transformation for the waves at t_I is then computed for the beach shape (the combination of the shoreline and the beach profile). From these transformed waves, erosion or accretion is calculated over a time Δt. These modify the beach shape and produce a new beach shape at $(t_I+\Delta t)$. The whole process is then repeated for the next wave condition on the new beach shape, stepping the computation forward to a final time $t_F=N\Delta t$. Because of the many repetitions of the calculations, a 1-D model needs to use relatively simple wave transformation computations. Normally, the simplest relationships of Ch. 7 are used. It is possible to use more complex wave refraction calculations, but this will affect the computational time. The added value of such extended calculations needs to be carefully evaluated in light of the uncertainties in the data and computations.

14.4.2 Wave Diffraction

Diffraction calculations are lengthy and need to be simplified before they can be included in a 1-D model. Parallel offshore breakwaters have been discussed in Perlin and Dean (1983), Hanson and Kraus (1989) and other publications related to the **GENESIS** model. Dabees and Kamphuis (1998) and Dabees (2000) discuss these diffraction routines in **ONELINE** extensively. As an example, of the reasoning and simplifications that is needed for the development of simplified refraction-diffraction computations, the refraction-diffraction relations near a groin are presented here in some detail.

Goda (1985) calculates wave diffraction, using the directional spreading of the incident waves (Ch. 3 and 7). He assumes that an obstruction simply blocks out a portion of the energy of incoming directional wave spectrum. Using Goda's method and some additional assumptions, simple expressions for refraction-diffraction behind a groin may be developed. In Fig 14.4, the incoming wave ray at a structure of length S_s makes an angle α_s with the structure and an angle θ with respect to the shadow line for the mean wave direction (AO). Goda (1985) assumes that all energy in the directional spectrum for which $\alpha_s<\alpha_m$ is blocked by the structure and removed from the spectrum. That removes half of the incident wave energy along the shadow line, and using Eq. [10] of Table 2.2 we find that H along the shadow line is $\sqrt{2}H_i = 0.71H_i$, where H_I is the incident wave height. Thus along the shadow line, the diffraction coefficient $K_d=0.71$. We can also relate the wave energy reaching a point P to the angle θ and regression analysis yields

$$K_d = 0.71 - 0.0093\,\theta + 0.000025\,\theta^2 \quad for \quad 0 \geq \theta \geq -90 \qquad (14.4)$$

$$K_d = 0.71 + 0.37 \sin\theta \quad for \quad 40 \geq \theta > 0 \tag{14.5}$$

$$K_d = 0.83 + 0.17 \sin\theta \quad for \quad 90 \geq \theta > 40 \tag{14.6}$$

Breaking wave height may be estimated by

$$H_{sb} = K_d \ H_{srb} \tag{14.7}$$

where H_{srb} is the wave height due to shoaling, refraction and breaking, without diffraction. Since the calculated values of K_d reduce the wave heights behind the structure, the breaking angle will also be reduced. The effect of wave diffraction on wave breaking angle was tested for ranges of S_s, α_s, T_p, d_c, d_b and H_b and the breaking angle adjusted for diffraction (α_{bd}) was found to be a simple relationship with K_d

$$\alpha_{bd} = \alpha_b K_d^{0.375} \tag{14.8}$$

Equation 14.8 is valid both inside and outside the shadow zone. Inside the shadow zone, however, a further decrease in breaking angle, resulting directly from wave diffraction, must be taken into account. The wave from the end of the groin according to Eq. 14.8 will have a breaking angle of

$$\alpha_{bo} = \alpha_b \ (0.71)^{0.375} = 0.88\alpha_b \tag{14.9}$$

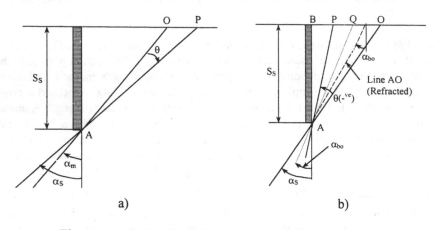

Figure 14.4 Refraction-Diffraction Definitions near a Groin

We assume that the wave ray from the end of the groin, makes landfall halfway between the shadow line AO and the line AQ, which makes an angle of α_{bo} with the groin. Since the breaking angle at the structure is zero, a simple proportionality ratio may be introduced so that for

$$\theta < 0 \quad and \quad \frac{PB}{S_s} < \frac{1}{2}\{\tan \alpha_S + \tan(0.88\alpha_b)\} \tag{14.10}$$

the adjusted breaking angle is

$$\alpha_{bd} = \alpha_b \, K_d^{0.375} \left[\frac{2\,PB}{S_s \{\tan \alpha_S + \tan(0.88\alpha_b)\}} \right] \tag{14.11}$$

14.5 Analytical Computation of Shore Morphology

14.5.1 Simplifications and Assumptions

The equations will now be simplified to obtain analytical solutions for simple boundary conditions. Such simple solutions give a quick and inexpensive impression about shoreline response, and can be used to benchmark numerical models (Ch. 13).

The shoreline is assumed to be initially straight along the Y-axis (x=0 everywhere) and a beach profile of constant shape is assumed to slide seaward during accretion or landward during erosion over a depth d_c. Accretion and erosion therefore cause a change in x.

The local shoreline orientation with respect to the original straight shoreline (the Y-axis) at any time is dx/dy and an effective local breaking angle, with respect to the rotated shoreline, may be defined as

$$\alpha_e = \alpha_b - \frac{dx}{dy} \tag{14.12}$$

Substituting Eq. 14.12 into Eqs. 12.33 or 12.37, results in

$$Q = q \sin 2\alpha_e = q \sin 2\left(\alpha_b - \frac{dx}{dy}\right) \tag{14.13}$$

where q is a collection of terms defined by Eq. 12.33 or 12.37 and Eq. 14.3. To solve the morphology problem analytically, α_e is assumed to be small so that

$$\sin 2\alpha_e \quad \rightarrow \quad 2\,\alpha_e \tag{14.14}$$

and Eq. 14.13 becomes

$$Q = 2q\left(\alpha_b - \frac{dx}{dy}\right) \tag{14.15}$$

Equation 14.15 may be differentiated to yield

$$\frac{dQ}{dy} = -2q\,\frac{d^2x}{dy^2} \tag{14.16}$$

Substitution into Eq. 14.1, assuming $q_0=0$ yields the diffusion equation

$$\frac{dx}{dt} - \underline{D}\,\frac{d^2x}{dy^2} = 0 \tag{14.17}$$

where from Eqs. 14.1, 14.15 and 14.16

$$\underline{D} = \frac{2q}{d_p} = \frac{Q}{\alpha_b\,d_p} \tag{14.18}$$

If Q is expressed in $[m^3/hr]$, then $\underline{D}$ is in $[m^2/hr]$.

Pelnard-Considère (1956) solved this diffusion equation for three simple boundary conditions: a barrier that interrupts the alongshore transport completely, a bypassing barrier and an instantaneous release of sand on a beach. Le Méhauté and Brebner (1960) also discuss analytical solutions of this equation. A later discussion of those and other analytical solutions may be found in Larson et al (1987). They treat several examples of sand supply through beach nourishment and river discharge and they solve shoreline evolution by groins, detached breakwaters and seawalls. Dean and Yoo (1993) also uses the concept to develop design criteria artificial beach nourishment.

This 1-D analytical solution is also known as a 1-Line analytical solution, because the whole profile moves as a unit so that it can be represented by a single contour line. The concept may be extended to include profiles that change shape. Willis (1978) proposes a profile that rotates. Bakker (1968) postulates a 2-Line analytical solution (Fig. 14.5), derived specifically to calculate the effect of groins on a beach. Essentially, two 1-Line models are stacked vertically and connected through the cross-shore exchange of sediment between them, which according to Bakker is linearly related to the difference between the existing beach profile and its equilibrium shape. Bakker solves these equations for zero net sediment input from external sources in the cross-shore direction. Other analytical solutions of the 1-D equations are by Le Méhauté and Soldate (1978) who included refraction and diffraction.

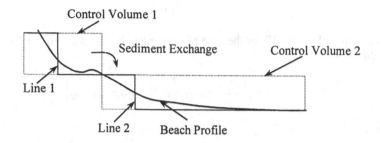

Figure 14.5 Two-Line Model (after Bakker, 1968)

14.5.2 Complete Barrier Solution

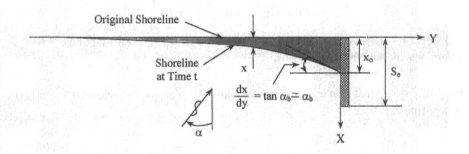

Figure 14.6 Analytical Solution for a Complete Barrier

The boundary condition governing a complete barrier located at y=0 (Fig. 14.6) is

$$Q_o = 0 \qquad (14.19)$$

Substitution into Eq. 14.15 yields

$$\left[\frac{dx}{dy}\right]_o = \alpha_b \cong \tan \alpha_b \tag{14.20}$$

The general solution for x in Eq. 14.17 is a function of y and t. We define the variable

$$u = \frac{y}{\sqrt{4\underline{D}\,t}} \tag{14.21}$$

The solution of Eq. 14.17 for the case of a complete barrier with Eq. 14.19 as boundary condition is

$$x = \sqrt{\frac{4\underline{D}\,t}{\pi}} \tan \alpha_b \left[e^{-u^2} - u\sqrt{\pi}\; erfc\,(u) \right] \tag{14.22}$$

where

$$erfc\,(u) = 1 - erf\,(u), \tag{14.23}$$

Erf (u) is the error function of u and erfc(u) is the complementary error function of u. Numerical approximations may be found in Abramowitz and Stegun (1965) and in spreadsheet and software packages.

We will denote the term in square brackets as

$$F(u) = \left[e^{-u^2} - u\sqrt{\pi}\; erfc\,(u) \right] \tag{14.24}$$

and Fig. 14.7 shows erf(u), erfc(u) and F(u). These are all even functions; function values are the same for (u) and (-u).

Eq. 14.22 shows that x, at any y, increases with $\sqrt{t}$. The surface area of the accretion at any time is

$$A_s = \frac{Q\,t}{d_p} \tag{14.25}$$

Since Q and d_p are constant and A_s is a product of x and y, then x at any y also increases with $\sqrt{t}$.

At the structure: y=0 and u=0, which means

$$x_o = \sqrt{\frac{4D\,t}{\pi}}\ \tan \alpha_b \tag{14.26}$$

and Eq. 14.22 becomes

$$x = x_o\ F(u) \tag{14.27}$$

Assuming $\tan \alpha_b \approx \alpha_b$, substitution of Eq. 14.18 into 14.26 yields

$$x_o = 2\sqrt{\frac{Q\,t}{\pi\,d_p}}\ \tan \alpha_b \tag{14.28}$$

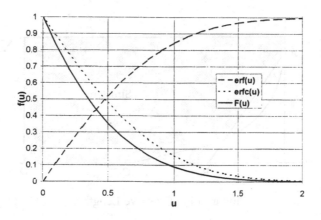

Figure 14.7 Error Function

14.5.3 Bypassing Barrier Solution

Unless a structure is infinitely long, the boundary condition (Eq. 14.19) will sooner or later become invalid and sediment will begin to bypass the structure. The time required to fill the structure may be calculated by setting

$$x_o = S_e \tag{14.29}$$

where S_e is the effective length of the structure. Equations 14.26 and 14.28 yields

$$t_{full} = \frac{\pi S_e^2}{4\underline{D}\tan^2 \alpha_b} = \frac{\pi d_p S_e^2}{4Q \tan \alpha_b} \qquad (14.30)$$

Once t_{full} is reached, Eq. 14.19 is no longer valid and the new boundary condition at the barrier is

$$x_o = S_e \qquad (14.31)$$

which yields the solution

$$x = S_e \, erfc(u) \qquad (14.32)$$

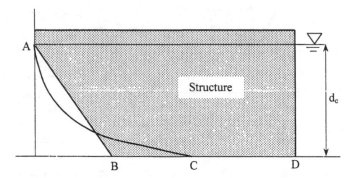

Figure 14.8 Effective Length

Effective length of the structure is the distance from where the beach profile intersects d_c to the end of the structure and depends, therefore, on the definition of the beach profile (Fig. 14.8). If it is assumed to be a simple slope (AB), then S_e=BD. If a beach profile is assumed (AC) then S_e=CD. But the position of point C is very sensitive to the choice of d_c, making the value of S_e difficult to determine. In view of the other simplifying assumptions, we will assume an average beach slope AB when calculating S_e.

The rate at which sediment bypasses the barrier is calculated as

$$Q_{by} = 2q\,\alpha_b \left[1 - \sqrt{\frac{t_{full}}{t}} \right] = Q \left[1 - \sqrt{\frac{t_{full}}{t}} \right] \tag{14.33}$$

Equations 14.22, 14.32 and 14.33 assume that the structure is long enough so that S_e is positive. For short groins, S_e will be negative and the equations will not work. The method is therefore only applicable to long groins, jetties and breakwaters.

Example 14.1[⊗] Analytical Calculation

For the conditions

$$\alpha_b = 3°; \quad Q = 10^6 \ m^3/yr: \quad S_e = 400\,m; \quad d_p = 5\,m$$

Equation 14.15 yields

$$q = \frac{Q}{2\alpha_b} = \frac{10^6}{2(3x\pi/180)} = 9.54 \cdot 10^6 \ m^3/yr$$

From Eq. 14.18

$$\underline{D} = \frac{Q}{\alpha_b d_p} = \frac{2q}{d_p} = 4.0 \cdot 10^6 \ m^2/yr$$

Equation 14.21 yields

$$u = \frac{y}{\sqrt{4\underline{D}t}} = \frac{y}{4000\sqrt{t}}$$

From Eq. 14.28

$$x_o = \left[\frac{4 \cdot 10^6}{\pi\ 5} 0.0524\,t \right]^{1/2} = 115.5\sqrt{t}$$

and from Eqs. 14.21 and 14.27 the shoreline may be calculated as

$$x = 115.5\sqrt{t}\ F\left(\frac{y}{4000\sqrt{t}} \right)$$

and t_{full} may be calculated using Eq. 14.31

$$t_{full} = (400/115.5)^2 = 12\ yrs.$$

After 12 years, shoreline position y is calculated with Eq. 14.32. The details of accretion and bypassing are presented in Table 14.1 and Fig. 14.9.

Table 14.1 Accretion Against Structure (m)

Time	Distances (m)							Q_{by}
(yrs)	0	500	1,000	2,000	5,000	10,000	20,000	(m³/yr)
1	116	91.2	70.6	39.7	4.0	0.0	0.0	0
4	231	206	182	141	57.7	8.1	0.0	0
10	365	340	315	270	162	58	3.4	0
12	400	374	350	304	192	77	6.7	0
25	400	389	377	354	287	188	59	307,000
50	400	392	384	367	319	244	123	510,000

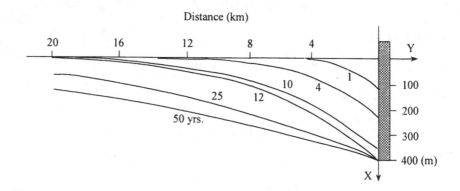

Figure 14.9 Example Analytical Solution

Analytical solutions have many obvious limitations. They calculate approximate accretion for one (average) wave condition and contain many simplifying assumptions. Yet they are often used in preliminary analysis for large structures.

If the sand accretes against the structure as in Fig. 14.9, the beach on the other side of the structure must be eroding (Ch. 12). A first estimate of the erosion pattern is the accretion pattern of Fig. 14.9, flipped about both the x and y-axes. This is obviously not correct because diffraction around the structure affects the erosion, modifying the pattern extensively, but some distance away from the structure, this approximation is reasonable.

14.6 Numerical Solutions

14.6.1 Basics

For practical problems, the equations, the input wave conditions and the boundary conditions cannot normally be simplified sufficiently for analytical solutions to be valid. In that case, Eq. 14.1 and 14.3, together with the bulk sediment transport rate equations of Ch. 12 are solved numerically.

The 1-D program called **ONELINE** and the N-Line program called **NLINE** will provide the examples for this chapter. **ONELINE** uses Eqs. 12,37, 14.1 and 14.3. The discussion below involves one single wave condition (one combination of H, T and α) but in normal 1-Line computations the procedure includes many such wave conditions, or a complete wave climate with many incident wave conditions (Fig. 14.3).

First, the shoreline is discretized into a series of sections of finite length as shown in Fig. 14.10. If the shoreline curves slowly, the wave angle, α_b, may be defined with respect to a y-axis, which is either the direction of the shoreline trend (average shoreline direction) or the original shoreline. If a strongly curved original shoreline is simulated, each shoreline section will be subjected to a different α_b for the same incident wave as in Fig. 14.11.

Calculation of sediment transport rate, using Eqs. 12.37 and 14.3 takes place at the ends of the sections and shoreline position is calculated with Eq. 14.1 at the middle of each section. The computation uses finite difference techniques and is stepped forward in time using increments of Δt. Finite difference methods are explained in many standard texts, such as Abbott (1979) and Hoffman (1992). The simplest finite difference scheme to program is the Explicit Finite Difference Scheme in which every new value of Q and x at a new time (t+Δt) is computed explicitly from the known values of Q and x at a previous time t. However, the explicit scheme easily becomes unstable (the errors grow to infinity). The stability condition is

$$\left[\frac{Q}{\alpha \, d_F} \cdot \frac{\Delta t}{(\Delta y)^2} \right] < \frac{1}{2} \tag{14.34}$$

and it only permits time steps of the order of hours, before computation becomes unstable. This is inefficient for many M and L-type calculations, which cover years to centuries.

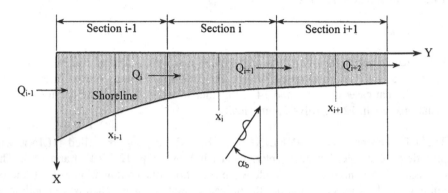

Figure 14.10 Discretized Shoreline for 1-D Model
(Shoreline Trend Option)

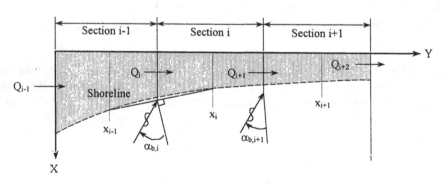

Figure 14.11 Discretized Shoreline for 1-D Model
(Shoreline Sections Option)

14.6.2 Implicit Finite Difference Scheme

Implicit Finite Difference Schemes are more difficult to program, but do not pose the same limitations on Δt. In such a scheme, the new values of Q and x at time ($t+\Delta t$) all along the shoreline are calculated simultaneously. Any implicit method, however, is based on (linear) matrix algebra and therefore Eqs. 12.37 and 14.1 must be expressed as first order equations in x and Q.

Equation 14.1 may be written in finite difference form, for each of the N sections. One possible form is

$$\frac{x_i^* - x_i}{\Delta t} = -\frac{1}{(d_p)_i} \left\{ \frac{Q_{i-1}^* - Q_i^*}{\Delta y} - q_o \right\} \tag{14.35}$$

where the * indicates calculated values at the new time ($t+\Delta t$). Eq. 14.35 may be written as

$$- A_i \, Q_i^* + x_i^* + A_i \, Q_{i-1}^* = S_i \tag{14.36}$$

where the left side contains the unknown values at time ($t+\Delta t$).

$$A_i = \frac{\Delta t}{(d_p)_i \, \Delta y} \tag{14.37}$$

and where the right side consists only of values known at time t

$$S_i = x_i + A_i \, \Delta y \, q_o \tag{14.38}$$

At the section ends Q is either specified (as a boundary condition) or calculated using Eqs. 12.37 and 14.3. The implicit solution could assume a small effective breaking angle, as discussed in Section 14.5 for the analytical solutions. However, this simplest linearization, which will be called the *small angle* option hereafter (SA), is only valid for both small incident breaking wave angles and small shoreline orientation angles from the mean shoreline trend. It yields uncertain results for finite values of α_b and dx/dy.

Following the example of Perlin and Dean (1978), it is possible to expand the term (sin $2\alpha_e$) so that α_e no longer needs to be small.

$$\sin \, 2\alpha_e = \sin \, 2\,(\alpha_b - dx/dy) = \sin \, 2\,(\alpha_b - \alpha_s) \tag{14.39}$$

where α_s is the shoreline angle, defined as

$$\alpha_s = \tan^{-1} (dx/dy) \tag{14.40}$$

Equation 14.39 may be expanded as

$$
\begin{aligned}
\sin 2\alpha_e &= \sin 2\alpha_b \ \cos 2\alpha_s - \cos 2\alpha_b \ \sin 2\alpha_s \\
&= \sin 2\alpha_b \ \cos 2\alpha_s - 2\cos 2\alpha_b \ \cos \alpha_s \ \sin \alpha_s \\
&= \sin 2\alpha_b \ \cos 2\alpha_s - 2 \ \cos 2\alpha_b \ \cos \alpha_s \left[\frac{dx}{ds} \right]
\end{aligned}
\tag{14.41}
$$

where

$$ds = \sqrt{dx^2 + dy^2} \tag{14.42}$$

Equation 14.41 may be written as a finite difference equation

$$
\begin{aligned}
\sin 2\alpha_e &= (\sin 2\alpha_b \ \cos 2\alpha_s) \\
&\quad - 2(\cos 2\alpha_b)(\cos \alpha_s) \left[\frac{\phi(x_i^* - x_{i-1}^*)}{ds} + \frac{(1-\phi)(x - x_{i-1})}{ds} \right]
\end{aligned}
\tag{14.43}
$$

For a completely implicit scheme $\phi=1$; a completely explicit scheme would have $\phi=0$. **ONELINE** sets $\phi=0.5$, taking into account some of the effects of both the old and new values. Shoreline angle, α_s, is assumed to be relatively small in the derivation of Eq. 14.43, but the assumption is not very restrictive. Nevertheless, **ONELINE** always keeps this angle small.

For the actual sediment transport rate Eqs. 12.37 and 14.3 may combined as

$$Q_i = (c_3)_i \ \sin^{0.6} 2\alpha_e \tag{14.44}$$

where

$$(c_3)_i = 7.3 C_Q \ (H_{sb})_i^2 \ (T_p)_i^{1.5} \ (m_b)_i^{0.75} \ (D_{50})^{-0.25} \tag{14.45}$$

We retain C_Q as a calibration coefficient. **ONELINE** evaluates Eq. 14.44 as

$$Q_i = \frac{(c_3)_i \ \sin \ 2\alpha_e}{(\sin^{0.4} 2 \alpha_e)} = (c_4)_i \ \sin 2 \alpha_e \tag{14.46}$$

Combining Eqs. 14.43 and 14.46 results in

$$-B_i x_{i-1}^* + Q_i^* + B_i x_i^* = R_i \qquad (14.47)$$

where the unknowns are on the left side of the equation and

$$B_i = \frac{(c_4)_i}{ds} \cos 2\alpha_b \cos \alpha_s \qquad (14.48)$$

and where all the values known from the previous time step are collected on the right side

$$R_i = (c_4)_i \ [\sin 2\alpha_b \cos 2\alpha_s - \cos 2\alpha_b \sin 2\alpha_s] \qquad (14.49)$$

If the boundary conditions are stated as known values of Q, then Eqs 14.36 and 14.47, expressed simultaneously for each of the N shoreline sections along with the boundary conditions result in a tri-diagonal matrix.

$$
\begin{bmatrix}
1 & & & & & & & & & \\
-A_1 & 1 & A_1 & & & & & & & \\
& -B_2 & 1 & B_2 & & & & & & \\
& & -A_2 & 1 & A_2 & & & & & \\
& & & -B_3 & 1 & B_3 & & & & \\
& & & & & & \ddots & & & \\
& & & & & -B_{N-2} & 1 & B_{N-2} & & \\
& & & & & & -A_{N-1} & 1 & A_{N-1} & \\
& & & & & & & -B_{N-1} & 1 & B_{N-1} \\
& & & & & & & & -A_N & 1 & A_N \\
& & & & & & & & & 1
\end{bmatrix}
\begin{bmatrix}
Q_1^* \\ x_1^* \\ Q_2^* \\ x_2^* \\ Q_3^* \\ \vdots \\ Q_{N-1}^* \\ x_{N-1}^* \\ Q_N^* \\ x_N^* \\ Q_{N+1}^*
\end{bmatrix}
=
\begin{bmatrix}
R_1 \\ S_1 \\ R_2 \\ S_2 \\ R_3 \\ \vdots \\ R_{N-1} \\ S_{N-1} \\ R_N \\ S_N \\ R_{N+1}
\end{bmatrix}
$$

$$(14.50)$$

The blanks indicate zeros. The first and last lines are the boundary conditions; R_1 and R_{N+1} are the sediment transport rates at the ends of the model. This matrix must be solved for each time step, to determine the new Q and x values. All values of R and S are based on values known from the previous time step.

14.6.3 Boundary Conditions

Boundary conditions connect the model to the outside world and are defined from outside the model. The boundaries of the computation must therefore be far enough away from any changes within the model, so that the boundary conditions are not affected by these changes. A usual set of boundary conditions is

$$Q_1 = R_{1,o} \; ; \quad Q_{N+1} = R_{N+1,o} \tag{14.51}$$

where $R_{i,o}$ is the sediment transport rate calculated for the original shoreline. Another boundary possible condition is the complete barrier.

$$Q_i = 0 \tag{14.52}$$

This condition describes the effect a structure or geological formation that is long enough to prevent any sediment from passing it. In time, however, as beach accretes against a structure, Eq. 14.52 will become invalid and a bypassing condition needs to be specified. **ONELINE** uses a bypassing expression that is based on the exponential beach profile shape of Eq. 12.15. The distance to the seaward end of the active profile (where the active profile intersects d_c) may be computed as

$$x_c = \left(\frac{d_c}{A_p} \right)^{3/2} \tag{14.53}$$

At any time, the structure has an effective length S_e defined as

$$S_e = S_s - x_s \tag{14.54}$$

where S_s is the structure length and x_s is the accumulation of sediment against the structure. It is assumed that rate of sediment bypassing is related to the active beach profile above d_c that extends beyond the end of the structure (A_2 in Fig. 14.12).

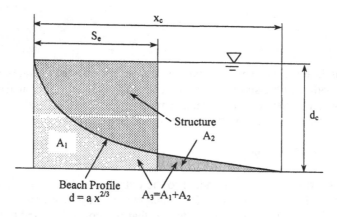

Figure 14.12 Definitions for Bypassing Computation

Integration of the profile leads to the bypassing sediment transport rate

$$\frac{Q_{by}}{Q} = \frac{A_2}{A_3} = \left(1 - \frac{A_1}{A_3}\right) =$$

$$1 - \left\{\frac{S_e d_c - 3/5 A_p S_e^{5/3}}{x_c d_c - 3/5 A_p x_c^{5/3}}\right\} \quad for \quad 0 < S_e < x_c \tag{14.55}$$

and

$$Q_{by} = 0 \quad for \quad S_e \geq x_c; \qquad Q_{by} = Q \quad for \quad S_e \leq 0 \tag{14.56}$$

where Q is the rate at which the sediment arrives at the structure. When the groin is filled $S_e=0$ and all the sediment bypasses the groin. Equations 14.53 to 14.56 are boundary conditions, but they can also be used to represent structures within the model.

The above boundary conditions define Q values. The rate of change of Q, (dQ/dy) could also be specified – for example, dQ/dy=0 indicates that Q at the boundaries is always equal to the ambient actual sediment transport rate. Similarly, the boundaries could be placed at a location where x is calculated. The last condition was used in **ONELINE** to simulate slowly eroding headlands.

14.6.4 Beach Slope

If a specific profile like in Eq. 12.15 is used in the computation, it should also be used to define the beach slopes for sediment transport rate and wave transformation. For the breaker criteria Eqs. 7.31 and 7.32 we need the slope immediately offshore of the breaker.

We can differentiate Eq. 12.15

$$m_x = \frac{d(d)}{dx} = \frac{2}{3} A_p \, x^{-1/3} \tag{14.57}$$

and this allows us to compute the beach slope at breaking

$$m_b = \frac{2}{3} A_p \, x_b^{-1/3} = \frac{2}{3} A_p^{3/2} \, d_b^{-1/2} \tag{14.58}$$

The average slope over the whole profile is

$$\overline{m} = \frac{d_c}{x_c} = \frac{d_c}{d_c^{3\,2}} A_p^{3/2} = \frac{A_p^{3/2}}{d_c^{l\,2}} \qquad (14.59)$$

14.6.5 Large Shoreline Curvatures

The development of Eq. 14.43 permits reasonably large shoreline orientation angles with respect to the original shoreline. For reasonably straight beaches, it will therefore permit computation with respect to a straight Y axis or shoreline trend, i.e. defining one single value of α_b for the whole beach for each wave condition, as in Fig. 14.10. In **ONELINE** this is called the (Oneline) *Shoreline Trend* option (OLST).

If the original shoreline is strongly curved, however, dx/dy (related to a straight line shoreline trend) will become large enough that the OLST option can no longer be used. This is the case, particularly, for L-type problems that cover large distances and long durations. It is possible to define the original (curved) shoreline, rather than a straight shoreline trend as the base line for the computation as in Fig. 14.11. This means the existing shoreline is discretized and α_b is calculated for each shoreline section. Equations 12.37 and 14.3 are used to calculate Q even though the equations assume an infinitely long, straight shoreline. A simple smoothing function helps to introduce the effects of the adjacent shoreline sections. This variation of **ONELINE** is called the (Oneline) *Shoreline Sections* option (OLSS).

For the OLSS option, the distance between the original shoreline and the Y axis is still called x. If the original shoreline is used as baseline for the calculation, the computation proceeds implicitly as in Section 14.6.2 above, calculating the distance between the new shoreline and the original shoreline, called x' to distinguish it from x. The values of (dx'/dy) are zero at the start of such a computation. As the calculation proceeds, the values of (dx'/dy) are monitored and when they exceed a limit, so that the computation could become inaccurate, x' is added to x and the latest calculated shoreline becomes a new "original" shoreline.

14.6.6 Summary

Three distinct computational schemes were discussed:
- The Oneline Line Small Angle (OLSA) model, uses the small angle assumption that sin $(2\alpha_e)$ may be replaced by $(2\alpha_e)$ in an implicit scheme. This numerical method is more versatile than the analytical method, but has similar shortcomings for large angles of wave incidence or shoreline direction.
- The Oneline Shoreline Trend (OLST) model uses Eq. 14.46, a much better

expansion of the sin $(2\alpha_e)$ term. That removes the restriction that α_b needs to be small.

– The Oneline Shoreline Sections (OLSS) model uses the actual shoreline as a calculation base. This removes the restriction that (dx/dy) needs to be small and when (dx'/dy) becomes sufficiently large, the latest calculated shoreline becomes a new calculation base.

The above presentation on 1-D modeling may appear as if the problem is simplified too much. This may indeed be the case for some problems. Yet the 1-Line model has some surprising applications to more complicated problems. First, there is the possibility to link several of these 1-Line models into 2-Line or Multi-Line Models as discussed earlier and exemplified by Bakker (1968), Kamphuis and Johnson (1988) and Kamphuis and Dabees (2000). The several 1-Line models are linked by cross-shore transport functions, which are normally related to how far the existing profile is out of equilibrium with the incident wave condition. This theme is expanded in Section 14.8. The 1-Line model can also be linked in the alongshore direction with several other 1-Line Models, thus allowing for quite complex boundary conditions and quite different physical conditions throughout the calculation area. Another application is to match the 1-Line Model with an equally global cross-shore model such as **SBEACH** (Kraus and Hanson, 1990). It was also shown in Ch. 13 that the 1-Line model can be used as the calculation base for a numerical composite model.

Finally, the 1-Line Model will have a major role in L-type behavioral modeling (Ch. 13). The whole cross-shore zone is taken into account over large distances and long durations. It consists of:

– A dune, subject to wave and wind erosion,
– The "active" profile of a constant shape, such as in Eq. 12.15 which moves in the cross-shore direction and which is ideally modeled by a 1-Line (or Multi-Line) model. This profile reaches down to a closure depth.
– An offshore point on the continental shelf, which is morphologically at rest.
– A transition region that connects the active profile to offshore.

14.7 Examples of ONELINE

The most recent updates of the 1-D program **ONELINE** are found in Dabees and Kamphuis (1998) and Dabees (2000). The examples presented here come from these references. **ONELINE** was first benchmarked against analytical solutions. Figure 14.13 is a vector plot showing wave heights and directions, produced from the simplified refraction-diffraction analysis behind a single offshore breakwater subjected to waves at 0° angle of incidence. Figure 14.14 shows diffraction coefficients for waves with a 10° angle of incidence. Figure 14.15 presents morphology over 2 years of a constant wave condition, at 0° incidence. These figures and others were used to benchmark the program against theoretical and published results. For example, Fig 14.15 was extensively benchmarked against results in Silvester and Hsu (1997).

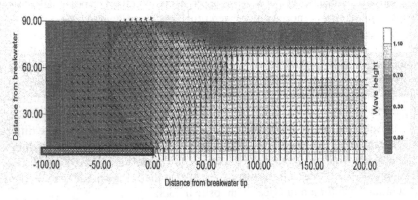

Figure 14.13 Wave Heights and Directions Behind Offshore Breakwater

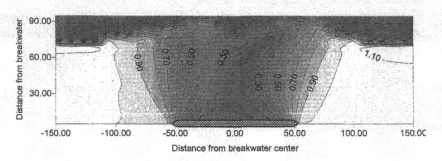

Figure 14.14 Diffraction Coefficients for 10° Angle of Incidence

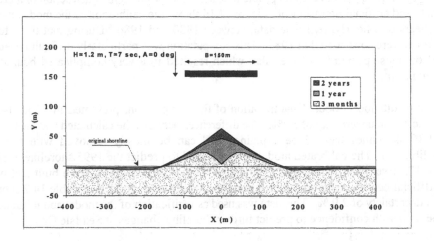

Figure 14.15 Accretion Behind Offshore Breakwater

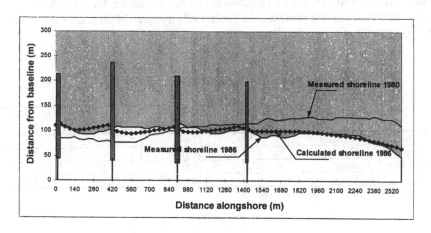

Figure 14.16 Calibration of Sea Isle City Model

Figure 14.16 shows part of the computational domain of a study of Sea Isle City, New Jersey. Here a coastal erosion problem near an inlet resulted in construction of a groin field and artificial nourishment. Figure 14.16 shows the calibration of the model, using 6 years of 4-hourly, real time data, between 1980 and 1986. During that time, four groins were constructed in 1983. This computation was particularly difficult because of the close proximity of the inlet, which resulted in a very complicated boundary condition[2].

The results look good. Quantification of the comparison, presented in Table 14.2, indicates an uncertainty of 3.2%. The differences between the calculated and measured 1986 shorelines can all be explained and can be mostly removed with further calibration. The calibrated model was then used to predict the 1995 shoreline in Fig. 14.17. Even though between 1986 and 1995 two further groins were built and two artificial nourishments were introduced, the results were good, and Table 14.2 shows an uncertainty of 2.9 %. This data was used as verification of the model and it can now be used with confidence to predict further shoreline changes at Sea Isle City.

Figure 14.18 shows the calibration of a model for Ras el Bar, on the Nile Delta, which eroded at 3 to 5 m/yr. at that location. The project involves a combination of groins, constructed in 1970 and offshore breakwaters, built in 1990. Model calibration for 2-hourly waves between 1986 and 1993 is realistic, even though the breakwaters were constructed in 1990. Figure 14.19 shows the model validation for the 1995 shoreline and the uncertainties are presented in Table 14.2. Figure. 14.20 shows the sediment transport rates that resulted in the accretion.

Table 14.2 Quantification of Errors

	Mean Shoreline Change (m)	σ (m)	Uncertainty (%)
Sea Isle City Calibration	27.6	0.88	3.2
Sea Isle City Verification	32.8	0.94	2.9
Ras El Bar Calibration	23.8	1.57	6.7
Ras El Bar Verification	32.4	1.86	5.7

$\sigma = \sigma(\text{predicted} - \text{observed}) = \sigma(\text{model} - \text{prototype})$

---

2. The boundary was so close to the model area that Q at the boundary was modified by the changes within the model area. It was therefore not a true boundary condition, which would have remained unaffected by what occurs within the model.

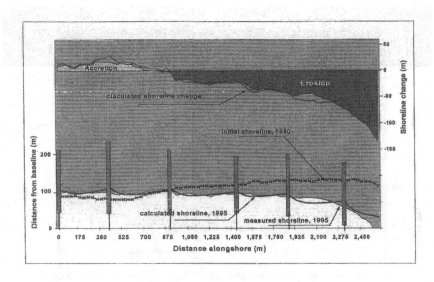

Figure 14.17 Final Results and Verification of Sea Isle City Model

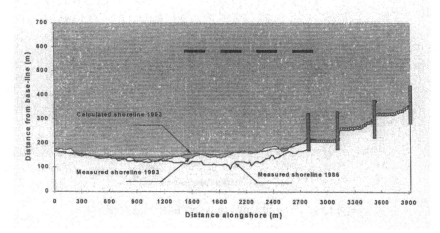

Figure 14.18 Calibration of Ras El Bar Model

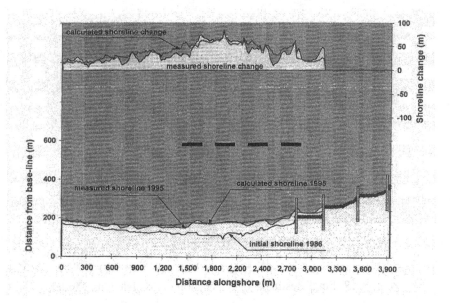

Figure 14.19 Verification of Ras El Bar Model

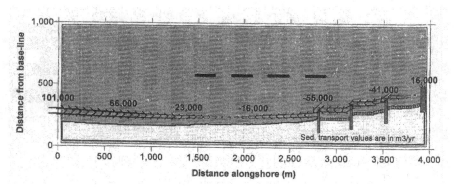

Figure 14.20 Sediment Transport at Ras El Bar

14.8 Examples of NLINE

An N-Line calculation is used when the constant beach profile as in 1-Line calculations, produces incorrect solutions. The basic calculation scheme for the program **NLINE** (Dabees, 2000; Kamphuis and Dabees, 2000) is shown in Fig 14.21. Figure 14.22 depicts the capability of **NLINE** to deal with profile changes as a result of cross-shore sediment transport rate. Alongshore sediment transport rate distribution behind a single offshore breakwater is shown in Fig. 14.23. Figure 14.24 shows the salient growth behind three offshore breakwaters. The last calculation was for a 1-year, 3-hourly wave climate on Lake Michigan. The mean wave parameters were H_s=1.1 m, T_p=4.6 sec and the maxima were H_s=5 m and T_p=9.1 sec. The incident wave angles varied about a mean angle of approximately 10°. The computation involved 3 breakwaters (6 simultaneous refraction-diffraction patterns). The model covered 2.4 km of shoreline at 20 m section lengths and at 10 contour lines, resulting in 1200 calculation cells. The whole computation required 10 minutes on a PC (Pentium III).

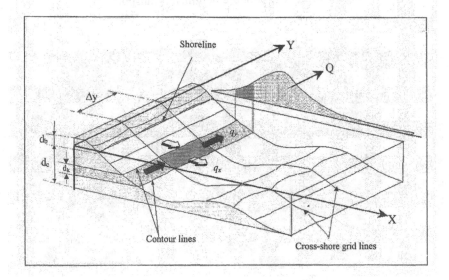

Figure 14.21 N-Line Model

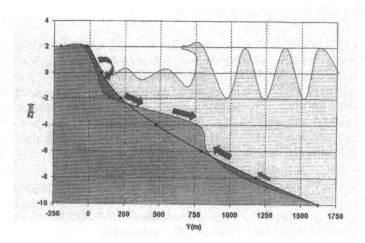

Figure 14.22 Cross-Shore Sediment Transport

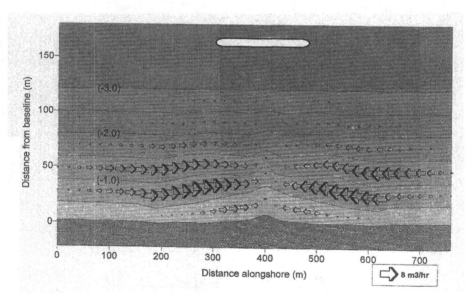

Figure 14.23 Longshore Current Distribution

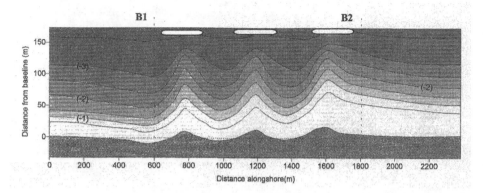

Figure 14.24 Salient Formation

15. Shore Protection

15.1 Introduction

Coastal engineering and management in the past consisted of providing protection against shore erosion and flooding. Life in coastal areas was a continuous battle of man against the sea and all possible methods were mustered to take part in this battle. When large machines were developed, man gained the upper hand in this battle and today with the help of machinery (dredges, earthmovers, cranes, concrete plants, etc.), man can live relatively safely near the sea. The sea still wins some battles (Ch. 1), but the coast is highly valued as living and recreational space (Ch. 10) and therefore major economic resources are available to ensure safety from the sea.

But precisely the countries that can afford to provide optimum protection are also most interested in the environment and quality of life. There is a desire to leave the coast as natural as possible and as a result, coastal management policy has become ambivalent. On the one hand, we want to keep the coast natural. On the other hand, we want the protection and the amenities that go with coastal living and recreation.

Critics of shore protection will say that all shore protection is temporary – so why build it and interfere with nature, which eventually will have its own way? On a geological scale, protection is not even temporary, but neither is the coastal system we are trying to protect. On an engineering time scale (Ch. 1) protection is indeed temporary. Even the very large protection systems such as the combination of dunes and sea dikes protecting the shore of the Netherlands require constant watchfulness, repair and changes in management techniques. But "temporary" with respect to shore protection is long enough to be of benefit for most applications. In any case, economic considerations (Ch. 10) decide if a coast should be protected. Particularly with the increase in tourism everywhere and demand for a lifestyle that includes the sea, it is unlikely that countries will permit their highly valued shorelines erode.

There does not appear to be a middle road with shore protection. If we do nothing, the shore will become ugly and dangerous through erosion and in time it will not be accessible. We do not want that. But to resist the sea successfully, shore protection must be massive and will often be ugly. Perhaps we also do not want that. A seawall, for example, discussed in Section 15.4, must be massive to withstand direct wave action. It cannot be replaced by pretty terraces and paving stones interspersed with pots of flowers[1], much as we might prefer that.

Given the necessity of shore protection, we should do it right. Unfortunately, there are few guidelines on how to build shore protection and any existing guidelines suffer from either too much simplification or too much generalization. As a result, much shore protection is built without adequate knowledge or appropriate design.

This chapter briefly discusses considerations for the design of coastal protection. Three questions that need to be asked are:
–   Do we want (or need) shore protection?
–   What are the available alternatives?
–   How can we implement protection and leave the coast as natural and attractive as possible?

The present discussion will focus on how certain shore protection schemes function best and on the impact of the protection methods, rather than on the details of their structural design.

The key concept in coastal protection design is *integration*. We saw in Ch 11 and 12 that nothing should be done within a littoral cell, without thinking about how it affects the rest of the cell, which by definition is part of the same system. Many times in practice, however, not even the neigboring properties are considered. On many shorelines, property owners simply look after their own interests. They hire their own consultants to design and build their own version of a shore protection theme, without regard to continuity with the adjacent properties, which are either unprotected, or covered by equally arbitrary, non-contiguous protection schemes (Section 10.7). The offending property owners are not necessarily callous individuals or economically motivated businesses. In many cases they are local, regional and federal governments making piecemeal decisions that only concern their own jurisdiction. Projects often

---

1. This was actually proposed in a round table discussion as a possible erosion protection for a section of the northwest shore of Lake Ontario (maximum wave conditions: H_s=6 m, T_p=10 sec.).

stop at regional or jurisdictional boundaries, rather than at system boundaries, as discussed in Ch 10.

Only integrated protection schemes can be ultimately successful. Such schemes must consider the cross-shore movements of the shoreline and the alongshore and cross-shore movement of quantities of sand. And as we saw in Ch 10, the schemes must also be designed and managed within a biological, ecological, judicial, political and sociological context.

15.2 Sediment Movement

In Ch. 11 and 12 we distinguished between alongshore and cross-shore sediment transport. That distinction must be clear in our mind, because most protection schemes do not function well with too much cross-shore sediment movement. In particular if the main cause of shoreline recession is systematic movement of sand offshore, the design of protection becomes difficult.

Incident wave angle is probably the most important ingredient in determining sediment movement, since it determines alongshore sediment transport rates and cross-shore sediment transport patterns. Wave angle was discussed in Ch. 11, 13 and 14 and we defined effective wave angle α_e as the angle between the breaking wave and the shore direction. This is an instantaneous angle that can be used to predict alongshore sediment transport rate at any time. When designing shore protection, we are more concerned with longer-term effects over the lifetime of a project. We normally define a morphology angle α_m as the long-term average effective angle. It is the angle between the long-term average angle of the wave climate and the long-term average beach orientation. This is also called the beach-forming angle and is expected to represent the overall beach-shaping forces[2]. Along with α_m goes a net long-term sediment transport rate Q_m (the morphology sediment transport rate). For a long, stable beach, α_e can vary in time and can be positive or negative, but α_m and Q_m are constant. The variation in α_e defines the fluctuations of the beach around its mean position. For most beaches, requiring protection, α_m is small ($< 10°$). If α_m increases along a section of beach, then Q_m increases which means more sediment is transported out of the section than into it. This causes a net deficit and results in erosion and shoreline recession (Ch. 12).

---

2. This approach implicitly assumes that design can be based on Q_{net}, which was shown to be incorrect in Ch. 11, when substantial sediment is transported in both directions.

15.3 Groins

Groins are structures that are perpendicular, or almost perpendicular to the shore. They were discussed briefly in Ch 11. An individual groin interrupts the sediment transport as shown in Fig. 11.18 forming accretion (and a beach) on its updrift side, and erosion (and damage to the shore) downdrift. Figure 11.18 shows shoreline change with respect to α_m but variation of α_e in time can produce large fluctuations about the basic pattern of Fig. 11.18. The effect of bi-directional sediment transport is shown in Figs. 11.19 and 11.20.

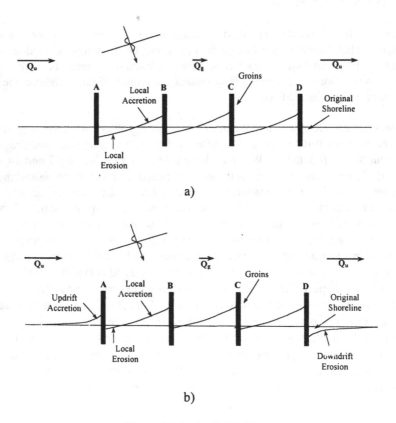

a)

b)

Figure 15.1 Groin Field

The purpose of a groin field (series of groins) is to divide a shoreline into short sections that can re-orientate themselves with respect to the incoming waves (Fig 15.1a). Over the long term, α_m within the groins will be less than for the original beach, causing Q_g through the groin field to be less than Q_u outside the groin field. Groins, therefore change the alongshore sediment transport rates. This will result in accretion updrift of the groins and within the groin field and erosion downdrift (Fig 15.1b). For each beach section, α_e will cause substantial temporal swings around this mean beach orientation (Fig 15.2).

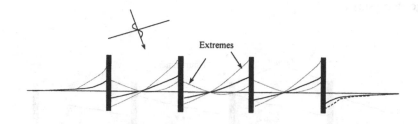

Figure 15.2 Extreme Beach Orientations

The length and spacing of the groins is based on the mean shoreline orientation (Fig. 15.1) and the extreme orientations (Fig 15.2). It is particularly important that the groins are placed well back into the existing shore to prevent the waves from flanking the groins (breaking through around the landward end of the structure). Flanking will normally result in deep scour trenches, landward of the groins and will compromise their stability.

Because the sediment transport rate past the groins (Q_g) is less than the rate in unprotected area outside the groin field (Q_u), such a groin field will act like a wide, single groin and cause local accretion, updrift and local erosion downdrift as in Fig. 11.18. This is shown in Fig. 15.1b. The erosion-accretion process will continue until all the groins are filled to capacity, so that they bypass all the sediment that arrives from updrift. In the time that it takes to fill the groins, however, extensive damage can be caused downdrift of the groins. Combining the groin construction with artificial beach nourishment as in Fig 15.3, providing the sand for the filling of the groin field and the updrift accretion area from elsewhere, can prevent such damage. That is a common method to integrate a groin field into its surroundings.

Cross-shore sediment transport, however, can rapidly add or remove sediment from the groin field. When offshore sediment motion resulting from high water levels and storm surge empties a groin field of sand and removes the accretion volumes collected updrift of the groins, downdrift erosion depicted in Figs. 11.8 and 15.1 will begin to take place. If the offshore movement of sand is severe, the shore will erode back far enough that the groins will flank, and the shore behind the groins will be damaged. Obviously, when the erosion is a result of a steep beach and foreshore, causing a net offshore motion of sand, groins will not help. Artificially filling the groins will also not work when there is a possibility of large temporary offshore transport rates or when there are large fluctuations in mean water level, such as along the Great Lakes, or in areas of large storm surge.

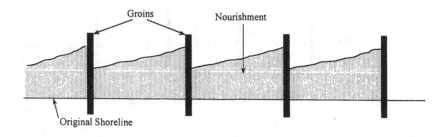

Figure 15.3 Groin Field with Nourishment

Thus, groins can only be applied in areas where erosion is a result of predominantly alongshore sediment transport, when erosion results from an increase in α_m (or Q_m) with distance along the shore. The re-orientation of the shoreline in the sections will decrease the sediment transport rate through the section. It is clear from Figs. 15.1 to 15.3 that the incident wave angles cannot be too large for groins to be effective, otherwise they would need to be either very long, or very closely spaced. And protection by groins is not effective when there are large long-term water level fluctuations. The method has therefore a very *restricted* window of application. The fact that the use of groins is so ubiquitous reflects a general misunderstanding about their functioning.

Damage by the groin field to the surrounding shore is a function of the rate of sediment bypassing. A filled groins system creates little damage. When the groin field is not filled, long, high groins will stop all sediment transport for a long time and cause much damage. Shorter, lower groins will cause less damage but will still affect the surrounding shore, until they are filled to capacity. Groins also generate offshore current as in Fig. 15.4. These currents move sediment offshore and can be a hazard to bathers. Most groins are short and will only obstruct the beach section where sediment transport takes place primarily by beach drifting (Fig. 11.16). That is the area where the largest grain sizes are found. Thus the currents in Fig. 15.4 will move primarily larger beach material offshore. This sediment is moved toward and perhaps over any alongshore bars to an area where normally, only finer sediment is moved. The wave action there has difficulty moving these large grains and returning them back to the shore. Thus a groin field can act as a sediment pump, moving coarse sediment to deeper water. Downdrift of the groins, the sediment gradation will then become finer and the shore will be less stable until the coarse sediment can finally come back to shore. This process of local decrease in grain size can cause additional erosion downdrift of a groin field and increase the extent of the downdrift damage out of all proportion to the groin sizes[3].

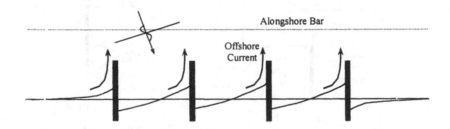

Figure 15.4 Offshore Currents near Groins

Some additional design considerations for groins are:
- Groins are mostly constructed out of armor stone or sheetpile.
- To minimize downdrift erosion, their height should only be just enough to contain the design beach profile.
- Their length and spacing are a function of α_m and the fluctuations of α_e about α_m.
- A wave climate that is not predominantly in one direction can produce much

---

3. In one case on Lake Huron, a single 30 m long groin caused rapid damage to more than 1 km of downdrift shoreline.

different erosion-accretion patterns (Ch. 11).
–   Groins impact the surrounding environment and habitat (Ch 10).
–   A discontinuity will arise where the groin field meets the surrounding area. To
    minimize damage to the adjacent downdrift areas, sometimes the end groins are
    shortened to form a transition. However, erosion-accretion around a groin field is
    a function of the complete groin system and not of the individual end groins. The
    difficulty in designing the ends of a groin field pleads for integral shore protection
    design.

15.4 Seawalls

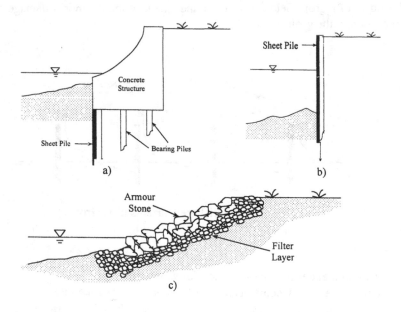

Figure 15.5 Typical Seawalls

A seawall is a protection wall, built along to the shore[4]. It is the protection method of choice for locations where further shore erosion will result in excessive damage, for example, when roads or buildings are about to fall into the water. Seawalls are designed to form the defining line of demarcation, dividing sea and land. Extreme examples of seawalls are the immense dikes built along the North Sea coasts. Most seawalls are, however, much smaller and many seawalls are close to vertical. They range from steel sheetpile walls to monolithic concrete barriers, to rubble mound structures, to brick or block walls to gabions (wire baskets filled with rocks). Typical examples may be found in Fig. 15.5

An attractive feature about seawalls is that their impact on the alongshore sediment transport is small. They do not result in the accretion-erosion patterns as in Fig. 11.18.

The primary design condition for seawalls is that they are stable and structurally sound. They are located at the top of the shore and will be out of reach of the water during good times (at low water). Sometimes they may even be covered with layers of beach sand. During times of stress (at high water), however, they will be exposed to direct wave action. Since seawalls are usually built as a last resort, most seawalls are continually under severe stress. The waves will attack the structure, move sand offshore and alongshore away from the structure. The wave action reflected off the seawall causes disturbed water near the wall that can promote deep scour holes immediately offshore of the seawall. The disturbed flows and scour areas can be dangerous and the scour may even excavate the supporting sand from under the structure, compromising the stability of the wall.

Water levels control the design environment for seawall design. High water levels allow higher waves to come closer into shore, subjecting the structure and its foreshore to high forces and high rates of erosion. Very high water levels will cause waves to overtop the seawall resulting in erosion at the back of the structure. Trapping of water behind the seawall, may cause drainage problems resulting in erosion and structural instability. The design of a seawall is not simple. Unfortunately, most seawall projects are installed by small contractors. There is little or no design and often, the most noticeable impact is a rapid destruction of the seawalls and the surrounding area during subsequent high water levels, storm surge and waves. Areas with long-term water level fluctuations, such as the Great Lakes are particularly vulnerable to cycles with periods of destruction of seawalls, followed by periods of lower water, when many new seawalls are built that are quite sound (until the next cycle of high water).

---

4. Sometimes the term revetment is used. A revetment normally refers to similar, but lighter protection built along rivers and small lakes.

Additional design considerations for seawalls are:
–   They are dangerous during times of high water and storm. People on or near the
    structure may be injured or swept out to sea.
–   For near-vertical structures, there will be much overtopping, sending salt water
    spray inland, resulting in accelerated corrosion.
–   They form a physical barrier to cross-shore movement of people and wildlife.
–   The ends of a seawall are difficult to design. Since the seawall actually defines
    where the shore-sea interface shall be in the shore section it protects, a
    discontinuity will form between the structure, which does not move, and the
    surrounding shore, which continues to recede. There will also be local accelerated
    erosion, damaging the adjacent shore. To prevent undermining and flanking of the
    seawall at its ends, the structure needs to be built well back into the existing shore.
       The difficulty in designing the ends of the structures and preventing erosion
    damage to adjacent properties again pleads for integral shore protection design.

In spite of the shortcomings, properly designed seawalls may be the only way to protect
shore property, particularly against damage by high water levels. They should be
integral with the system in which they are placed, taking into account their own
structural integrity and their environmental impacts (Ch. 10). Design alternatives
should always be considered and in many tourist areas, seawalls have been replaced by
offshore breakwaters, artificial nourishment or both.

15.5 Headlands

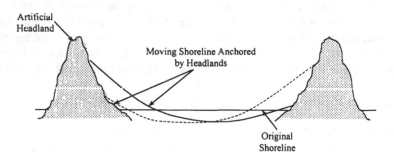

Figure 15.6 Artificial Headlands

When headlands occur naturally along a shore with some sand, they will contain pocket beaches. It is possible to emulate this on a smaller scale with artificial headlands as in Fig 15.6. This is really a scaled up version of a groin system and its larger size can withstand extensive cross-shore transport of sediment during periods of high water and storm surge. It is very versatile in that shore directions can be varied much more with an appropriate combination of landfill, headland construction and beach nourishment (Fig 15.7). The approach has been used extensively, for example, along the Toronto shore where attractive multi-purpose projects host parks, wildlife areas, marinas and bathing beaches, where originally there was an eroding, more-or-less straight bluff shoreline. In Toronto, both the headlands and the beaches were built up of clean excavation and construction debris produced by the nearby city. Silvester and Hsu (1997) discuss the shapes of headland-controlled beaches in detail. Clearly, major structural units as in Fig. 15.7 must be carefully integrated with the surroundings. Downdrift erosion is a major consideration and hence such large structures can only be used if Q_{net} is small or erosion can be readily mitigated.

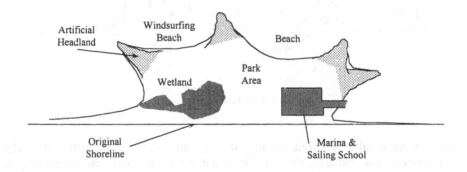

Figure 15.7 Innovation with Artificial Headlands

15.6 Offshore Breakwaters

Offshore breakwaters (Fig. 15.8) have been used as beach protection, particularly in tourist areas, where seawalls and groins are not attractive alternatives. They can be used in areas with substantial cross-shore transport.

Offshore breakwaters intercept much of the incident wave energy, resulting in reduced wave action behind the structures. The waves enter through the breakwater gaps and then diffract as they travel toward the shore (Figs 7.9 to 7.11). The diffracted waves change the beach shape from a relatively straight shore to an attractively curved shoreline with salients or tombolos. A salient is an accretion formation that does not reach the breakwaters; a tombolo is attached to a breakwater. In general, breakwaters that are longer or placed close to shore form tombolos. Salients form when the breakwaters are further from shore and there are substantial gaps between the breakwaters. Silvester and Hsu (1997) survey some rules of thumb to determine if salients or tombolos are likely to form. CUR (1997) presents a detailed review of applications of offshore breakwaters.

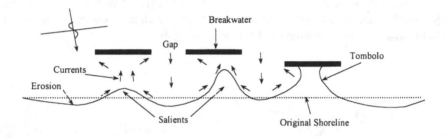

Figure 15.8 Offshore Breakwaters

Salients are usually preferred, because they do not block the currents behind the breakwaters, thus enhancing water quality in the swimming areas. However, they are essentially an unstable beach form between a straight beach and a tombolo. Small changes in conditions can convert a salient into a tombolo, which means that incident wave and water level conditions must be more or less constant in order to produce salients.

The diffracted wave crests and currents in the diffraction zone behind the breakwaters shape the salients and tombolos. The currents are forced by the mass transport of water from the waves entering through the breakwater gaps and by the wave height gradients along the shore as a result of the wave diffraction. Beach material to form the salients and tombolos is swept from adjacent areas of the original beach, causing areas of local erosion, within the project, as well as outside it. Combination of these structures with artificial nourishment is ideal. The artificial nourishment prevents the erosion and the structures serve to keep the artificial nourishment in place.

The design of beaches, using offshore breakwaters is quite complex and also uncertain in the case of salients. A major complication is introduced by waves overtopping the breakwaters. Mass transport by the waves coming over the breakwater decreases the currents shown in Fig 15.8 or may even reverse them. Thus, a particular layout may form tombolos, if the breakwaters are high, but lower breakwater crests may result in salients. Since wave overtopping is closely related to water levels, the morphology is very sensitive to water level fluctuations. For that reason, applications of offshore breakwaters, particularly to form salients are mainly found in areas where the water level fluctuations are small, such as along the Mediterranean Sea. Many designs have been implemented in Japan, but most of these are tombolo designs, which are less sensitive to water levels. To maintain salients may require periodic redistribution of the sand behind the breakwaters. This can be readily done with land-based equipment or a small dredge that can operate behind the breakwaters, safe from large waves.

The currents behind offshore breakwaters can be dangerous to swimmers, during storm periods. Because the waves behind the breakwaters are benign, people are not aware of the strong currents, which are a function of the large waves outside the breakwaters. Careful lifeguard patrol during storms must keep people away from areas of strong current activity, such as near the ends of the structures and off the tips of the salients.

15.7 Artificial Nourishment

The principles of artificial nourishment were discussed extensively in Section 11.3.4. The concept is based on simulating natural dune-beach formations. The present discussion will focus on some design details. The artificially placed material has a profile that is different from the stable profile and it has a limited length (along the shoreline). No matter what the constructed plan shape of the nourishment is, it will spread out (diffuse) and tend toward a straight or slowly curving shoreline as in Fig. 15.9. Its center of mass will also move in the direction of net sediment transport (advection). In addition, the nourishment will tend toward a stable profile shape in the cross-shore direction.

Since artificial nourishment emulates nature itself, it is environmentally the most friendly protection alternative. It has the least impact on adjacent properties and the environment, and instead of harming the surroundings, a beach fill will benefit adjacent eroding properties. Only when depth needs to be maintained at the adjacent properties, such as in a navigation channel, or when the sand added to the system threatens valuable habitat, does the diffusion and advection of a beach fill present a problem.

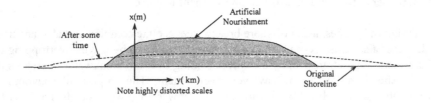

Figure 15.9 Artificial Beach Nourishment

Although artificial nourishment has been practiced for many years, the technology is still very much intuitive. It is important to understand why erosion takes place at the site. If it is through greed or ignorance (Figs. 11.5 to 11.8), it is possible to use a correcting artificial nourishment. It essentially attempts to rectify what ignorance destroyed. If it is a slow, systematic erosion, as in Fig. 11.9, an artificial nourishment will be subjected to the same erosion. The design is then not only concerned with how much sand to place, but also with how often it needs to be replenished. Artificial nourishment in most areas then becomes a beach maintenance solution, based on annual cost/benefit figures. If the site erodes more rapidly, as a result of offshore conditions, such as a locally steeper shoreline or a convergence of wave energy, the artificially placed fill will also be subjected to the same conditions and will not perform well.

Reviews of artificial nourishment projects in various countries may be found in Schwarz and Bird (1993), Stauble and Kraus (1993), and Hamm et al (1998). The placement method is a function of the equipment used. In general, because of the large volumes of sand required, beaches are nourished by hydraulic fill from dredges. Some nourishments have been executed by placing sediment on the shoreface, in the breaking zone or seaward of the breaker bars. The material is then placed in 5 to 10 m of water (Nourtec, 1997). Placement is easy in that case, since hopper-suction dredges can come over the fill areas, so that no rehandling of the material is required. In such shoreface nourishment the sand does not redistribute itself very much and essentially forms an offshore sandy reef that protects the shore. Only a small portion of the offshore material will come onshore and little additional recreational beach area is created[5]. The new offshore mass of sand will

---

5. Unless there is substantial alongshore transport, which will be deposited behind the offshore sand bar.

prevent further beach erosion, because the waves break further offshore and the beach slope to deep water is substantially decreased.

Since a major objective of most artificial nourishment schemes is to provide protection as well as additional recreational beach, most nourishments are placed as beach fills sometimes in combinations with shoreface nourishment. Beach fill normally requires rehandling of the sand so that it can be placed by pipeline dredge and perhaps be reshaped by land-based earthmoving equipment. The onshore sand is usually placed with a steep seaward slope. The wave action on such a fill will shape the most seaward part of the fill mass into a beach profile. During this adjustment period and at any later time, when other beach material, further landward is redistributed, fine grain sizes will be winnowed out of the mass of sand and lost to deep water, until the grain size distribution of the remaining sand mass is similar to the native distribution. Once the fill has been re-adjusted by the waves to form a beach profile, a steep scarp may have formed at the top of the beach.

Both diffusion and advection of the beach material will decrease with grain size. James (CERC, 1984) developed relationships between grain size and fill effectiveness. If the average grain size of the fill is smaller than the size of the native material, more fill needs to be placed than can be expected to stay. The nourishment sand contains relatively more fine material than the native sand and to produce a volume of sand with the same grain size gradation as the native sand requires a larger volume of nourishment sand. The finer nourishment sand that does not fit the distribution will be winnowed out and lost. Nourishment sand of larger mean diameter than the native sand will armor the beach, because there is a relative excess of coarse material in the grain size distribution. If the fill material is more uniform in size than the native material, a larger volume of fill is also needed to reproduce the native gradation. As a result, less uniform or smaller nourishment material, requires larger nourishment volumes and earlier renourishment.

Dean and Yoo (1993) offer another explanation for the relationship between required fill volume and grain size. They assume that a stable beach profile (solid line in Fig 15.10) is represented by Eq. 12.15. If the native material is used as nourishment, the beach profile is simply shifted to seaward over a horizontal plane, as in Fig 14.2. Since A_p in Eq. 12.15 increases with D, both the beach profiles and the representative beach slopes will become steeper with grain size. Thus fill material that is coarser than the native sand will result in a steeper profile (dotted line in Fig. 15.10) that intersects the existing profile. Fill material that is finer than the native sand produces a flatter profile thast does not intersect with the existing profile (dashed line in Fig. 15.10). The volumes of beach fill needed to effect a certain nourishment width W_n are clearly defined in the case of intersecting profiles. For non-intersecting profiles, it is not clear how far out from shore the nourishment material will migrate and a large portion of the fill material is needed to supply the offshore part of the profile. Since a thin layer of sand in deep water provides no protection and since the success of artificial beach nourishment is usually considered to be the visible dry beach remaining after nourishment, it is obvious that non-intersecting profiles must be avoided. Hence, the nourishment material should ideally be coarser than the native material.

Unfortunately, the most readily available source of nourishment sand is usually offshore sand, which is considerably finer and more uniform than the native beach material.

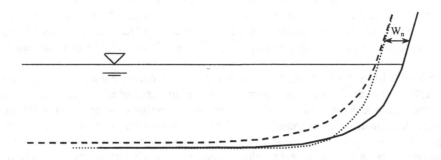

Figure 15.10 Intersecting and Non-Intersecting Profiles

In plan, nourishment projects are always of limited length and the angle of wave approach increases over the ends of the project. Although wave refraction decreases the incident wave angle over the ends, beach orientation, dx/dy increases, resulting in a larger effective angle α_e. Thus sediment transport rate, Q, is increased causing the nourishment to stretch out. Dean and Yoo (1993) use a Pelnard-Considère-type analytical (diffusion) solution, as was done in Ch. 14, to calculate the planform evolution of a beach fill. The "diffusion coefficient" is modified with distance to account for the process changes over the ends of the nourishment. Similar to the analytical diffusion solution in Ch 14, a 1-Line numerical model that can take into account the different incident wave angles at different locations in the project will provide a more general solution.

Other aspects of design of artificial nourishment are:
–   Where will the nourishment material come from and is there sufficient material?
–   The end effects discussed above, along with lower unit costs for placing large volumes of dredged material lead to the general impression that long beach fills are

more effective than short ones.
- Schwarz and Bird (1990) show that fill length has only a modest influence on longevity of a beach nourishment.
- Dean and Yoo (1993) use the diffusion equation to show that the portion of material remaining on the shore for a simple rectangular beach fill is inversely proportional to the project length.
- There is the whole question of longevity of a fill.
- When assessing the amount of fill material left in a project, one must carefully distinguish between real fill losses and apparent losses, which are reversed, as in Fig 11.1.
- The sand volume in a fill is normally assumed to decrease exponentially with time after placement. This does not seem to be true at Norderney (Kunz, 1993).
- Dean and Yoo (1993) show the fill volume remaining on the beach is proportional to $\sqrt{t}$, where t is the time after placement.
- Several authors state that the longevity of a project is a function of individual storms, but beach fills at Ocean City, USA that were exposed to storms of totally unexpected severity seems to disprove this.

It is obvious that the combination of artificial nourishment with structures such as groins or offshore breakwaters will help contain the fill material. Structures also provide an opportunity to use beach fills in areas, which would never be stable with artificial nourishment alone. Examples are Hilton Head (Bodge et al, 1993) and Norderney (Kunz, 1993).

Water levels are a very important design parameter in determining the stability and longevity of a beach fill. On maritime shores, the water levels are changed by periodic storm surges, which are known to result in major damage. On reservoirs and lakes, periodic high water levels will cause much damage to a nourishment project and it is not clear if artificial nourishment is even possible with large water level fluctuations. Finally, long term water level rise resulting from eustatic sea level rise, isostatic rebound and global warming need to be taken into account.

A beach is biologically relatively unproductive. There are indications that any benthic communities covered by a beach fill re-establish quite quickly after nourishment. The surrounding ecosystem , however, will need to be carefully considered (Ch 10).

Further information about design of artificial nourishment projects may be found in the many papers published in the proceedings of the coastal zone conferences, the international conferences on coastal engineering, and in CUR (1987), CUR (1997), Simm (1996) and NRC (1995).

15.8 Water Levels

We need to remind ourselves at the end of this chapter that all shore protection is very sensitive to water level and its fluctuations. In the end, it is the most important design consideration and the major cause of destruction of shore property and shore protection schemes.

16. Problems

16.1 Introduction

These problems form an integral part of the book. Without actually applying the material presented in the first 15 chapters, you will not really be able to understand the details. The problems presented in this chapter also attempt to simulate real working environments and real life situations.

Problem 1.1 Preparation

Purpose: To provide the basis for relevant problems.

It is impossible to develop exercises that are relevant to all the readers. Since much design information pertains to hydrographic charts, you are asked at this time to locate three sites in your own area. Use appropriate hydrographic charts to locate a substantial beach - *Site B*, a marina or small craft harbour - *Site M,* and a shallow water shore section (with a long, shallow shelf offshore of the site) - *Site S.* Some of the problems in this chapter will refer to your sites B, M and S in order to make the problems relevant.

Problem 1.2 Proposal

Purpose: To simulate a typical design environment.

Your company is asked to produce a technical proposal for a project. Your knowledge is insufficient at the present time, but you have information available to you (this book and perhaps other references). The proposal is due in one week.

a) Form a small "company" (2 or 3 people) and appoint a Chief Executive officer (CEO) who functions as spokesperson and is responsible for the success of the project.

b) Your company has been asked to submit a proposal to redesign the breakwater for a 50% expansion at Site M. Write a proposal containing at least:
- Your approach to the problem.
- (Very) preliminary design – in other words, what is the expected outcome?
- A description of the data you will need.
- A description of the system(s) that need to be considered
  - for the direct influence of the waves.
  - for any sediment transport.
  - for any environmental considerations such as pollution, habitat, etc.
- A description of the Quaternary (particularly Holocene) geology of the area.
- Recommendations regarding the design tools you will use (such as models).
- The proposed cost of this redesign (not the cost of the project!).

At this stage, your report will not be a real design proposal. No matter. The secondary purpose of this report is to provide a base line to which you can refer later. You can use it to gauge your progress. You can also use it to find the larger picture again when you become involved in the details of the later chapters.

Deliverables: The proposals will be presented orally. Your CEO will have 10 minutes for the presentation and the competing companies will discuss your proposal with you for 10 minutes[1].

---

1. It is assumed throughout these problems that you are part of a class with other similar groups – "competing companies" in this case.

16.2 Water Waves

Problem 2.1 Basic Wave Calculations

Purpose: To use the wave tables and develop your first design tool.

A wave has a height of 2.0 m in a depth of water of 20 m. The wave period is 8 sec.

a) Use the wave tables to calculate:
    - wave length, L
    - velocity of propagation, C
    - energy density, E
    - group velocity, C_G
    - wave power, P.
b) at 12 m below the water surface, calculate:
    - the maximum values of orbital velocities, u and w
    - the pressure fluctuation due to the wave.
c) Using, Eq. 2.17, write a program or spreadsheet to calculate L and C at water depths of 100, 60, 40, 30, 20, 10, 5 and 2 m.
d) Extend the program or spreadsheet to calculate the quantities in Items a) and b).

Note: Make sure you do items a) and b) by wave table and calculator *before* you set up the program or spreadsheet.

Deliverables: Wave table solutions to a) and b) and one working program or spreadsheet with solutions.

Problem 2.2 Wave Reflection

In a model test with simple, regular (monochromatic) waves a wave probe is moved very slowly perpendicular to the shore. The output of such the slowly moving probe is shown in Fig. P-2.2 and the envelope of this signal means the same as Fig. 2.14.

From this record deduce:

a) Wave length
b) Incident wave height

c) Reflection coefficient
d) The depth of water is 0.2 m; calculate:
- horizontal component of orbital motion at the bottom under the antinode
- vertical component of orbital motion there at a depth of 0.1 m
- the difference between mean water level and still water level.

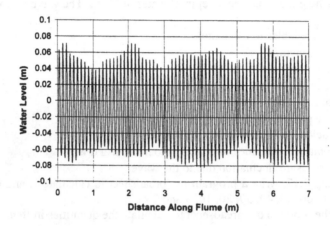

Figure P2-2 Measured Reflection Envelope

16.3 Short-Term wave Analysis

Problem 3.1 Analysis of Fig 3.4

Purpose: To trace the steps in the examples of Ch. 3.

The wave data for Fig 3.4 are provided as **F3-4.dat**[®].

a) Plot the wave record, using a spreadsheet and compute σ.
b) Use **WAVAN**[®] to obtain the distribution of individual wave heights, and the significant wave height and average wave period.
c) Check to see if the wave height distribution is Rayleigh.

d) Use **WAVAN**$^{\circledR}$ to determine the wave spectrum.
e) For the recording, what is:
   - The highest frequency that is computed correctly?
   - the Nyquist frequency?
   - Spectral bandwidth?
   - T_p, T_1 and T_2?
f) Use the calculated values of H_{mo} and T_p to plot the associated Jonswap and the PM spectra. (Use a spreadsheet).
g) Calculate $\overline{H_{0.1}}$ for the record.
h) Calculate the expected maximum wave height for 3 hrs of waves.

Deliverables: A 3-page report, discussing your findings and your experience with **WAVAN**$^{\circledR}$. All detailed information should be in appendices.

Problem 3.2 Analysis of Collected Wave Data

Purpose: To practice short-term wave analysis and to learn about the Jonswap and PM wave spectra.

Eight irregular wave data files are provided - **WD-01.dat** to **WD-08.dat.**

- **WD-01.dat**$^{\circledR}$ to **WD-03.dat**$^{\circledR}$ were collected during a breakwater test at the Queen's University Coastal Engineering Laboratory.
- The National Water Research Institute of Canada provided **WD-04.dat**$^{\circledR}$ and **WD-05.dat**$^{\circledR}$. The data were collected by a wave tower at the west end of Lake Ontario.
- Rijkswaterstaat in the Netherlands provided **WD-06.dat**$^{\circledR}$ to **WD-08.dat**$^{\circledR}$. They were collected during a field experiment on the North Sea shore at Egmond, Netherlands.

For one (or more) of these recordings, answer the questions of Problem 3.1.

Deliverables: A 3-page report, discussing your findings. All detailed information should be in appendices. Each group in the class should analyze a different data set.

Problem 3.3 Rayleigh Distribution

A 10 minute wave record was found to contain 100 waves, which were distributed as
in Table P3-1

a) Is this a Rayleigh Distribution?
b) What is H_s?
c) What is the estimated maximum wave height over 3 hours?

Table P3-1 Wave Height Distribution

H (m)	No of Waves	H (m)	No of Waves
< 0.60	15	1.8-1.99	2
0.6-0.79	15	2.0-2.19	2
0.8-0.99	7	2.2-2.39	2
1.0-1.19	17	2.4-2.59	3
1.2-1.39	5	2.6-2.79	1
1.4-1.59	9	2.8-2.99	1
1.6-1.79	20	3.0-3.19	1
Total			100

Problem 3.4 Zero Crossing Analysis

Zero down-crossing analysis of a 10 minute wave record results in $\sigma_z = 0.71$ m and

$$P = e^{-\left(\frac{H}{2.06}\right)^{2.11}}$$

There are 58 waves in the record

a) Are the wave heights Rayleigh distributed? (Explain your answer)
b) What is the average of the highest 1% of the waves?
c) What is the highest wave in 1 hour?
d) What is the highest wave in 10 years?

e) The wave record was obtained using a pressure gauge located in 30 m of water, 3
    m below the surface. For the significant wave height - what were
    –   the pressure fluctuation at the gauge?
    –   the maximum horizontal velocity at the gauge?
    –   the velocity at the bottom?

Problem 3.5 Wave Spectrum

The wave spectrum in Figure P3-5 was measured in 30 m of water. This spectrum
may be found digitally in **WS-01.dat**[®]. The incident wave angle in 30 m is 25
degrees with respect to the shoreline. A water intake is located near the bottom in
10 m of water.

a) Calculate the pressure fluctuation on this intake, caused by the waves
    represented by Fig. P3-5.
b) Calculate the velocity of propagation and the group velocity at the intake.
c) Calculate the size of the water particle orbits at mid-depth (5m) above the
    intake.
d) What is the probability of exceedence of $H = 2.1$ m in 30 m of water?
e) If the spectrum is representative of a 3 hour segment of a single storm, what is
    the estimated maximum wave height that occurred in those 3 hours.

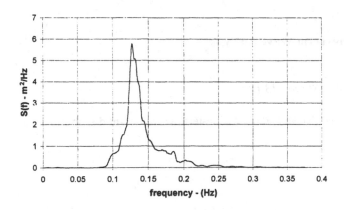

Figure P3-5 Wave Spectrum

Problem 3.6 Laboratory Record

A laboratory wave record was analyzed by zero crossing analysis. Figure P3-6 shows the Weibull graph using $\alpha = 2$ of the zero crossing wave heights. The digital version may be found in **P03-6.dat**[⊗].

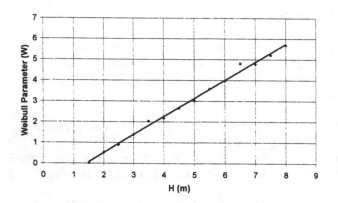

Figure P3-6 Laboratory Wave Height Distribution

a)   Is the wave height distribution a Rayleigh Distribution? Comment.
b)   Estimate σ_z from Fig. P3-6.
c)   What is the average of the highest 1% of the waves?
d)   Estimate the maximum wave in a wave train of 2000 waves.

16.4 Long-Term Wave Analysis

Problem 4.1 Station 13 Data

Purpose: To derive and use a long-term wave height distribution.

The file **P04-1.xls** contains wave data for Station 13, near the East end of Lake Ontario (Fig. P4-1). The data are given as bivariate distributions (hours of occurrence vs H and T) for waves from eight major directions (N, NE, E, SE, S, SW, W, NW) for the years 1964 to 1983 (20 years). We will not be able to do a POT analysis and hence the data will not quite be statistically independent. That is, however, the format in which much wave data are provided. Use only the data with wave heights greater than 1.5 m, and consider this as grouped data.

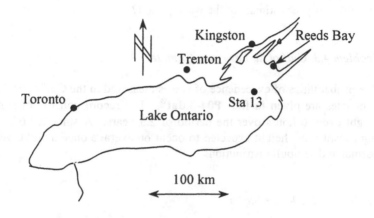

Figure P4-1 Lake Ontario

a) Predict the significant wave height expected on average at the site once in 20, 50, 100 and 200 years, using the Log Normal, Gumbel and Weibull distributions.

b) Determine the local relationship between T and H.

Deliverables: A three page report plus appendices, showing among others:
  – Assumptions you made.
  – Graphs of data and the fitted functions (as in Figures 4.4 and 4.6).
  – Output of the regression analyses for the distributions.
  – A table similar to Table 4.5.
  – A graph like Fig 4.11.
  – Discussion of your results.

Problem 4.2 North Sea Wave Climate

The probabilities of exceedence of waves measured in the North Sea 600 km off
Aberdeen are given in file **P04-2.dat**[⊗]. The original record from the POT analysis
consists of significant wave heights every 6 hours over the period of one year -
λ=44.

a) What is the significant wave height expected to occur on average once in 50 and
   100 years?
b) What is your estimate of the wave period?

Problem 4.3 Gulf of St. Lawrence Climate

The probabilities of exceedence of waves measured in the Gulf of St. Lawrence near
Sept Isles are given in file **P04-3.dat**[⊗]. The record consists of significant wave
height every 6 hours over the period of 2.5 years. Assume λ= 68. Calculate the
significant wave height expected to occur on average once in 1000 years, using log
normal and Weibull distributions.

Problem 4.4 50-year Storm

A 5 year 'Peak Over Threshold' wave analysis yielded the distribution in **P04-
4.dat**[⊗]. Assume λ=19. What is the wave height of the once in 50 year storm?

16.5 Wave Hindcasting

Problem 5.1 Very Simple Wave Hindcast

Calculate the significant wave height and the peak period that will approach Site B
and Site M after a wind of 18 m/s that has blown for 4 hours and 10 hours from the
direction of longest fetch.

a) Are the waves fetch-limited, or duration limited?
b) If duration limited, what is the effective fetch?
c) What is the maximum possible wave condition you can expect?

Problem 5.2 Simple Wave Hindcast

Purpose: In **P04-1.xls**® you were given hindcast wave data for Station 13 at the east end of Lake Ontario. You will now hindcast some extreme waves for this station, using simple Jonswap hindcasting method and compare the results with the data in file.

The longest fetch to Station 13 is 240 km from the southwest and the maximum hourly wind speeds recorded at three sites around the lake - Toronto, Trenton and Kingston are 25, 24 and 23 m/s (see Fig. P4-1). The National Building Code of Canada estimates the wind speed to be 23, 26 and 28 m/s, with return periods of 10, 30 and 100 yrs respectively. Assume $R_T = R_L = 1$.

a) Use this information to give your best estimate of the highest significant wave heights at Lake Ontario Station 13.
b) What would be the highest waves from the north and the west, for which the fetches are 35 km and 180 km?
c) Compare your results with **P04-1.xls**®.

Deliverables: One report: What did you do? What does it mean? What are the limitations? etc. Obviously, this report will be better if it answers some "what if" questions. This will require that you make several hindcast calculations, representing several scenarios. Therefore use computer programs or spreadsheets to answer questions in a), b) and c), as well as such questions as:
 – What if the wind speed were 10 % greater or less?
 – What if the fetches were 10 % greater or less?

Problem 5.3 WAVGEN and Shallow Water

Purpose: To do simple hindcasting and to develop a hindcast tool for shallow water.

a) Use **WAVGEN**® to calculate H_s and T_p for a wind speed of 24 m/s, a fetch length of 400 km and durations of 6 hours and 60 hours. Check your results with Fig. 5.3.

b) Use **WAVGEN**® to compute H_s and T_p at Urk on the IJssel Lake in the Netherlands. The fetch is 60 km, the wind speed is 20 m/s. What storm duration is needed to make this a fetch-limited problem?

c) The IJssel Lake, however, has an average depth over the fetch of 5 m. Use Fig. 5.4 from the text, **WAVGEN**® and Eqs. 5.12 to 5.14 to develop a program to calculate waves in shallow water and then re-calculate wave conditions at Urk. Compare your results with b).

16.6 Storm Surge

Problem 6.1 Storm Surge at Reeds Bay

Determine the storm surge at Reeds Bay, at the East end of Lake Ontario (Fig. P4-1) for the wind conditions in Problem 5.2. Assume that the profile to the site is as in Table P6-1. The profile starts in the middle of Lake Ontario. Assume the storm surge to be zero there.

Table P6-1 Offshore Profile

Section	1	2	3	4	5	6	7	8	9	10
Length (km)	24	56	3	4	2	22	6	3	0.3	1.5
d_{avg} (m)	125	75	45	35	25	36	22	7.5	3.5	0.5

Problem 6.2 Storm Surge and Waves

Hay Bay may be schematized as a narrow body of water with a profile as in Table 1. A 50 knot wind blows directly up the bay for 4 hours. This storm is accompanied by a pressure drop of 3 kPa.

Table P6-2 Offshore Profile

Section	1	2	3
Length (km)	5	5	1.5
d_{avg} (m)	7	4	5

a) Estimate the wave height and period at the end of the bay? (1 knot = 0.5 m/s).

b) What is the storm surge at the end of the bay? – Assume storm surge is zero at the beginning of the bay.

c) Give your best estimate of the period of oscillation of the bay.

Problem 6.3 Storm Surge and Waves at Site S

Determine the maximum wind conditions at Site S and calculate the maximum storm surge and the waves at the shore.

16.7 Wave Transformation

Problem 7.1 Wave Refraction and Breaking

Purpose: To practice simple refraction-shoaling-breaking calculation.

Assume that the waves in Table P7-1 occur in 20 m of water offshore of your Site M^2. (These are just the largest waves, which we will use by way of example). You have learned about wave refraction and shoaling, but only for simple situations, with regular contours that don't vary much along the shore. If it becomes more complicated, we need computer programs. On the other hand, many studies are based on using the simpler methods.

Table P7-1 Wave Bins

Wave	H (m)	T (sec)	α (°)	f (hrs/yr)
1	4.0	8	-8	23
2	4.5	8	-8	18
3	5.0	8	-8	12
4	5.0	9	-8	6
5	5.5	9	-8	3
6	6.0	9	-8	2
7	6.5	9	-8	1
8	3.5	8	12	13
9	4.0	8	12	6
10	4.5	8	12	1

---

2. If you know the actual wave climate at Site M, you should use it of course.

To determine the inshore wave climate, we need to transform each of the "wave bins" (combinations of H, T, α and f in 20 m of water) into the related values near the shore. Assume that the depth contours are parallel to the shore.

For the waves in Table P7-1, calculate:
a) Wave height and angle of approach in 15, 10 and 5 m of water off Site M.
b) Breaking wave conditions.

This problem should be done by expanding the program or spreadsheet you developed in Problem 2.1. Such a program will allow you to repeat the calculation readily for Steps a) and b) and perform additional calculations. Alternatively, you could use the program **RSB**®. Make sure you understand what you are doing with this program and in any case, you should do one complete calculation using the wave tables and a calculator.

Deliverables: A 3-page report, discussing your calculations, the demonstrated wave transformation program and any limitations you foresee. All other material should be in appendices.

Problem 7.2. Wave Transformation

For the measured wave condition: H_s=2 m, T_p=7 sec, α=7° in d=6 m of water, calculate:
a) Deep water parameters H_o, T_o, α_o
b) Breaking parameters H_b, T_b, α_b, d_b (the beach slope is 1:40 and the contours may be considered parallel to the coast)

Problem 7.3. Wave Diffraction

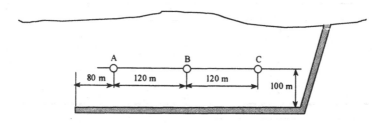

Figure P7-3 Wave Diffraction

For H_s=2.0 m, T_p=6.5 sec and α=15°, all observed in 20 m of water depth, calculate the maximum wave height to be expected at Points A, B and C behind the breakwater shown in Fig. P7-3. The breakwater is in 5 m of water and you may assume that the depth at points A, B and C is also 5 m. State all your other assumptions clearly.

16.8 Design

Problem 8.1 Probability of Failure

Purpose: To calculate probability of failure of an existing design.

The data file **P08-1.dat**[⊗] contains the results of a 4-year wave hindcast on Lake Winnipeg. The total number of Peak-over-Threshold events in 4 years was 68.

a) What is the wave height with a return period of 200 yrs?

A large dam has been built in 20 m of water using 8 tonne stone for armoring against the wave action.

b) What is the overall Γ for deterministic design if we use the T_R=200 yr wave ?

Assuming the design life of the structure is 50 years

c) What size of stone is needed to produce P_L=0.04, using the PIANC expression for a target P_F of 0.1?
d) What is the lifetime probability of failure of the 8 tonne armor layer for the 50 yr wave, using both the Level I calculation of Table 9.9 and the simple Level II design of Table 9.11 with
   – the uncertainty in wave height σ'_H=0.15
   – the uncertainty is stone mass σ'_M=0.3.

Deliverables: A report (three pages plus appendices) that digests your work. Do not just present your work.

Problem 8.2 Vertical Breakwwater

Purpose: To design a vertical breakwater.

This is a lot of work and you should organize a "company" to divide the work. Appoint a CEO who writes the final report and correlates the various aspects of the work. Meet about the results so everyone understands what was done.

The hindcast wave conditions are provided in **P08-2.dat**. The hindcast is for 15.7 yrs and λ=38.5. The design wave approaches the shore at an incident angle of 14°. To provide the appropriate draft for the ships, this structure must be located in a depth of 4.0 m below low water (usually below Chart Datum - CD). To get the design water depth, you will need to take into account water level fluctuations and local storm surge. The highest tides are 2.3 m above CD and the storm surge can be +0.3 m and –0.1 m. The foreshore slope near the breakwater consists of sand (N=20) and has a slope of 0.02. The design life of the structure is 50 years. For our initial design (before a model study) we will use $(H_{s,b})_{max}$ for a 50 year return period with the Burcharth and Sorensen coefficients for a target P_F=0.01.

Assume that the breakwater is built parallel to the shore, as in Fig. P7-3. Your design will be a composite structure consisting of a caisson, placed on top of a 2 m high berm of rock. Make appropriate additional assumptions for any information not given.

a)   Determine the necessary design depth of water at the structure.
b)   Determine if the design is for breaking waves.

Use the hydrostatic or the Goda and Minikin methods, whichever is appropriate; the waves immediately behind the breakwater resulting from overtopping waves should be less than 0.4 m high.

c)   Design the caisson for several conditions as was done in Tables 9.2, 9.3 and 9.4 and choose the best design.
d)   Check geotechnical stability for each design alternative.
e)   What would the design look like, if we used H_s as design wave height, instead of $(H_b)_{max}$.
f)   Calculate probability of failure (P_f) against sliding for the design in c) for the uncertainties ($\sigma'=\sigma/\mu$): σ_R'=0.25 and σ_H'=0.1 (see Table 9.11). Since S is mainly due to the dynamic wave force/unit length of structure, which is proportional to H^2, then σ_S' will be 0.2.

g) Calculate the required caisson width (B_v) to produce $P_f = 5\%$.
h) How thick a sheet of ice would cause the breakwater to slide?

Deliverables: A report (three pages plus appendices and tables of results). Digest your work and draw appropriate conclusions.

Problem 8.3 Vertical Breakwater at Site M

Purpose: To design a vertical breakwater at site M.

This problem is the same as Problem 8.2, but for your site M. Use the local wave climate, or else the climate in **P08-1.dat**®.

Deliverables: Same as Problem 8.2.

Problem 8.4 Vertical loading dock on Gulf of St. Lawrence

Purpose: To design a vertical loading facility.

A vertical loading dock needs to be built in 7 m of water. The wave climate is represented by **P04-3.dat**®. Assume that m=0.03 and $T_p=3.8H_s^{0.55}$. The design life of the structure is 50 years. For our initial design (before a model study) we will use $(H_{s,b})_{max}$ for a 50 year return period with the Burcharth and Sorensen coefficients for a target $P_F=0.01$.

For your design, assume that the dock is built parallel to the shore direction. Your design will be a composite structure consisting of a caisson, placed on top of a 1.5 m high berm of rock.

Make appropriate assumptions for any information not given and answer all questions in Problem 8.2.

Deliverables: Same as Problem 8.2.

Problem 8.5 Rubble Mound Breakwater

Purpose: To design a rubble mound breakwater.

Use the same company as in Problem 8.2, but select a different CEO. Remember to divide the work.

a) Design a conventional rock breakwater for the conditions of Problem 8.2, using
    - Hudson Formula
    - Van der Meer Formula
    Determine all dimensions – crest elevation, crest width, etc. and cost.
    What would the crest elevation be if it stops $R_{2\%}$?
    Assume:
    - Cost of Armor is $ 80 per tonne
    - Cost of core and underlayers is $35.- per tonne.
    - What happens if the armor stone cost becomes $55.- per tonne
b) Design the breakwater using Tetrapods and compare cost if concrete units are cast at $ 93 per tonne.
c) Design a berm breakwater, using readily available stone:
    - $D_{50} = 0.6$ m
    - $D_{90} = 1.2$ m
    - $D_{15} = 0.3$ m
    - $P_R = 0.2$

Deliverables: As in Problem 8.2

Problem 8.6 Rubble Mound Breakwater at Site M

Purpose: To design a rubble mound breakwater at Site M.

This problem is the same as Problem 8.5, but for your site M. Use the local wave climate, or else the climate in **P08-1.dat**[⊗].

Deliverables: Same as Problem 8.2.

16.9 Coastal Management

Problem 9.1 Expansion at Site M

Purpose: To study what needs to be done to expand a marina in your area.

Prepare a proposal for a 50 % expansion of Site M. This is a companion piece to Problem 1.2. You will need to divide the work. Set up a group, appoint a spokesman, etc. Determine:

a) need,
b) alternatives to the present site,
c) what agencies need to be involved (Table 10.10),
d) what legislation needs to be satisfied (Table 10.11),
e) what are the conflicts (Tables 10.6 and 10.7),
f) what are the regulations regarding development near your shoreline,
g) physical impact,
h) environmental impact,
i) necessary mitigation.

Deliverables: A report addressing the above items and summarizing the opportunities and problems with the project (digest your work).

Problem 9.2 Facilities at Site B

Purpose: To study what needs to be done to increase the facilities at the beach site in your area. There are two definite proposals on the table

a) One developer wants to build a recreational park, complete with merry-go-rounds and Ferris wheels.
b) Another plans to build a multi-purpose facility, having housing, shops and a 2000 seat theater.
c) It is probably not possible to simply say "no" to all development, because the municipal government sees much income from these ventures. Argumentation to stop any or all development will need to be well-founded, citing alternatives, impacts, mitigation plans, etc.

Prepare documentation for expansion of Site B, including construction of one beach house, parking facilities and the setting up of a sailing club at the site. Carefully review the two development ideas. Determine:

a) need,
b) alternatives to the present site,
c) what agencies need to be involved (Table 10.10),
d) what legislation needs to be satisfied (Table 10.11),
e) what are the conflicts (Tables 10.6 and 10.7),
f) what are the regulations regarding development near your shoreline,
g) physical impact,
h) environmental impact,
i) necessary mitigation.

Deliverables: A report addressing the above items and summarizing the possibilities and problems with the project (digest your work).

Problem 9.3 Development of Property

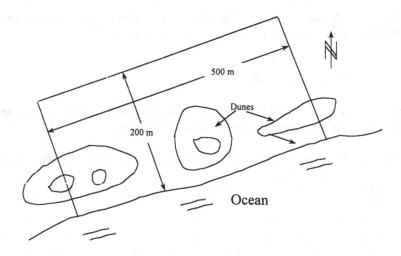

Figure P9-3 Development Site

A developer has acquired the property shown in Fig. P9-3 and asks for some initial advice (pre-feasibility report) on how to develop the property. She wants a 100 room hotel with appropriate parking. Water supply, sewage disposal must be accomplished within the site boundaries. She also want a facility for small boats. The nearest road runs parallel to the beach and 1 km to the north.

Your report discusses:
a) Basic guidelines for development of this property.
b) Data needed before any design can be made.
c) A list of relevant agencies and legislation.
d) Conflicts
e) Problems that you foresee

16.10 Sediment Transport and Morphology

Problem 10.1 Potential Sediment Transport Rate

Calculate potential sediment transport rate for:
a) Problem 5.1; the wave has a deep water angle of 15°.
b) Problem 7.2.
c) Problem 7.3.

Problem 10.2 Potential Sediment Transport Rate

Assume that the waves in 20 m of water offshore of Site B are the same as in Table P7-1[3]. Calculate

a) Sediment transport rates along the shore in both directions.
b) Gross and net sediment transport rates.

Problem 10.3 Accretion

To calculate long-term sediment transport, it is necessary to divide a long-term wave climate into deepwater "bins" of H, T, α and f. (H is wave weight, T is wave period, α

---

3. If you know the actual wave climate at Site B, you should use it of course.

is wave angle and f is frequency of occurrence). For this problem consider only one single bin ($H_{s,o}$ = 1.5 m, T_p = 9 s and α_o = 12°) and assume that is the representative wave to describe the morphology process. Assume also that the actual sediment transport rate equals the potential rate. Referring to Fig. P10-3: m=0.02, S_s=200 m, d_c=4 m and D_{50}=0.25 mm:

a) How long will it take for this wave to "fill up" the structure, i.e., when will the structure begin to bypass sand?
b) Calculate the accretion near the structure from t=0 to t=2t_{full}.
c) What is the accretion at a point 2000 m updrift of the groin at the time the groin begins to bypass?

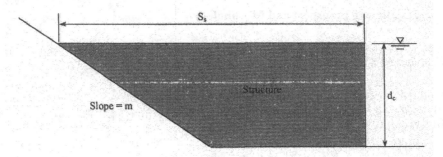

Figure P10-3 Accretion at a structure

Problem 10.4 Sediment Transport in two Directions

Alongshore sediment transport may be summarized by two wave conditions: H_{so}=1.2 m, T_p=6 sec, α_o=10° with C_Q (ratio between Q_a and Q_p)=0.4, and H_{so}=0.9 m, T_p=7 sec and α_o=-14° with C_Q=0.3. The structure and the beach profile are as shown in Fig. P10-3 with m=0.03, S_s=350 m, d_c=5 m and D_{50}=0.22 mm. For these simplified conditions, answer the questions of Problem 10.3.

Problem 10.5 Sea Level Rise

A shore consists of D_{50} = 0.35 mm sand and the beach is backed by a dune that is 4.3 m high. The wave climate on this profile is **P04-1.xls**[®]. Calculate the recession to be expected from a 0.5 m rise in sea level.

Problem 10.6 Northeaster Storm

An "Atlantic Northeaster" storm blows from 70° Azimuth (East of true North) for 3 days with an average wind speed of 14 m/s. A straight beach, backed by 6 m high dunes overlies a geological formation that is nearly horizontal, about 6.5 m below water. The formation extends about 6 km out from shore. Beyond 6 km, the shore slopes at m=0.01. The grain size of the sand on the beach is 0.28 mm. (All this is obviously a simplification of reality).

a)   What are the wave height, period and direction over this 6.5 m deep shelf?
b)   How much shoreline recession is caused by this storm?
c)   Estimate the alongshore sediment transport rate.

16.11 Modeling

Problem 11.1 Physical Models

You have a laboratory that has two major facilities. A 100 m long wave flume, 5 m wide, capable of generating waves with the following characteristics:

$$H_s \leq 0.18\,m$$
$$0.5 \leq T_p \leq 2.6\,sec$$
$$0.5 \leq d \leq 1.3\,m$$

and a 30 by 50 m wave basin with the same wave generation capabilities. For the rubble mound tests, we have model armor stone with mean mass of 0.1 kg, 0.18 kg, 0.32 kg and 0.62 kg, 0.23 and 0.59 kg tetrapods, and 0.17 and 0.48 kg dolos.

Design model studies for Problems 7.3, 8.1, 8.2, 8.3, 8.4, 8.5, 8.6, 9.1 and/or 9.2.

Deliverables: A report for each model study describing scales selected, other scales resulting from this selection, problems foreseen and quality of the results.

Problem 11.2 Numerical Models

Your beach site B is eroding and you must design a numerical model study to evaluate the installation of groins along the whole beach. Your model must determine the optimum groin lengths and spacings. Assume that the site is subjected to the wave climate in **P04-1.xls**®. The direction SW of **P04-1.xls**® is at an angle of $-10°$ with respect to Site B.

Deliverables: A report describing the design process, numerical methods, any problems and quality of the results.

16.12 Comprehensive Problems

Problem 12.1 Design Analysis

Purpose: To do a comprehensive analysis and design

Your company is asked to analyze the design of a small craft harbor at Jordan Station on Lake Ontario in light of new information. The site is shown in Fig. P12-1a and the breakwater in Fig P12.1b. You are also asked to determine the design wave height for a tower to measure waves offshore of the site.

a) What are the 10 most important pieces of information you will need in order to check the design of the breakwater? Give a short description of each.
b) What tools would you use to check the breakwater design? Give a short description of each.
c) Wind records at St Catharines indicate that the maximum hourly wind speed off the lake is 56 km/hr. Estimate the deep-water significant wave height, period and direction for the largest waves that can be generated by a 14-hour storm, based on the 56 km/hr maximum hourly wind speed.
d) Waves were measured near the site in 20 m of water. During one large storm, the wave spectrum shown in Fig. P12-1c was measured (available as **WS-**

02.dat[⊗]). The direction of the waves was 30° Azimuth (30° East of North). Determine the following parameters, based on this measured wave.

-   The characteristic (or significant) wave.
-   The expected value of H_{max}, if the duration of the measurement is 20 minutes.
-   H_{max} for the 6 hours, if the record represents 6 hours.
-   What does the peak at f=0.07 Hz represent?
-   What is the maximum pressure fluctuation at the bottom in 20 m of water during 6 hours?
-   To design the cable to hold the wave recording tower in place, it is necessary to calculate the wave-generated velocities and accelerations all along the cable. Just calculate the maximum wave-generated velocities at mid-depth (10 m) in 6 hours.
-   What is the wave energy density of this wave climate?
-   What is its wave power?
-   What was the deep water wave angle for that wave?
-   What are the breaking depth, wave height, period and direction? Assume m = 0.03.
-   What is the potential sediment transport rate for this wave? Assume D = 0.2 mm.

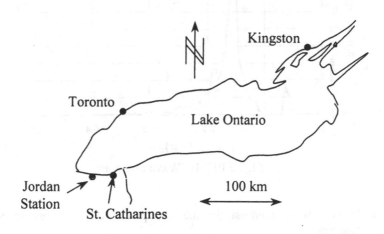

Figure P12-1a   Jordan Station

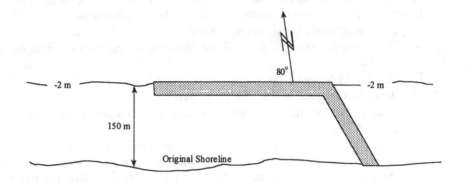

Figure P12-1b Jordan Station Breakwater

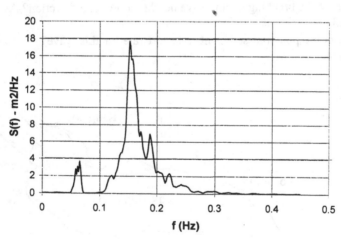

Figure P12-1c Wave Spectrum

e) A 35 year wave hindcast for the site produced the following long-term relationship

$$Q = \exp\left[-\left\{ \frac{H_s - 1.56}{0.34} \right\}^{1.12} \right]$$

This equation is based on an average of 93 measurements per year ($\lambda=93$). A wave measuring tower must be designed offshore of the site (in deep water) with a 25 year life span and a lifetime probability of failure (P_L) of 1 %. What should the design wave height be?

f) What is the design wave height at the breakwater?
g) Design a conventional breakwater section. Sketch the section and explain the fine points of your design. Use deterministic design methods to determine the armor stone size using
   - the Hudson formula
   - Van der Meer's formula.
h) Repeat g) using the PIANC approach.
i) Design the breakwater as a berm breakwater with 600 kg armor stone.
j) The isostatic rebound for the area is 0.75 mm/yr. What effect does that have on your design?
k) Comment on the effect of global climate change on your design.
l) Design a hydraulic model study to determine the stability of the conventional breakwater you designed in g). The study is to be carried out at a laboratory with the following limitations

$$H_s < 0.2 \text{ m}$$
$$T_p < 3 \text{ sec}$$
$$d < 1.5 \text{ m}$$

m) The Ontario Department of Tourism has invited your company to discuss the impact on the environment by the structure in Fig. P12-1b. What will be the five most important points in your report?
n) Five years after construction of the harbor, the accretion on the east side of the harbor is as in Fig P12-1d. The hydrographic charts indicate that the depth of the active profile, $d_p= 7$ m. Assume that S_e for the harbor is 110 m.
   - How much sand will be bypassing the harbor after 20 years?
   - How far will the shoreline have moved seaward in 20 years at a sewagetreatment plant, 700 m East of the harbor?

Deliverables: An extensive and detailed report, outlining approach, calculations and results, and discussing difficulties and problems.

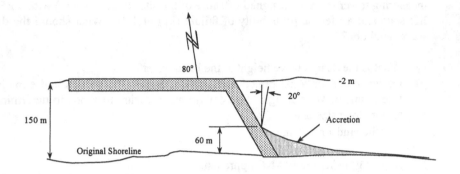

Figure P12-1d   Accretion

Problem 12.2 Design of Breakwater with Parapet Wall

Purpose: Design of a composite structure.

A natural armor stone breakwater must be designed. The 50 yr offshore wave used for the design has $H_{s,0} = 4$ m, $T_p = 9$ s and $\alpha_o = 40°$ with respect to the shoreline. With an offshore slope of 1:50 and a breakwater front slope of 1:1.5, what armour stone size is required for a conventional breakwater design

a)   at a section located in 2 m depth of water?
b)   at a section located in 15 m depth of water?

The breakwater will be topped off with a parapet wall to form a walkway (Fig P12-2).

c)   Design the parapet wall as best you can.
d)   Derive all the pertinent dimensionless ratios and the scales for a model to determine the forces on the parapet wall.
e)   If in a certain test the measured force on the wall in the model is 10 N/m, what is the prototype force?

Deliverables: As in Problem 12.1

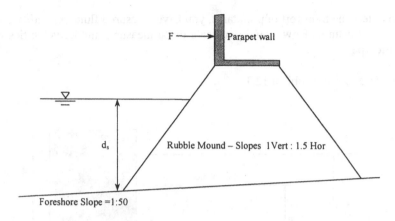

Figure P12-2 Breakwater with Parapet Wall

Problem 12.3 Vertical Breakwater Design

Purpose: Design check of a vertical breakwater.

For the vertical breakwater section in Figure P12-3, the tidal range is 1.0 m and the maximum storm surge (setup) is 0.25 m. The once in 100-year wave condition is H_s=4.5 m, T_p=8.5 s

a. Is the structure stable against sliding and overturning?
b. You have built a model of this structure to scale 36 and you want to measure the pressures in the front face using pressure transducers. Derive the pressure scale.
c. You also want to measure the total wave force at the structure by placing the vertical caisson on ball bearings and measuring the forces needed to keep the model in place. Derive a force scale.
d. Your model is 2 m wide (across a 2 m wide wave flume). You can buy three force meters with the following capacity ranges

1 to 100 N

10 to 1000 N

100 to 10000 N

Which force meter would you buy?

e.   To determine transport of pollutants, you have measured fluid velocities in front of the structure. How would you scale those measured model velocities up to prototype.

Deliverables: As in Problem 12.1

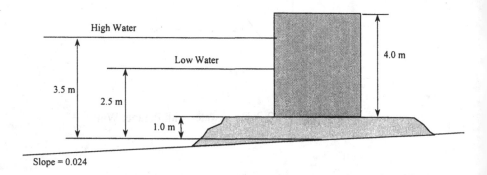

Figure P12-3 Vertical Breakwater

References

Abbott, M.B. (1979), *"Computational Hydraulics – Elements of the Theory of Free Surface Flow"*, Pitman Press.

Abbott, M.B. and D.R. Basco (1989), *"Computational Fluid Mechanics – An Introduction for Engineers"*, Longman Press.

Abbott and Price (1994), *"Coastal, Estuarial and Harbour Engineers' Handbook"*, Chapman and Hall.

Abernethy, C.L. and G. Gilbert (1975), "Refraction of Wave Spectra", Rep. INT 117, *Hydraulics Research Wallingford.*

Abramowitz, M and I. Stegun (1965), *"Handbook of Mathematical Functions"*, Dover Press.

Airy, G.B. (1845), "Tides and Waves", *Encyclopedia Metropolitana.*, Vol 192, pp 241-396.

Allsop, T.R. and Cohen, S.J. (1986), "CO_2 Induced Climate Change and its Potential Impact on the Province of Ontario"", *Proc. Conf. on Climate and Water Management, Am. Met. Soc*, Boston, pp 285-290.

Ang, A. H-S, and W.H. Tang (1984), *"Probability Concepts in Engineering Planning and Design, Volume II - Decision, Risk and Reliability"*, John Wiley and Sons.

ASCE (1974), *Int. Symp. on Ocean Wave Measurement and Analysis*, New Orleans.

Bakker, W.T. (1968), "The Dynamics of a Coast with a Groin System", *Proc. 11th Int. Conf. on Co. Eng.*, ASCE, London, pp 492-517.

Basco, D.R. (1983). Surfzone currents. *Co. Eng.*, Vol. 7, pp 331-517.

Bascom, W. (1964), *"Waves and Beaches"*, Doubleday-Anchor.

Battjes, J.A. (1974), "Surf Similarity Parameter", *Proc. 14th Int. Conf. on Co. Eng.*, ASCE, Copenhagen, pp 69-85.

Bendat, J.S., and Piersol, A.G. (1966), *"Measurement and Analysis of Random Data"*, Wiley.

Benjamin, J.R., and Cornell, C.A. (1970), *"Probability Statistics and Decision for Civil Engineers"*, McGraw-Hill.

Berkhoff, J.C.W. (1972), "Computation of Combined Refraction-Diffraction", *Proc. 13th Int. Conf. on Co. Eng.*, ASCE, Vancouver, pp 471-490.

Bird, E.C.F. (1984), *"Coasts"*, 3rd Ed., Basil Blackwell Press.

Bird, E.C.F. (1985), *"Coastline Changes, a Global Review"*, Wiley.

Bird, E. C. F. (1993), *"Submerging Coasts: The Effects of a Rising Sea Level on Coastal Environments"*, Wiley.

Black, K.S., D.M. Patterson and A. Cramp (1998), "Sedimentary Processes in the Intertidal Zone", *Geological Society (London)*, Special Pub. 131.

Bodge, K.R., E.J. Olson and C.G. Creed (1993), "Performance of Beach Nourishment at Hilton Head Island, S.C.", Beach Nourishment Engineering and Management Considerations, *Proc. Co. Zone '93*, ASCE, New Orleans, pp 16-30.

Booij, N., L.H. Holthuysen and R.C. Ris (1996), "The Swan Wave Model for Shallow Water", *Proc. 25th Int. Conf. on Co. Eng.*, ASCE, Orlando, pp 668-676.

Boussinesq, J. (1872), "Theorie des Ondes et de Remais qui se Propagent le long d'un Canal Rectangulaire Horizontal, en Communiquant au Liquide Contenu dans ce Canal des Vitesses Sensiblement Paralleles de la Surface au Fond". *J. Math. Lionvilles*, 17:55.

Bouws, E., H. Gunther, W. Rosenthal and C.L. Vincent (1985), "Similarity of the Wind Wave Spectrum in Finite Depth of Water, Part 1", *J. Phys. Oceanography*, Vol 17 (C1), pp 1288-1295.

Bouws, E., H. Gunther, W. Rosenthal and C.L. Vincent (1985), "Similarity of the Wind Wave Spectrum in Finite Water Depth", *J. Geoph. Res.*, Vol 90, C1, pp 975-986.

Brampton, A.H. (1977), "A Computer Method for Wave Refraction", Rep. IT-22, *Hydraulics Research Wallingford.*

Bretschneider, C.L. and R.O. Reid (1953), "Change in Wave Height due to Bottom Friction, Percolation and Refraction", *Proc. 34th Annual Mtg, Am. Geoph. Union*, pp 34-37.

Bretschneider, C.L. (1958), "Revisions in Wave Forecasting; Deep and Shallow Water", *Proc. 6th Int. Conf. on Co. Eng.*, ASCE, pp 17-24.

Bretschneider, C.L. (1959), "Wave Variability and Wave Spectra for Wind Generated Waves", Beach Erosion Board Tech. Memo No. 113, *U.S. Army Corps of Eng.*

Briand M.H.G. and J.W. Kamphuis (1990), "A Micro Computer Based Quasi 3-D Sediment Transport Model", *Proc. 22nd Int. Conf. on Co. Eng.*, ASCE, Delft, pp 2159-2172.

Briand, M.H., and J.W. Kamphuis (1993), "Waves and Currents on Natural Beaches: A Quasi 3-D Numerical Model", *Co. Eng.*, Vol 20, No 1, pp 101-134.

Briand, M.H., and J.W. Kamphuis (1993a), "Sediment Transport in the Surf Zone: A Quasi 3-D Numerical Model", *Co. Eng.*, Vol 20, No 1, pp 135-150.

Broker, I. B., R. Deigaard and J. Fredsoe (1991). Onshore/Offshore Sediment Transport and Morphological Modelling of Coastal Profiles. *Proc. Co. Sediments '91*, ASCE, Seattle, pp 643-657.

Bruun, P. (1954), "Coast Erosion and the Development of Beach Profiles", Beach Erosion Board Tech. Memo No. 44, *U.S. Army Corps of Eng.*

Burcharth, H.F. (1992), "Reliability Evaluation of a Structure at Sea", Short Course on Reliability Design, *24th Int. Conf. on Co. Eng.*, ASCE, Venice, Ch 21.

Burcharth, H.F and J.D. Sorensen (1998), "Design of Vertical Wall Caisson Breakwaters Using Partial Safety Factors", *Proc. 26th Int. Conf. on Co. Eng.*, ASCE, Copenhagen, pp 2138-2151.

Carter, R.W.G (1988), "*Coastal Environments*", Academic Press, London.

Cartwright, D.E., and Longuet-Higgins, M.S. (1956), "The Statistical Distribution of the Maxima of a Random Function", *Proc. Royal Soc.* A, Vol 237, pp 212-232.

CCREM (1978), *Proc. Shore Management Symp.*, Canadian Council of Resources and Energy Ministers, Victoria.

CC-SEA (1991), *Proc. Symp. on Canadian Environmental Regulations and Issues in the Shore Zone*, CC-SEA, Mt. Ste Marie.

CERC (1977), "*Shore Protection Manual*", Co. Eng. Res. Center, U.S. Corps of Eng., Vicksburg.

CERC (1984), "*Shore Protection Manual*", Co. Eng. Res. Center, U.S. Corps of Eng., Vicksburg.

Chakrabarti, S.K. (1994), "*Offshore Structure Modeling*", Advanced Series on Ocean Engineering, Vol 9, World Scientific Publishing, Singapore.

Chapman, L.J. and D.F. Putnam (1984), "*The Physiography of Southern Ontario*", 3rd Ed., Queen's Printer of Ontario, Toronto.

Clancy, R.M., J.E. Kaitala and L.F. Zambreski (1986), "The Fleet Numerical Oceanography Center Global Spectral Ocean Wave Model", *Bull. Am. Met. Soc*, Vol 67, No 5, pp 498-512.

Clark R.H and N.P. Persoage (1970), "Some Implications of Crustal Movement in Engineering Planning", *Can. J. of Earth Sc.*, Vol 7, pp 628-633.

Cohen, S.J. (1986), "Impacts of CO_2 Induced Climate Change on Water Resources in the Great Lakes Basin", *Climate Change*, Vol 8, pp 135-153.

Cohen, S.J. (1987), "Sensitivity of Water Resources in the Great Lakes Region to Changes in Temperature, Precipitation, Humidity and Wind Speed", *Proc. IUGG/IAHS Symp. on the Influence of Climate Changes and Climatic Variability on Hydrological Regime and Water Resources,* pp 35-49.

Crombie, D. (1992), *"Regeneration",* Final Report, Royal Commission on the Future of the Toronto Waterfront, Queen's Printer of Ontario, Toronto.

Cunge,J.A., F.M. Holly and A. Verwey (1980), *"Practical Applications of Computational River Hydraulics",* Pitman Press.

CUR (1990), *"Manual on Artificial Beach Nourishment",* CUR Rep 130, Balkema Press, Rotterdam.

CUR (1997), *"Beach Nourishment and Shore Parallel Structurest",* CUR Rep 97-2, CUR, Gouda.

CUR/CIRIA. (1991), *"Manual on the Use of Rock in Coastal and Shoreline Engineering",* Balkema Press, Rotterdam.

Dabees, M.A. and J.W. Kamphuis (1998), "ONELINE, a Numerical Model for Shoreline Change", *Proc. 26th Int. Conf. on Co. Eng.,* ASCE, Copenhagen, pp 2668-2681.

Dabees, M.A. and J.W. Kamphuis (1999), "Beach Evolution Modelling", *Proc. Canadian Co. Conf. '99,* (CCSEA), Victoria, pp 29-42

Dabees, M. A. and J.W. Kamphuis (2000), "N_LINE: Efficient Modelling of 3-D Beach Change", *Proc. 27th Int. Conf. on Co. Eng.,* ASCE, Sydney, (to be published).

Dabees, M. (2000), " Efficient Modeling of Beach Evolution", Ph.D. Thesis, *Queen's Univ.*

Davies, A.G., J.S. Ribberink, A. Temperville and J.W. Zyserman (1997), "Comparison between Sediment Transport Models and Observations made in Wave and Current Flows above Plane Beds", *Co. Eng.,* Vol 31, pp 163-198.

Dean, R.G. (1974), *"Evaluation and Development of Water and Wave Theories for Engineering Application",* Special Rep. No. 1, Co. Eng. Res. Center, U.S. Army Corps of Eng.

Dean, R.G. (1977), "Equilibrium Beach Profiles, U.S. Atlantic and Gulf Coasts", Tech Rep No 12, *U. Delaware,* Newark.

Dean, R.G. (1983), "Principles of Beach Nourishment", *Handbook of Co. Processes and Erosion,* P. Komar (Ed), CRC Press.

Dean, R.G. and R.A. Dalrymple (1984), *Water Wave Mechanics for Engineers and Scientists,* Prentice Hall – republished as Advanced Series on Ocean Engineering, Vol 2, World Scientific Publishing, Singapore (1992).

Dean, R.G. and C. Yoo (1993), "Predictability of Beach Nourishment Performance", Beach Nourishment Engineering and Management Considerations, *Proc. Co. Zone '93*, ASCE, New Orleans, pp 86-102.

Defant, A. (1961), *"Physical Oceanography, Volume II"*, Pergamon Press.

De Vriend, H.J., M. Copabianco, T. Chesher, H.W. de Swart, B. Latteux and M.J.F. Stive (1993), "Approaches to Long-Term Modelling of Coastal Morphology: a Review", *Co. Eng.*, Vol. 21, pp 225-269.

De Vriend, H.J., J. Zyserman, J. Nicholson, J.A. Roelvink, P.Pechon and H.N Southgate (1993a), "Medium-Term 2DH Coastal Area Modelling", *Co. Eng.*, Vol 21, pp 193-224.

Dingemans, M. W. (1997), *"Water Wave Propagation over Uneven Bottoms"*, Advanced Series on Ocean Engineering, Vol 13, World Scientific Publishing, Singapore.

Donelan, M.A. (1980), "Similarity Theory Applied to the Forecasting of Wave Heights, Periods and Directions", *Proc. Canadian Co. Conf. '80*, CC-SEA, Burlington, pp 47-61.

Donelan, M.A., J. Hamilton and W.H. Hui (1985), "Directional Spectra of Wind Generated Waves", *Phil. Trans., Roy. Soc., London*, Vol A 113, pp 509-562.

Donnelly, P, (1991), "Shore Management (A Conservation Authorities Perspective)", *Proc. Symp. on Canadian Environmental Regulations and Issues in the Shore Zone*, CC-SEA, Mt. Ste Marie, pp 138-166.

Dronkers, J.J. (1964), *"Tidal Computations in Rivers and Coastal Waters"*, North Holland Publishing Company.

Dyer, K. (2000), "Intertidal Mudflats: Properties and Processes", Special Issue, *Continental Shelf Research*, (to be published)

Ebersole, B.A. (1985), "Refraction-Diffraction Model for Linear Water Waves", *J. Waterways, Port, Coastal and Ocean Eng.*, ASCE, Vol 111, pp 939-953.

Ebersole, B.A. and R.A. Dalrymple (1980), "Numerical Modelling of Nearshore Circulation", *Proc. 17th Int. Conf. on Co. Eng.*, ASCE, Sydney, pp 2710-2725.

Fleming,, C.A., K.L. Philpott and B.M. Pinchin (1984), "Evaluation of Coastal Sediment Transport Techniques, Phase I: Implementation of Alongshore Sediment Transport Models and Calibration of Wave Hindcasting Procedures", Rep. C2S2-10, *Can. Co. Sed. Study*, Nat. Res. Co. of Canada.

Forrester, W.D. (1983), *"Canadian Tidal Manual"*, Canada Fisheries and Oceans, Ottawa.

Fredsoe, J. and R. Deigaard (1992), *"Mechanics of Coastal Sediment Transport"*, Advanced Series on Ocean Engineering, Vol 3, World Scientific Publishing, Singapore.

Gerstner, F. (1809), Theorie der Wellen, Abh. d. k. bohm. Ges. d. Wiss., reprinted in, *Ann. der Physik*, No. 32, 412-442.

Goda, Y. (1970), "Numerical Experiments on Wave Statistics with Spectral Simulation", *Port and Harbour Res. Inst.*, Japan, Vol 9, No 3.

Goda, Y. (1985), *Random Seas and Design of Maritime Structures*, U. Tokyo Press.

Goda, Y., (1992), "Uncertainty of Design Parameters from Viewpoint of Extreme Statistics", *J. Offshore Mechanics and Arctic Eng.*, ASME, Vol 114, pp 76-82.

Goda, Y., P. Hawkes, E. Mansard, M.J. Martin, M. Mathiesen, E. Peltier, E. Thompson, G. Van Vledder (1993), "Intercomparison of Extremal Wave Analysis Methods Using Numerically Simulated Data", *Proc. 2nd Int. Symp. on Ocean Wave Measurement and Analysis*, ASCE, New Orleans, pp 963-977.

Godin, G., (1972), "*The Analysis of Tides*", U. Toronto Press.

Gourlay, M.R. (1978), "Wave Generated Currents", Ph.D. Thesis, *U. Queensland*, Brisbane.

Hall, K.R. (1993), "*Design Guidelines for Berm Breakwaters*", Canada Fisheries and Oceans, Small Craft Harbors Branch, Ottawa.

Hallermeier, R.J. (1981), "A Profile Zonation for Seasonal Sand Beaches from Wave Climate", *Co. Eng.* Vol 4, pp 253-271.

Hanson, H. and Kraus, N.C. (1989). Genesis: Generalized Model for Simulating Shoreline Change. CERC Report 89-19, *US Corps of Eng.*, Vicksburg.

Hamm, L., H. Hanson, M. Copabianco, H.H. Dette, A. Legucha and R. Spanhof (1998), "Beach Fills in Europe – Projects, Practices and Objectives", *Proc. 26th Int. Conf. on Co. Eng.*, ASCE, Copenhagen, pp 3060- 3073.

Harker, J.M. and Mortson, L.D. (1995), "Environmentally Sound Hazard Management: Great Lakes - St Lawrence River System Shorelines", *Proc. Canadian Co. Conf. '95*, CC-SEA, Halifax, pp 431-447.

Hasselmann, K., T.P. Barnett, E. Bouws, H. Carlson, D.E. Cartwright, K. Enke, J.A. Ewing, H. Gienapp, D.E. Hasselmann, P. Kruseman, A. Meerburg, P. Muller, D.J. Olbers, K. Richter, W. Sell and H. Walden (1973), "Measurements of Wind-Wave Growth and Swell Decay during the Joint North Sea Wave Project (JONSWAP)", *Deut. Hydrog. Zeitschrift*, Supp. A8, No 12.

Hedges, T.S. (1976), "An Empirical Modification to Linear Wave Theory", *Proc. Inst. of Civil Eng.*, Vol 61, pp 575-579.

Herbich, J.B. (1989), "*Handbook of Coastal and Ocean Engineering*", Gulf Publishing, Houston.

Hoffman, J.D. (1992), "*Numerical Methods for Engineers and Scientists*" McGraw-Hill.

Holthuysen,L.H., N. Booij and T.H.C. Herbers (1989), "A Prediction Model for Stationary, Short-Crested Waves in Shallow Water with Ambient Currents", *Co. Eng.*, Vol 13, pp 23-54.

Horikawa, K. (1978), *"Coastal Engineering"*, U. Tokyo Press.

Horikawa, K. (1988), *"Nearshore Dynamics of Coastal Processes"*, U. Tokyo Press.

Houston, J.R. (1996), "International Tourism and U.S. Beaches", *Shore and Beach,* Vol. 64, No. 2, pp 3-4.

Houston, J.R. (1995), "The Economic Value of Beaches," CERC 95-4, Co. Eng. Res. Center, *US Army Corps of Eng.,* pp 1-4.

Hughes, S. (1993), *"Physical Models and Laboratory Techniques in Coastal Engineering"*, Advanced Series on Ocean Engineering, Vol 7, World Scientific Publishing, Singapore.

Hunt, J.N. (1979), "Direct Solution of Wave Dispersion Equation", *J. Waterway, Port, Coastal and Ocean Eng.*, ASCE, Vol 105, WW4, pp 457-459.

Ippen, A. (1966), *"Estuary and Coastline Hydrodynamics"*, McGraw-Hill.

Isobe, M. (1985), "Calculation and Application of First-Order Cnoidal Wave Theory", *Co. Eng.*, Vol 9, p 309-325.

Ivicsics, L. (1980), *"Hydraulic Models"*, Water Resources Publications, Fort Collins.

Janssen, T.T. (1999), "A Study of the Behavior of Long Wave Motion Driven by Short Wave Groups", M.Sc Thesis, *Delft U. of Technology*, Delft.

Johnson, H.K. and J.W. Kamphuis (1988), "N-line Model for a Large Initially Conical Sand Island", *Proc. Symp. Mathematical Modelling of Sediment Transport in the Coastal Zone.* IAHR, Copenhagen, pp 275-289.

Kamphuis, J.W. (1975), *"The Coastal Mobile Bed Model"*, CE Rep. No 75, *Queen's U.*, Kingston.

Kamphuis, J.W. (1983), "On the Erosion of Consolidated Clay Material by a Fluid Containing Sand", *Can. J. of Civil Eng.* (CSCE), Vol 10, pp 213-231.

Kamphuis, J.W. (1985), "On Understanding Scale Effect in Coastal Mobile Bed Models", *Physical Modelling in Coastal Engineering*, R. Dalrymple (Ed), Balkema Press, Rotterdam.

Kamphuis, J.W., M.H. Davies, R.W. Nairn and O.J. Sayao (1986), "Calculation of Littoral Sand Transport Rate" *Co. Eng.,* Vol. 10, pp 1-21.

Kamphuis, J.W. and Kooistra, J. (1986a), "On Volume Erosion and Scale Effects for Hydraulic Models of Large Conical Islands", *Proc. Symp. Scale Effects in Modelling Sediment Transport Phenomena*, IAHR, Toronto, pp 10-23.

Kamphuis, J.W. (1987), "Recession Rate of Glacial Till Bluffs", *J. Waterway, Port, Coastal and Ocean Eng.,* ASCE, Vol 113, pp 60-73.

Kamphuis, J.W. and M.R. Larson (1987), "A Mobile bed Scale Model Series under Attack by Constant, Regular and Irregular Waves", *Proc. Co. Sediments '87*, ASCE, New Orleans, pp 154-174.

Kamphuis, J.W. (1990), "The Influence of Sand on the Erosion of Cohesive Material", *J. of Hydraulic Res.* (IAHR), Vol 28, pp 43-53.

Kamphuis, J.W. (1991), "Alongshore Sediment Transport Rate", *J. Waterway, Port, Coastal and Ocean Eng*, ASCE, Vol 117, pp 624-640.

Kamphuis, J.W. (1991a), "Wave Transformation", *Co. Eng.*, Vol 15, pp 173-184.

Kamphuis, J.W. (1991b), "Incipient Wave Breaking", *Co. Eng.*, Vol 15, pp 185-203.

Kamphuis, J.W. (1991c), "Alongshore Sediment Transport Rate Distribution", *Proc. Co. Sediments '91*, ASCE, Seattle, pp 170-183.

Kamphuis, J.W. (1991d), "Physical Modeling", *Handbook of Coastal and Ocean Engineering*, J. Herbich (Ed), Gulf Publishing, Vol 2, Ch 21, pp 1049-1066.

Kamphuis, J.W. (1991e), "Peaks in the Cross-Shore Distribution of Littoral Drift Rate", *Proc. 24th Congress of International Association of Hydraulic Research*, (IAHR), Madrid, pp B477-488.

Kamphuis, J.W. (1993), "Effective Modelling of Coastal Morphology", *Proc. 11th Australasian Conf. on Coastal and Ocean Eng.*, Inst. of Eng. of Australia, Sydney, pp 173-179.

Kamphuis, J.W. (1995), "Composite Modelling - an Old Tool in a New Context", *Proc. 26th Congress of International Association of Hydraulic Research*, (IAHR), London, Vol 2, pp 230-235.

Kamphuis, J.W. (1996), "Physical Modeling of Coastal Processes", *Advances in Coastal and Ocean Engineering*, P. Liu (Ed), World Scientific Publishing, Singapore, Vol 2, pp 79-114.

Kamphuis, J.W. (1996a), "Depth-Limited Design Wave", *Proc. 25th Int. Conf. on Co. Eng.*, ASCE, Orlando, pp 221-232.

Kamphuis, J.W. (1998), "Long Waves in Flume Experiments", *Proc. 26th Int. Conf. on Co. Eng.*, ASCE, Copenhagen, pp 1154-1167.

Kamphuis, J.W. (1999). "Marketing Uncertainty", *Proc, 5th Int. Conf. on Co. and Port Eng. in Developing Countries (COPEDEC)*, Capetown, pp 2088-2099.

Kamphuis, J.W. (1999a), "Coastal Modelling for the Next Millenium", *Proc. Can. Coastal Conf. '99*, (CCSEA), Victoria, pp 13-28.

Kamphuis, J.W. (2000), "Coastal and Estuary Processes", *Hydraulic Modeling*, ASCE Manual 97, Ch 9.

Kamphuis, J.W. (2000a), "Designing with Models", Keynote Address, *Proc. 27th Int. Conf. on Co. Eng.*, ASCE, Sydney, (to be published).

Kamphuis, J.W. (2000b), "Physical Modelling – Renewed Importance", Ch 12, *International Handbook of Coastal Engineering,* Academic Press, (To be published)

Katopodi, I, and Ribberink, K. (1992), "Quasi-3D Modelling of Suspended Sediment Transport by Currents and Waves", *Co. Eng.*, Vol 18, pp 83-110.

King, C.A.M. (1972), *"Beaches and Coasts"*, 2nd Ed., Arnold Press.

Kinsman, B. (1965), *"Wind Waves"*, Prentice-Hall.

Kirby, J.T. and R.A. Dalrymple (1986), "An Approximate Model for Non-Linear Dispersion in Monochromatic Wave Propagation Models, *Co. Eng.*, Vol 9, pp 545-561.

Kolberg, M (1995), "Addressing the Hazards and Assessing the Impacts: Great Lakes - St Lawrence River System Shorelines", *Proc. Canadian Co. Conf. '95*, CC-SEA, Halifax, pp 495-510.

Komar, P.D. (1983), *"A Handbook of Coastal Processes and Erosion"*, CRH Publications.

Komar, P.D (1998), *"Beach processes and sedimentation"* (2nd edition)", Prentice-Hall.

Korteweg, D.J. and G. de Vries (1895), "On the Change of Form of Long Waves Advancing in a Rectangular Canal, and on a New Type of Long Stationary Wave", *Phil. Mag.*, 5 Ser., 39: 422-443.

Kraus, N. and S. Harikai (1983), "Numerical Model of the Shoreline Change of Oarai Beach", *Co. Eng.*, Vol 7, pp 1-28.

Kraus, N.C. and H. Hanson (1990), "Decoupled Numerical Model of 3-D Beach Change", *Proc. 22nd Int. Conf. on Co. Eng.,* ASCE, New York, pp 2173-2185.

Kraus, N.C, M. Larson and D.L. Kriebel (1991), "Evaluation of Beach Erosion and Accretion Predictors", *Proc. Co. Sediments '91*, ASCE, Seattle, pp 572-587.

Kunz, H. (1993), "Sand Losses from Artificially Nourished Beach Stabilized by Groynes", Beach Nourishment Engineering and Management Considerations, *Proc. Co. Zone '93*, ASCE, New Orleans, pp 191-205.

Langhaar, H.L. (1951), *"Dimensional Analysis and Theory of Models"*, John Wiley and Sons.

Larson, M., Hanson, H., and Kraus, N.C. (1987), "Analytical Solutions of the One-Line Model of Shoreline Change", CERC Report 87-15, *US Corps of Engineers,* Vicksburg.

Le Méhauté, B. and A. Brebner (1960), "An Introduction to Coastal Morphology and Littoral Processes" CE Rep. 14, *Queen's U.,* Kingston.

Le Méhauté B. (1976), *"An Introduction to Hydrodynamics and Water Waves"*, Springer-Verlag.

Le Méhauté, B. and M. Soldate (1978), "Mathematical Modelling of Shoreline Evolution", *Proc. 16th Int. Conf. on Co. Eng.*, ASCE, Hamburg, pp 1163-1179.

Le Méhauté B. and D.H. Hanes (1990), *"The Sea"*, Vol 9, Parts A and B, Wiley Interscience.

Liu, P.L.F. (1990), "Wave Transformation", *The Sea*, (B. LeMehaute and D. Hanes, Ed.) Wiley, Vol 9A, Ch 2, pp 27-63.

Longuet-Higgins, M.S. (1952), "On the Statistical Distribution of Heights of Sea Waves", *J. of Marine Res.*, Vol. XI, No 3, pp 13-22.

Longuet-Higgins, M.S. (1970), "Longshore Currents Generated by Obliquely Incident Sea Waves", *J. Geophysical Res.*, Vol. 75, pp 6778-6801.

Louisse, C. and A.J. Kuik (1990), "Coastal Defence Alternatives in the Netherlands", *Proc. 22nd Int. Conf. on Co. Eng.*, ASCE, Delft, pp 1862-1875.

Madsen, H.O., Krenk, S. and Lind, N.C. (1986), *"Methods of Structural Safety"*, Prentice Hall.

Mansard, E.P.D. and E.R. Funke (1980), The Measurement of Incident and Reflected Spectra, Using Least Squares Method", *Proc. 17th Int. Conf. on Co. Eng.*, ASCE, Sydney, pp 154-172.

Marchuk, G.I. and Kagan, P.A. (1984), *"Ocean Tides"*, Pergamon.

Martins, R.(1989), *"Recent Advances in Physical Modelling"*, R. Martins (Ed), Kluwer Academic Publishers.

Mathiesen, M., P. Hawkes, M.J. Martin, E. Thompson, Y. Goda, E. Mansard, E. Peltier, G. Van Vledder, (1994), "Recommended Practice for Extreme Wave Analysis", *J. of Hydraulics Res.*, IAHR, Vol 32, pp 803-814.

McCowan, J. (1891), "On the Solitary Wave", *Phil. Mag.* Series 5, 32, pp 45-58.

McCowan, J. (1894), "On the Highest Wave of Permanent Type", *Phil. Mag.*, Ser. 5, 38, pp 351-357.

McKeen, P (1995), "Ontario's Natural Heritage, Environmental Protection and Hazard Policies: Great Lakes - St. Lawrence River System Shorelines", *Proc. Canadian Co. Conf. '95*, CC-SEA, Halifax, pp 609-620.

Miche, R. (1944), "Mouvements Ondulatoires des Mers en Profondeur Constante et Decroisante", *Ann. des Ponts et Chaussees*, pp 25-78, 131-164, 270-292, 369-406.

Miles, J.W. (1957), "On the Generation of Surface Waves by Shear Flow", *J. Fluid Mechanics*, Vol 3, pp 185-204.

Mitsuyasu, H. (1980), "Observations of the Power Spectrum of Ocean Waves using a Cloverleaf Buoy", *J. Phy. Oceanog.*, Vol 10, No 2, pp 286-296.

Moore, B. (1982), "Beach Profile Evolution in Response to Changes in Water Level and Wave Height", M.S.. Thesis, *U. Delaware*, Newark.

Moulton, R. and D. Cuthbert (1987), "Man and Nature in the Shore Zone", *Proc. Canadian Co. Conf. '87,* CC-SEA, Quebec, pp 19-30.

Muir Wood, A.M. and C.A. Fleming (1981), *"Coastal Hydraulics",* 2nd Ed., MacMillan.

Munk, W.H. (1949), "The Solitary Wave Theory and its Application to Surf Problems". *Ann. New York Acad. of Science,* 51:376-424.

Murthy, T.S. (1984), *"Storm Surges",* Bull 212, Canada Fisheries and Oceans, Ottawa.

National Research Council (1979), *"Carbon Dioxide and Climate: A Scientific Assessment",* National Academy Press, Washington.

National Research Council (1983), *"Changing Climate",* National Academy Press, Washington.

National Research Council (1987), *"Responding to Changes in Sea Level",* National Academy Press, Washington.

National Research Council (1995), *"Beach Nourishment and Protection",* National Academy Press, Washington.

Neumann, G., and Pierson, W.J. (1966), *"Principles of Physical Oceanography",* Prentice-Hall.

Nicholson, J., I. Broker, J.A. Roelvink, D. Price, J.M. Tanguy and L. Moreno (1997), "Intercomparison of Coastal Area Morphodynamic Models", *Co. Eng.,* Vol 31, pp 97-123.

Nourtec (1997), "Innovative Nourishment Techniques Evaluation", *Rijkswaterstaat (RIKZ),* The Hague.

Osaza, H. and A.H. Brampton (1980), "Mathematical Modelling of Beaches Backed by Seawalls", *C. Eng.* Vol 4, pp 47-64.

Open University - U.K. (1989), *"Waves, Tides and Shallow-water Processes",* Pergamon Press.

Péchon, P., F. Rivero, H. Johnson, T Chesher, B. O'Connor, J.M. Tanguy, T. Karambas, M. Mory and L. Hamm (1997), "Intercomparison of Wave-Driven Current Models", *Co. Eng.,* Vol 31, pp 199-215.

Peck, R.B., W.E. Hanson and T.H. Thorburn (1974), *"Foundation Engineering",* Wiley.

Pelnard-Considère, R. (1956), "Essai de Theorie de l'Evolution des Formes de Rivage en Plages de Sable et de Galets", *4-ieme Journees de l'Hydraulique, Les engeries de la mer,* Question III, Rapport No. 1.

Penney, W.G. and A.T. Price (1952), "The Diffraction Theory of Sea Waves and the Shelter Afforded by Breakwaters", *Phil. Trans. Royal Soc., A,* Vol 224 (882), pp 236-253.

Perlin, M. and R.G. Dean (1978). "Prediction of Beach Platforms with Littoral Controls", *Proc. 16th Int. Conf. on Co. Eng.*, ASCE, Hamburg, pp 1818-1838.

Perlin, M. and R.G. Dean (1983). "A Numerical Model to Simulate Sediment Transport in the Vicinity of Coastal Structures", Report MR-83-10, *U.S. Army Corps of Eng.*

Phillips, O.M. (1957), "On the Generation of Waves by Turbulent Wind", *J. Fluid Mech.* Vol 2, pp 417-445.

Phillips, O.M. (1958), "The Equilibrium Range in the Spectrum of Wind-Generated Waves", *J. Fluid Mech.*, Vol 4, pp 426-434.

PIANC (1992), "*Analysis of Rubble Mound Breakwaters*", Supplement to Bulletin 78/79.

Pierson, W.J. and L. Moskowitz (1964), "A Proposed Spectral Form for Fully Developed Wind Seas on the Similarity Theory of S.A. Kitaigorodskii", *J. Geoph. Res.*, Vol 69, No 24, pp 5181-5190.

Pilarczyk, K.W. (1990), "*Coastal Protection*", Balkema Press, Rotterdam.

Pilarczyk, K. W. and R.B. Zeidler (1996), "*Offshore breakwaters and shore evolution control*", Balkema Press, Rotterdam.

Radder, A.C. (1979), "On the Parabolic Equation Method for Water Wave Propagation", *J. Fluid Mech.*, Vol. 95, Part 1, pp 159-176.

Rakha, K.A. and J.W. Kamphuis (1995), "Numerical and Hydraulic Model Study of Wave Decay on a Shelf Beach", *Co. Eng.*, Vol 24, pp 275-296.

Reid, R.O. (1990), "Tides and Storm Surges", Ch 9 in *Handbook of Coastal and Ocean Engineering*, Gulf Publishing.

Resio, D.T. and C.L. Vincent (1977), "Estimation of Winds over the Great Lakes", *J. Waterway, Port Coastal and Ocean Eng.*, ASCE, Vol 103, No 103, pp 265-285.

Roelvink, J.A. (1991), Modelling of Cross-Shore Flow and Morphology. *Proc. Co. Sediments '91*, ASCE, Seattle, pp 603-617.

Roelvink, J.A. and H.I. Broker (1993), "Cross Shore Profile Models", *Co. Eng.*, Vol 21, pp 163-191.

Ross, D.A. (1977), "*Introduction to Oceanography*", Prentice-Hall.

Rye, H.(1977), "The Stability of Some Currently Used Wave Parameters", *Co. Eng.*, Vol 1, No 1, pp 17- 30.

Sarpkaya, T. and M. Isaacson (1981), "*Mechanics of Wave Forces on Offshore Structures*", Van Nostrand-Reinhold, New York.

Sato, S. and N. Mitsunobu (1991), "A Numerical Model of Beach Profile Change due to Random Waves", *Proc. Co. Sediments '91*, ASCE, Seattle, pp 674-687.

Sawaragi, T (1995), *"Coastal Engineering: Waves, Beaches, Wave-Structure Interactions"*, Elsevier.

Schwab, D.J., J.R. Bennett, P.C. Liu, and M.A. Donelan (1984), "Application of a Simple Wave Prediction Model to Lake Erie", *J. Geoph. Res.*, Vol 89, C3, pp 3586-3592.

Schwarz, M.L. and E.C.F. Bird (1990), Special Issue No.6, *J. of Co. Research.*

Sharp, J.J. (1981), *"Hydraulic Modelling"*, Butterworths.

Shen, H.W. (1990), "Introductory Remarks", *Movable Bed Physical Models*, (H.W. Shen, Ed), Kluwer Academic Publishers, pp 1-12.

Silvester, R. (1974), *"Coastal Engineering"*, Elsevier.

Silvester, R. and J.R.C. Hsu (1997), *"Coastal Stabilization"*, Advanced Series on Ocean Engineering, Vol 14, World Scientific Publishing, Singapore.

Simm, J. (1996), *"Beach Management Manual"*, Construction Industry Research and Information Association (CIRIA), Rep 154, London.

Sorensen, R. M. (1993), *"Basic Wave Mechanics for Coastal and Ocean Engineers"*, Wiley.

Stauble, D.K and N.C. Kraus (1993), *"Beach Nourishment Engineering and Management Considerations"*, Proc. Co. Zone '93, ASCE, New Orleans.

Steetzel, H.J. (1987), "A Model for Beach and Dune Profile Changes near Dune Revetments", *Proc. Co. Sediments '87*, New Orleans, ASCE, pp 87-97.

Steetzel, H.J. (1990), "Cross-Shore Transport during Storm Surges", *Proc. 22nd Int. Conf. on Co. Eng.*, ASCE, New York, pp 1922-1934.

Stive, M.J.F. (1986), "A Model for Cross-shore Sediment Transport", *Proc. 20th Int. Conf. on Co. Eng.*, ASCE, Capetown, pp 1550-1564.

Stive, M.J.F. and J.A. Battjes (1984), "A Model for Offshore Sediment Transport", *Proc. 19th Int. Conf. on Co. Eng.*, ASCE, Houston, pp 1420-1436.

Stokes, G.G. (1847), *"On the Theory of Oscillatory Waves, Mathematical and Physical Papers"*, Cambridge U. Press, London, Vol. 1, 314-326.

Sullivan, J.D. and R.G.D. Davidson-Arnott (1995), "Hazard and Regulatory Standards: Great Lakes - St. Lawrence River System Shorelines", *Proc. Canadian Co. Conf. '95*, CC-SEA, Halifax, pp 799-814.

Svendsen, I.A. (1974), "Cnoidal Waves over a Gently Sloping Beach", *Inst. of Hydrodynamics and Hydraulic Eng.*, Denmark, Series paper 6.

Sverdrup, H.U. and Munk. W.H. (1947), "Wind, Sea and Swell: Theory of Relations for Forecasting", Pub No 601, *U.S. Navy Hydr. Off,* Washington.

Takahashi, S., K. Tanimoto and K. Shimosako (1994), "Dynamic Response and Sliding of Breakwater Caisson against Impulsive Breaking Wave Forces", *Japan Port and Harbour Res. Inst.*

Takahashi, S. (1996), "Lecture Notes - Short Course on Vertical Breakwaters", *25th Int. Conf. on Coastal Eng,* ASCE, Orlando.

Thoft-Christensen, P and M.B. Baker (1982), *"Structural Reliability Theory and its Applications"*, Springer Verlag.

Tolman, H.L. (1991), "A Third-Generation Model for Wind Waves on slowly Varying, Unsteady and Inhomogeneous Depths and Currents", *J. Phys. Oceanog.*, Vol 21, 6, pp 782-797.

Townend, I.H. (1994), "Coastal Management", Ch 22 in *Coastal, Estuarial and Harbour Engineers' Handbook*, (M.B. Abbott and W.A. Price, Ed), Chapman and Hall.

Tsai, C.P. and L.L. Tsong (1999), "Back-Propagation Neural Network in Tidal Level Forecasting", *J. Waterway, Port Coastal and Ocean Eng., ASCE*, Vol 125, No 4, pp 195-202.

Van der Meer, J.W. (1987), "Stability of Breakwater Armour Layers", *Co. Eng.*, Vol 11, pp 219-239.

Van der Meer, J.W. (1993), "Conceptual Design of Rubble Mound Breakwaters", Rep. 483, *Delft Hydraulics*, Delft.

Van Rijn, L.C. (1998), *"Principles of Coastal Morphology"*, Aqua Publications, Netherlands.

Walker, S. (1973), *"Wind and Strategy"*, Norton Press.

Wamdi (1988), "The WAM Model: a Third Generation Ocean Wave Prediction Model", *J. Phys. Oceanog.*, Vol 18, pp 1775-1810.

Watanabe, A., Riho, Y. and Horikawa, K., (1980), "Beach Profile and On-offshore Sediment Transport", *Proc. 17th Int. Conf. on Co. Eng.*, ASCE, Sydney, pp 1106-1121.

Watanabe, A. and Dibajnia, M. (1988), "Numerical Modelling of Nearshore Waves, Cross-shore Sediment Transport and Beach Profile Change", *Proc. Symp. Mathematical Modelling of Sediment Transport in the Coastal Zone* (IAHR), Copenhagen, pp 166-174.

Waterfront Regeneration Trust (1995), *Lake Ontario Greenway Strategy*, Queen's Printer of Ontario, Toronto.

Wiegel, R.L. (1964), *Oceanographical Engineering*, Prentice-Hall.

Willis, D.H. (1978), "An Alongshore Beach Evolution Model", Report HY-92, *National Research Council of Canada*.

Wind, H.G. (1987), *"Impact of Sea Level Rise on Society"*, Balkema Press, Rotterdam.

Yalin, M.S. (1971), *"Theory of Hydraulic Models"*, MacMillan Press.

Young, I.R. and L.A. Verhagen (1996), "The Growth of Fetch Limited Waves in Water of Finite Depth. Part 1. Total Energy and Peak Frequency", *Co. Eng.*, Vol 29, pp 47-78.

Author Index

Subject Index